Missiles for the Fatherland

Missiles for the Fatherland tells the story of the scientists and engineers who built the V-2 missile in Hitler's Germany. This is the first scholarly history of the culture and society that underpinned missile development at Germany's secret missile base at Peenemünde. Using mainly primary source documents and publicly available oral history interviews, Michael B. Petersen examines the lives of the men and women who worked at Peenemünde and later at the underground slave labor complex called Mittelbau-Dora. His research reveals a complex interaction of professional ambition, internal cultural dynamics, military pressure, and political coercion, which coalesced in the texture of life at the facility. The interaction of these forces made the rapid development of the V-2 possible but also contributed to an environment in which stunning brutality could be committed against concentration camp prisoners who manufactured the missile.

Michael B. Petersen received his doctorate from the University of Maryland in 2005. He is currently a historian with Science Applications International Corporation, where he writes on intelligence and space history. He has also worked for the Nazi War Crimes and Japanese Imperial Government Records Interagency Working Group (IWG) at the National Archives and Records Administration, where he contributed to a collection of essays on Japanese war crimes records held by the National Archives.

Cambridge Centennial of Flight

Editors:

John Anderson
Curator of Aerodynamics, National Air and Space Museum, and
Professor Emeritus, Aerospace Engineering, University of Maryland

Von Hardesty
Smithsonian Institution

The series presents new titles dealing with the drama and historical impact of human flight. The Air Age began on December 17, 1903, with the epic powered and controlled flight by the Wright brothers at Kitty Hawk. The airplane rapidly developed into an efficient means of global travel and a lethal weapon of war. Modern rocketry has allowed heirs of the Wrights to orbit the Earth and to land on the Moon, inaugurating a new era of exploration of the solar system by humans and robotic machines. The Centennial of Flight series offers pioneering studies with fresh interpretive insights and broad appeal on key themes, events, and personalities that shaped the evolution of aerospace technology.

Also published in this series:
Scott W. Palmer, *Dictatorship of the Air: Aviation Culture and the Fate of Modern Russia*

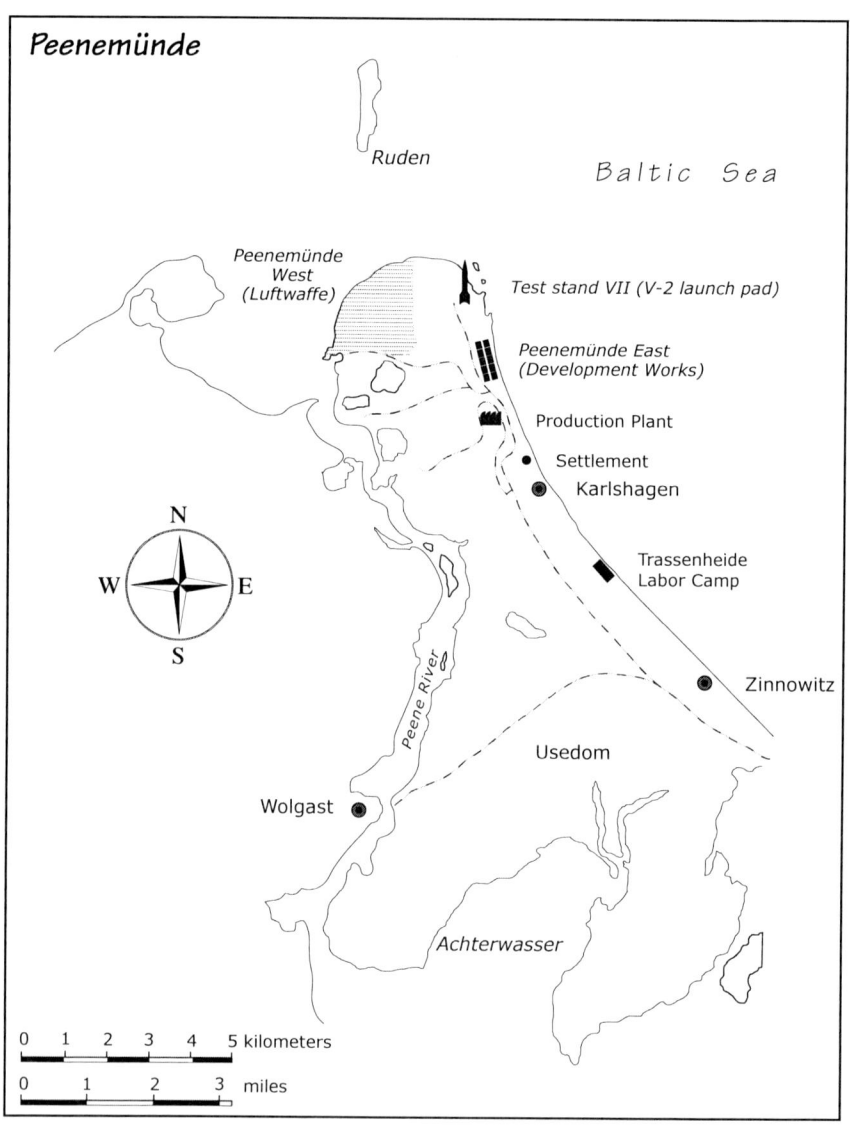

Peenemünde

Ruden

Baltic Sea

Peenemünde
West
(Luftwaffe)

Test stand VII (V-2 launch pad)

Peenemünde East
(Development Works)

Production Plant

Settlement

Karlshagen

Trassenheide
Labor Camp

N

W E

S

Zinnowitz

Peene River

Usedom

Wolgast

Achterwasser

0 1 2 3 4 5 kilometers

0 1 2 3 miles

Frontispiece: Map of Peenemünde missile base, Germany, 1943 (map by John Apinis).

Missiles for the Fatherland

Peenemünde, National Socialism, and the V-2 Missile

MICHAEL B. PETERSEN

CAMBRIDGE
UNIVERSITY PRESS

CAMBRIDGE UNIVERSITY PRESS
Cambridge, New York, Melbourne, Madrid, Cape Town,
Singapore, São Paulo, Delhi, Tokyo, Mexico City

Cambridge University Press
The Edinburgh Building, Cambridge CB2 8RU, UK

Published in the United States of America by Cambridge University Press, New York

www.cambridge.org
Information on this title: www.cambridge.org/9780521283403

First published 2009
First paperback edition 2011

A catalogue record for this publication is available from the British Library

Library of Congress Cataloguing in Publication data
Petersen, Michael B.
Missiles for the fatherland : Peenemünde, national socialism, and the V-2 missile /
Michael B. Petersen.
p.   cm. – (Cambridge centennial of flight)
Includes bibliographical references and index.
ISBN 978-0-521-88270-5 (hbk.)
1. V-2 rocket – History. 2. Germany. Heer. Heeresversuchstelle Peenemünde – History.
3. Rocketry – Germany – Peenemünde – History – 20th century. 4. National socialism and
technology.   I. Title.
UF535.G3P49   2008
623.4'51953094309044 – dc22   2008031150

ISBN 978-0-521-88270-5 Hardback
ISBN 978-0-521-28340-3 Paperback

Contents

List of Illustrations

Acknowledgments

This book would not have been completed without the support of many people and institutions. In particular, I owe a considerable debt to Jeffrey Herf, whose work on the history of Nazi Germany has done so much to shape our knowledge of that terrible era. His guidance and constructive criticism improved each draft that I handed him and pushed me to conceptualize this book in ways that addressed historical issues beyond the scope of its subject. Michael Neufeld, whose first-rate scholarly work inspired this project, generously commented on numerous drafts and offered the benefit of his extraordinary knowledge of the German missile program as well as several obscure but important sources necessary to complete this project. His expert advice is incorporated everywhere in this book. Thomas Zeller availed me of his remarkable intellectual gifts, his impressive knowledge of German technical archives, and his warm friendship. Marsha Rozenblit and John Lampe provided a great deal of friendly and helpful advice, especially at the earliest stages of this project. I also owe many others my gratitude. Martin Collins, John Delaney, David DeVorkin, Norm Goda, Wolf Grüner, Von Hardesty, Martina Hessler, Laura Hilton, Alex Roland, and Jill Stephenson read early versions of chapters presented at various conferences and talks. My friend Suzanne Brown-Fleming, a gifted historian in her own right, provided both intellectual and moral support at the most difficult stages of this process. I would also like to extend my appreciation to the University of Maryland, which supplied me with research and travel funding to carry out much of this work; the Smithsonian National Air and Space Museum for the very generous Guggenheim Fellowship that enabled me to complete a great deal of research and writing; and the United States Holocaust Memorial Museum.

My former colleagues on the Nazi War Crimes and Japanese Imperial Government Records Interagency Working Group at the National Archives, Richard Breitman, William Cunliffe, Norman Goda, and David Van Tassel, also deserve thanks for their intellectual support and archival guidance. Great thanks are also due to the staff at the Archives of the Smithsonian National Air and Space Museum's Garber Facility, Marilyn Graskowiak, Mark Kahn, David Schwartz, Paul Silbermann, and Larry Wilson. The staff was wonderfully accommodating to a researcher who spent more than his fair share of time requesting rolls of microfilm and taking up space in their rather cozy but rich archive. I quickly came to value their expertise and friendship. Finally, Eric Crahan, my excellent editor at Cambridge, was extraordinarily patient while waiting for me to complete this book.

In Germany, I had the benefit of expert advice from other notable historians, archivists, and curators. Andreas Heusler provided me with helpful research advice for German archives. Jens-Christian Wagner shared his unmatched knowledge of Mittelbau-Dora and the incumbent historical issues that are wrapped up in what Jeffrey Herf has called the "divided memory" of the postwar Germanys. Torsten Hess also kindly contributed his expertise on Mittelbau-Dora. Dirk Zache and Manfred Kanetzky of the *Historisches-Technisches Informationszentrum Peenemünde* paid me the great privilege of making available their rich collection of archival and museum holdings that are normally not open to the public. Their contribution is writ large across the pages that follow. Rolf-Dieter Müller and Berghard Ciesla, both at the *Militärgeschichtliches Forschungsamt* in Potsdam, also provided helpful advice. Frau Bärbel Galonska, archivist at the Stasi Archive in Berlin, went well beyond the call of duty to provide me with many important documents while regaling me with stories of her childhood outside of Peenemünde and stories of the fall of the Berlin Wall. Peggy Schmidt generously opened her wonderful apartment in the center of Berlin to me, giving me not only a comfortable couch to sleep on but also a trove of great memories that include trips to see the Berlin Philharmonic Orchestra perform and thrice-weekly jogs through the Brandenburg Gate.

My family, friends, and loved ones deserve all of the thanks that one person can muster. My parents, Hans Petersen and Theo Fellows, first inspired my love for history and nurtured it through many years while no doubt silently wondering if I would ever be able to get a "real" job. John Apinis always provided the best laughs when I needed them most. Thanks also to the two Charlies, Mingus and Parker, for their accompaniment, and of course, to my dog Rugby, who is a constant reminder that it's still fun to wrestle on the living room floor. Finally, this project would never have been completed if not for the love and encouragement of Heather Jacobsen. Her steadfast support and profound patience were constants throughout this process, and it is to her and my parents that this book is dedicated.

Introduction

The Community of Innovation and Culture
of Consent in the *Raketen-Stadt*

> It's a factory-state here, a City of the Future full of extrapolated 1930s swoop-facaded and balconied skyscrapers, lean chrome caryatids with bobbed hairdos, classy airships of all descriptions drifting in the boom and hush of the city abysses, golden lovelies sunning in roof gardens and turning to wave as you pass. It is the *Raketen-Stadt*.
>
> Thomas Pynchon, *Gravity's Rainbow*[1]

One of the twentieth century's most dramatic technological achievements occurred the afternoon of October 3, 1942. That day, a Saturday, was a clear and unseasonably warm one at Germany's supersecret missile base Peenemünde, on the picturesque island of Usedom on the Baltic coast. For just over three years, the Third Reich had been waging a cataclysmic war in Europe and around the world, and the nation's fortunes were beginning to take a turn for the worse. Since early 1942, Germany had been suffering from a massive, nearly continuous Allied bombing campaign over its cities. Fourteen hundred miles to east, the battle of Stalingrad raged as German troops tried desperately to dislodge Soviet defenders from that ruined city. At that moment, though, such concerns were only secondary to the work occupying the scientists and engineers at Peenemünde. Dozens of them checked and rechecked equipment, made final technical calculations, and prepared the gear that would measure the flight of the missile that currently sat on its launch table in the middle of Peenemünde's huge test stand. With final preparations complete, the ground crews retreated to their protective bunkers, and the countdown began. Just before 4:00 in the afternoon, twenty-five tons of thrust lifted the forty-six-foot-tall A-4 (or V-2) from its launch moorings and into the sky. The black and white missile accelerated rapidly until it hurtled through the air at nearly 3,500 miles per hour, cut off its thrust, slipped out of Earth's atmosphere, and then came careening back to the planet at over three times the speed of sound, landing five minutes later some 125 miles away in the Baltic Sea.[2]

---

[1] Thomas Pynchon, *Gravity's Rainbow* (New York: Viking Press, 1973), 674.
[2] Walter Dornberger, *V-2* (New York: Viking Press, 1954), 3–15.

After years of toil, the scientists and engineers at Peenemünde had carried out the first successful launch of the A-4. For the first time, humans had launched an object into space, an epochal achievement that they accomplished with virtually no previous practical knowledge and only a few years of theoretical experience. A mere six years earlier, neither the base at Peenemünde nor the plans for the A-4 existed. It was only in late 1936 that developers laid plans for this particular missile and began constructing the facilities for its development. Thus, the successful launch of the A-4 – a scientific and technical event of fundamental importance to the modern world – was one that Germany's brilliant missile specialists managed to pull off with amazing haste.

Though many of its ambitious developers argued after the war that they dreamed of nothing but spaceflight, this was no humanitarian project. The A-4's purpose was to terrify civilian populations by delivering, without warning, a high-explosive warhead to a target nearly 150 miles from its launch origin. German military strategists dreamed that it would so devastate enemy morale that foreign governments would have no choice but to sue for peace. In January 1944, the first of those missiles rolled off the assembly line over the emaciated bodies of thousands of prisoners of the Third Reich at the terrifying underground missile production facility known as Mittelbau-Dora. By the time of Dora's liberation at the hands of American soldiers in April 1945, Nazi Germany had rained nearly 2,200 missiles on London and Antwerp, and perhaps as many as 20,000 slave laborers at the Mittelbau-Dora camp complex were dead as a result of the conditions they endured to build the V-2.[3]

This book tells the story of life and work within the German missile program as it played itself out at the missile base at Peenemünde. A complex interaction of professional ambition, internal cultural dynamics, military pressure, and political coercion coalesced in the texture of life at the facility. The interaction of these forces made the rapid development of the A-4 possible, but it also contributed to an environment in which stunning brutality could be committed against concentration camp prisoners in the name of defending the Nazi state. The engineers and other missile specialists at Peenemünde, only some of whom were committed National Socialists, reacted to these pressures in a variety of ways. Most became passive facilitators of Nazi brutality, doing their duty in support of the Nazi war effort. Through their passivity, they legitimized the tendencies of a smaller group

---

[3] Manfred Bornemann and Martin Broszat, "Das KL Dora/Mittelbau," in *Studien zur Geschichte der Konzentrationslager* (Stuttgart: Deutsche-Verlags Anstalt, 1970), 154–198. This estimate includes the 1,500 prisoners killed by the British bombing raids on the neighboring town of Nordhausen on April 3–4, 1945, Dora prisoners deemed "unfit for work" and sent to the gas chambers at Auschwitz and Majdanek, and those who were murdered during the evacuation of the camp.

that manifested a more radical tendency, combining scientific and technological rationality with Nazi ideology in a way that served the dual goals of
producing weapons and persecuting perceived enemies of the state.

Understanding the ways in which the institution of Peenemünde was able
to enlist the unequivocal support of its members is also central to a deeper
comprehension of how major technological systems develop and reproduce
themselves, especially in the intensified atmosphere of war. This study moves
beyond the external functions of state financing and resource support to
examine how individuals within the program endowed their institution with
personal importance. Moreover, in the Nazi context, identification with the
institution's goals also meant that many scientists, engineers, and technicians
were willing to tolerate, even participate in, the regime's brutal excesses.
Though Peenemünde experienced the impact of Nazification as much as any
place in Germany, the reasons for its employees' complicity were not solely
or explicitly ideological. Rather, they are located in the quotidian pattern of
events taking place at the research station on the Baltic coast.

This book takes what appeared to those at Peenemünde as commonsense
beliefs and everyday, rational routines and shows that they were in fact
part of the process of what anthropologists call "enculturation," the steady,
relentless internalization of a particular set of group norms and ideals.[4] At
Peenemünde, technical specialists internalized a specific set of beliefs about
the importance of their work in a nation in the midst of a desperate war
for its very survival. This created and reinforced the group's own ideas as
a collective entity. In their new role as weapons specialists in the service of
the Third Reich, they came to see the concerns of outside groups as being
far less consequential than their own. The result was a narrowed technical
and patriotic vision that consented to some of the worst crimes of the Nazi
regime.

Missile developers at Peenemünde, however, were not solely united by
any explicit political or ideological program but rather by a shared belief in
a technological program. This set of ideas was characterized by cultural and
technological dynamics that could function across a broad spectrum of political ideologies, subtly reinforcing an individual's loyalty to any number of
political agendas. This, combined with their own active anticommunism, is
what made former German missile specialists so amenable to working for the
United States after the war. During the Nazi era, however, missile specialists
at Peenemünde also displayed a durable sense of loyalty to Hitler's regime.
In the context of a National Socialist government that pursued rearmament,
war, and total war as policy ends, the decisions of weapons engineers, whose
very work helped to both realize these goals and defend the system that set
them forth in the first place, were nothing if not conclusive statements about

---

[4] See Hugh Gusterson, *Nuclear Rites: A Weapons Laboratory at the End of the Cold War*
(Berkeley, CA: University of California Press, 1996).

their political sentiments toward the Nazi state. Peenemünde engineers and technicians did not just help supply Nazi Germany's war machine. They also contributed, in their own small way, to legitimizing the deeds of the government that made their work possible.

The issue of consensus and collaboration under Hitler is perhaps the most important and therefore most contentious issue in the historiography on Nazi Germany. In the 1980s, the effort to document the "history of every-day life" (*Alltagsgeschichte*) in Nazi Germany led historians to conclude that support for the Nazis emerged for many different reasons from many differing segments of society. Even so, the Nazis were successful in carrying out only those policies that the German population did not widely oppose. Though Nazism was a mass movement, only a minority of Germans took up the Nazi banner and its ideological causes. Those who did not were mostly passive onlookers or fellow travelers. This cleared the ground for the ideo-logical vanguard to establish increasingly radical policies. Fanatical Nazi ideas were most successful when German citizens had nothing against them and raised no protest; a failure to voice disapproval of National Socialist fanaticism amounting to a passive acceptance of it. A de facto consensus on certain issues moved people to docile toleration and cooperation.[5] More recent books have reexamined consensus for Nazi policy and have shown that even passive onlookers were in fact not so passive. The historian Robert Gellately, for example, has demonstrated the proactive participation of aver-age Germans in the policing of the Nazi state.[6] He also shows how a fluid but lasting consensus for Hitler developed within the first months of Hitler's regime and, through a combination of selective rewards and repression, remained until the end of the war.[7] Through all of this work, one thing has become clear: The Nazi regime carried out a colossal social, political, and cultural project in Germany that would not have been possible without the activism of a minority of the population coupled with the positive consent of the majority. That they were as "successful" as they were indicates that, one way or another, the Nazis were able to produce powerful social bonds between individuals and with the regime.

Personal happiness and a positive self-perception, therefore, had a deter-mining effect on what was possible in Hitler's Germany. The success of the A-4 endeavor is a case in point. This book revisits the historical traditions of *Alltagsgeschichte* by examining the texture of life at the Peenemünde

[5] Ian Kershaw, *Popular Opinion and Political Dissent in the Third Reich: Bavaria, 1933–1945* (Oxford: Oxford University Press, 1983). See also Detlev Peukert, "Alltag und Barberei: Zur Normalität des Dritten Reiches," in Dan Diner, ed., *Ist der Nationalsozialismus Geschichte? Zur Historisierung und Historikerstreit* (Frankfurt am Main: Fischer, 1987), 51–61.

[6] Robert Gellately, *The Gestapo and German Society: Enforcing Racial Policy, 1933–1945* (New York: Oxford University Press, 1990).

[7] Robert Gellately, *Backing Hitler: Consent and Coercion in Nazi Germany* (New York: Oxford University Press, 2001).

missile facility. The local practices in place at Peenemünde resocialized its employees from an aggregate of disparate individuals into a cohesive group that strongly identified with the same sets of social, political, and technical ideals. In becoming a part of the community of missile specialists at Peenemünde (a "Peenemünder"), individual specialists became firmly convinced that what they were doing was essential to the survival of their nation. The work was, in their eyes, a noble project. Despite whatever demographic differences that they might have had – there were, in fact, few – the basic practices at Peenemünde bound them together into a cohesive unit with a single mission. A distinct set of dynamic social and professional practices ensured their commitment to Peenemünde's goals, which were inextricably linked to the murderous government that sponsored them in the first place. Support for National Socialism, was, to borrow historian Alf Lüdtke's term, "co-produced" by the common practices of everyday life at the base.[8]

Moreover, with a few exceptions, much scholarship on Nazi Germany has asked why virtually no one resisted the murderous policies of the Nazi regime. Alternatively, historians and others have sought to understand how it was that Nazi perpetrators were able to overcome revulsion at crimes they committed in the name of the regime as well as the disillusionment that must have come along with these acts.[9] This work has been instructive but flawed. It makes a basic assumption that the perpetrators viewed what they were doing as immoral and criminal, or that they should have at least understood that it was wrong. Nazi criminals carried out their acts without feeling. Other forces were at work that enabled them to surmount their natural predilections that these were in fact immoral, illegal acts. Thus historians immediately constructed a framework that implicitly assumed the Germans understood that what they were doing was wrong; that they should have resisted such terrible acts. Historical actors therefore repressed their true feelings of revulsion and avoided moral introspection about their actions. This framework has been helpful, but it has not been entirely satisfying. Most often, it does not actually address the issue of personal dedication to the tasks confronting individuals. The work itself is merely a task to be

---

[8] Alf Lüdtke, *Eigen-Sinn: Fabrikalltag, Arbeitererfahrungen und Politik vom Kaiserreich bis in den Faschismus* (Hamburg: Ergebnisse Verlag, 1993), 332.

[9] Robert Lifton, *The Nazi Doctors: Medical Killing and the Psychology of Genocide* (New York: Basic Books, 1986); Hans Mommsen, "Die Realisierung des Utopischen: Die 'Endlösung der Judenfrage' im 'Dritten Reich,'" in *Der Nationalsozialismus und die deutsche Gesellschaft* (Reinbek bei Hamburg: Rowohlt, 1991), asks "Why did so many who participated in the events that led directly or indirectly to the extermination of the Jews fail to withdraw their contribution either through passive resistance or any form of resistance at all?" (p. 186). See also Christopher Browning, *Ordinary Men: Reserve Police Battalion 101 and the Final Solution in Poland* (New York: HarperCollins, 1992). Finally, though problematic, one of the benefits of Daniel Goldhagen's *Hitler's Willing Executioners: Ordinary Germans and the Holocaust* (New York: Knopf, 1996) was to ask whether or not Germans actually faced any dilemma at all in persecuting the Jews.

performed, not a possible source of binding energy or motivation. Instead, it is useful to recall that the problems confronting those who forcibly relocated Jews and other perceived enemies of the state, coordinated massive slave labor projects, developed the world's first ballistic missile, or, for that matter, executed the "Final Solution," were gargantuan. The success of these projects could only be counted on if those carrying them out were dedicated, conscientious, and motivated workers.[10] Repression, avoidance, and denial do not fully explain the willingness of individuals to carry out criminal acts.

The work by the Peenemünders to produce missiles for the Nazi regime as well their participation in the practice of slave labor have become the central points of controversy in historical discussions about Peenemünde generally. For nearly fifty years after the war, most histories of the German ballistic missile program were written by participants themselves or their supporters. The result was a narrative that both distanced their work from the regime that sponsored it and misrepresented or ignored their decisions about participation in the use of slave labor.[11] In the late 1980s, after the Justice Department's investigation of Arthur Rudolph, the Factory Director at the Peenemünde production plant and the slave labor factory at Mittelwerk, some journalists began scrutinizing the Nazi past of the former Peenemünders. This work was valuable for the documents it turned up, but unfortunately, it was similar to the earlier work in that it painted a crude, though very different, picture of life in the Third Reich and the missile specialists' place in it.[12] Thus, for nearly half a century, historians were left

---

[10] See Michael Thad Allen's work on the SS Economic and Administrative Main Office, *The Business of Genocide: The SS, Slave Labor, and the Concentration Camps* (Chapel Hill, NC: University of North Carolina Press, 2002); Michael Wildt's qualitative study of the *Reichssicherheitshauptamt* (Reich Security Main Office – RSHA) officer corps, *Generation des Unbedingten: Das Führerkorps des Reichssicherheitshauptamt* (Hamburg: Hamburg Edition, 2002); and Eric A. Johnson's study of the Krefeld Gestapo, *Nazi Terror: The Gestapo, Jews, and Ordinary Germans* (New York: Basic Books, 1999) for excellent efforts to surmount this tendency. Allen's book also contains a chapter dedicated to the effort to manufacture the A-4. In this chapter, he places ideology at the center of activities, missing, in my estimation, the connections between Peenemünde and Mittelbau-Dora, and therefore the other factors motivating work that resulted from this connection.

[11] An excellent example of this type of work is Walter Dornberger's *V-2, Der Schuss ins Weltall: Geschichte einer Grossen Erfindung* (Esslingen: Bechtle Verlag, 1952). Among the many examples of work written by other supporters of the Peenemünders, see Thomas Franklin (pseudonym for Hugh McInnish), *An American in Exile: The Story of Arthur Rudolph* (Huntsville, AL: Christopher Kaylor, 1987) and Marsha Freeman, *How We Got to the Moon: The Story of the German Space Pioneers* (Washington, DC: 21st Century Science Associates, 1994).

[12] See Tom Bower, *The Paperclip Conspiracy: The Hunt for Nazi Scientists* (Boston: Little, Brown, 1987) and Linda Hunt, *Secret Agenda: The United States Government, Nazi Scientists, and Project Paperclip, 1945 to 1990* (New York: St. Martin's Press, 1991). Another, less valuable book is Dennis Piszkiewicz, *The Nazi Rocketeers: Dreams of Space and Crimes of War* (Westport, CT: Praeger Press, 1995).

with a thoroughly incomplete understanding of one of the most significant technological endeavors of the twentieth century.

In 1995, Michael Neufeld addressed this oversight with his important book *The Rocket and the Reich: Peenemünde and the Coming of the Ballistic Missile Era*. His work is an account of the technological and administrative history of the German ballistic missile program. According to Neufeld, the A-4 was "the product of a narrow technological vision that obscured the strategic bankruptcy of the project."[13] Though immensely sophisticated, it was a weapon that had virtually no tactical or strategic value because it was wildly inaccurate and could only deliver a payload of one ton, much less than even a single American bomber. Administrators of the project inflamed the expectations of the regime and used the regime's polycratic struggles to establish the missile as Germany's best chance to win the war. Allied bombing raids provided the rationale for continued access to resources during the war, which were allocated at the expense of other more strategically valuable projects. The use of slave labor to mass produce the missile was, according to Neufeld, the Nazis' contribution to the program. In all, Neufeld shows that in the German context, such a huge technological leap forward would not have been possible without the megalomaniacal, even irrational, ambitions of National Socialism.

The complementary converse is also true. Although the grand designs of the Nazi regime were undoubtedly critical, such a task could also not have been accomplished without the willing identification of individual engineers and technicians with many of the same overblown ambitions. The social, cultural, and political fabric at Peenemünde inextricably bound the missile specialists to the goals of their institution and, through them, to the objectives of the regime itself. Neufeld necessarily focuses on the specialists' accomplishments as purely technological achievements, as ends themselves. This book examines the Peenemünders' accomplishments not as technological statements but as political and military ones. In less than a decade, missile specialists at Peenemünde carried out one of the twentieth century's most impressive technological achievements. Such a stunning feat could indeed not have taken place without the willing and active identification of the Peenemünders with the important work to which they were assigned. An important part of their connection with these goals was a willingness to set aside the priorities of all other groups and to engage in slave labor under some of the most horrific conditions in the Nazi empire. The process by

---

[13] Michael J. Neufeld, *The Rocket and the Reich: Peenemünde and the Coming of the Ballistic Missile Era* (Cambridge, MA: Harvard University Press, 1995), 274. In 1984, Heinz-Dieter Hölsken published the scholarly work *Die V-Waffen: Entstehung, Propaganda, Kriegseinsatz* (Stuttgart: Deutsche Verlags-Anstalt, 1984), but his work did not have access to the entire documentary record and fell prey to many of the myths about Peenemünde established after the war.

which the Peenemünders came to internalize such imperious ambitions is at the center of this study.

Thus the story of Peenemünde is one that sits at the intersection of the history of Nazi Germany and the history of spaceflight. The A-4 was undeniably developed to serve as a weapon, but it was nevertheless an important step in the development of spaceflight technology. In exploring the forces that led individual engineers, scientists, and technicians at Peenemünde into the arms of the Nazi regime, this book also describes the lived experiences of these people as the sun began to dawn on the space age. It explores the role that culture and society played in shaping the environment in which they worked, and how these factors in turn either helped or hindered their decision making. These issues, often overlooked in the general historiography on spaceflight, can provide a special insight into the whys and hows of successful space programs.

This book is arranged both chronologically and thematically. Chapter 1 examines the roots of rocket engineering in Weimar Germany. The central feature of rocketry in this period was the collection of amateur rocket societies that were dedicated to the idea of spaceflight. Perhaps the most important experimental facility was located in Reinickendorf, outside of Berlin, and had the impressive moniker *Raketenflugplatz Berlin* (Rocket Base Berlin). The members of the *Raketenflugplatz* were mostly unemployed engineers, technicians, manual laborers, and other enthusiasts who were fascinated by the idea of space travel. Many of them commonly cast their work as an assertion of German cultural and national interest. Radically new rocket technology was a statement of strength made by its practitioners on behalf of a nation that suffered so terribly in the wake of World War I. Moreover, the common practices and shared conditions on the shop floor at the *Raketenflugplatz* acted to bind its members together into a closely knit group that identified intensely with its work. When the German Army began its own in-house missile program and was able to co-opt the services of some of the amateur rocketeers, those few who joined the Army program began to fulfill their technological, economic, and nationalist interests, and the process by which their identities would be reshaped as rocket specialists in the service of the state had begun.

As the Army dedicated more and more resources to the work, it became clear that a new research facility was necessary. Chapter 2 examines the rise of Peenemünde and the framework within which Peenemünde's unique institutional culture would crystallize. Missile specialists were drawn into a close cooperative relationship with authorities within the Nazi regime through a combination of military decisions, professional aspirations, and demands for secrecy. The steadily strengthening Army made its commitment to missile technology clear. Frenetic rearmament in the 1930s gave the specialists a first-rate research facility on the Baltic coast that was the most closely guarded secret in the nation. The secrecy around this project had important implications for the formation of the engineers' group identity as missile specialists in the service of the Nazi state. It fostered a sense of community, privilege, loyalty, and an overriding sense of being observed by the authorities. This general atmosphere set the framework for all of their future efforts on behalf of the regime.

Chapter 3 analyzes the life and work of specialists inside the Peenemünde research station. Those who worked at the facility, which was something between an army base and a utopian social experiment, recalled their years there as some of the best of their lives. Engineers and scientists, most of whom would have been drafted into the Army to serve at the front if not for their work, were positively thrilled about being hired or assigned to Peenemünde. The development work, so profoundly advanced and playing about the edges of science fiction, was supremely exciting. Many of them bonded personally and professionally while making many major technological advances. The tasks at Peenemünde deeply satisfied many of their personal and professional goals. At the same time, engineers who designed and built the missile base made sure that the specialists were afforded spacious, comfortable housing for themselves and their families. Community life at Peenemünde was distinctly pleasant. Inhabitants of the small, enclosed settlement established tight bonds with each other by holding regular social events and partaking in the many leisure and recreation opportunities on their island base. These activities helped solidify their identification with each other and established the community of "Peenemünders," a group of professionally and personally like-minded people whose shared circumstances fostered close bonds of personal familiarity and professional friendship.

This work, however, was not entirely set in an apolitical, technocratic environment. It was clear to these Peenemünders, who owed their identities and professional lives to the Nazi regime, that their work was being carried out in order to defend the government that made their work possible. They were to develop and produce a powerful weapon for which there was no defense, and they were to do so as quickly as possible. That they were doing so for a regime that embarked on a war that engulfed the continent, openly persecuted Jews, homosexuals, and others, and enslaved foreign civilians was not a matter of particular concern for them. A number of them even embraced Nazi political and military goals. Those who were not necessarily committed Nazis still accepted the National Socialist rhetoric in which their work was cast. Their comfortable personal lives and profound professional satisfaction, all established within a framework of intense secrecy that tended to stunt the development of contrary positions, led to the nearly automatic adherence to Peenemünde's central mission of developing an unstoppable weapon that could be used to defend the Nazi state. Their concerns were central. Those of other groups paled in comparison.

This dynamic led the Peenemünders to participate in one of Nazi Germany's most heinous acts of cruelty. Chapter 4 examines the decision by

Peenemünde managers to employ slave labor in the mass production of the A-4. Specialists at Peenemünde actively sought out slave labor as a solution to the increasingly pressing labor shortages that were occurring across Germany, and they welcomed the contributions of the SS (*Schutzstaffel*) in this regard. Chapter 4 also analyzes the treatment of forced and slave laborers who worked at Peenemünde. An important dynamic established itself at the base, in which unskilled foreign labor suffered poor treatment, extremely arduous work, and impossible living conditions, while skilled labor, because of the its value for the project, enjoyed better treatment, easier work, and more comfortable housing. Those prisoners who were in a position to directly help the Peenemünders and their work received much better treatment than those who were involved in more menial construction and materials transport work. Peenemünde specialists made no efforts to alleviate the condition of those unfortunate laborers who were not lucky enough to possess the skills that would enable them to assemble a functional ballistic missile. This was a pattern that would be reflected, with much more catastrophic results, at the notorious slave labor facility Mittelbau-Dora. The Peenemünders' narrowed ethical outlook, a result of their strong identification with each other and the goals of their project, meant that the concerns of others barely weighed in the balance.

The result was ready accommodation to increasingly barbarous slave labor in the missile program in 1943 and after. Chapter 5 examines the actions of Peenemünde specialists who were engaged in mass production in the terrifying slave labor factory of Mittelwerk. The missile program's midlevel managers who carried out their tasks at Mittelwerk proved to be willing collaborators with the SS, which supplied labor for the factory and set the overall conditions for its use, because both groups strongly identified with the military and technical goals of the missile project. Former Peenemünde specialists assumed important positions in the factory in which they had to make daily decisions that directly affected the lives and well-being of slave laborers who worked on the shop floor. Their strong identification with the program's objectives, the major professional advances that they made in the move to Mittelwerk, and, it must be noted, a dramatically increased feeling of personal coercion to conduct the work successfully, combined to ensure the civilian specialists' utmost dedication to their production tasks. The same dynamic as at Peenemünde, in which management viewed skilled labor as a valuable commodity and treated it as such while not concerning itself with the fate of unskilled labor, rapidly took shape at Mittelwerk. The result was a dynamic in which decisions about human value were made based on criteria of function and skill; humanitarian considerations did not fit into the equation at all.

Chapter 6 shifts the focus back to the experts at Peenemünde. In the last eighteen months of the war, the missile program was buffeted by major bureaucratic conflict at the highest levels of the regime. The increased influence of the Armaments Ministry and SS, along with the Army's weakening influence, opened up gray areas of influence in which these organizations each sought greater control. However, these conflicts were attenuated by the close cooperation between individuals in these organizations at the level of middle and lower management. The Peenemünders' expertise made them irreplaceable, whereas their shared dedication to the program's goals made them willing collaborators with other organizations. This working arrangement was the model for the solution to the administrative conflicts at the top of the program.

Moreover, over the course of 1944 and early 1945, the missile specialists at Peenemünde worked furiously to reverse Germany's fortunes in the war. This was a period of immense technological creativity that was characterized by both a steady advance in missile technology and the development of new weapons that sometimes were no more than desperation projects borne of technological fantasy. In both cases, the scientists, engineers, and technicians at Peenemünde prosecuted their work with phenomenal effort. This chapter in part confirms Karl-Heinz Ludwig's influential thesis on *Selbstmobilisierung* (self-mobilization), the notion that engineers under the Nazis went far beyond the normal call of duty in their daily work.[14] The Peenemünders never flagged in their technical dedication to missile technology and, therefore, the regime that sponsored them. In this way, they made their own technological contribution to the cumulative radicalization that took place in Nazi Germany in the last months of the war. Their experience at Peenemünde, a place characterized by its utter secrecy, tightly knit community, fascinating work, and persistent political rhetoric, fully imbued them with the idea that their livelihoods depended entirely upon the continued service to the state, and they were bent on doing everything they could to ensure its survival.

In a recent essay, the historian Norbert Frei has argued that it is necessary to look at periods of "normalcy" under the Nazis and ask what kind of effect they had on the lives of regular Germans. He holds that "One must take into account collective feelings and subjective experiences which in part seemed to be more positive than was to be expected under the objective political circumstance of a dictatorship."[15] This book is an effort to do

---

[14] Karl-Heinz Ludwig, *Technik und Ingenieure im Dritten Reich* (Düsseldorf: Droste Verlag, 1974). Ludwig examined the sociopolitical conduct of engineers under the Nazis generally. See also Jeffrey Herf, *Reactionary Modernism: Technology, Culture, and Politics in Weimar and the Third Reich* (Cambridge, UK: Cambridge University Press, 1984) and Helmuth Trischler, "Self-Mobilization or Resistance? Aeronautical Research and National Socialism," in Monika Renneberg and Mark Walker, eds., *Science, Technology, and National Socialism* (New York: Cambridge University Press, 1994).

[15] Norbert Frei, "Peoples' Community and War: Hitler's Popular Support," in Hans Mommsen, ed., *The Third Reich Between Vision and Reality: New Perspectives on German History, 1918–1945* (New York: Berg, 2001).

exactly that. Consensus and collaboration under the Nazis was not achieved by the dynamic established because of an individual's or a group's repression and avoidance. Rather, the positive integration of individuals into a collective body that believed in the goals of the Nazi project was central to the success of Hitler's regime.[16] Like many Germans, those at Peenemünde shared some of the same goals as many of the most ardent members of the regime. Many of the megalomaniacal ambitions of the Third Reich would not have been as successful as they were any other way.

[16] Despite its problems, Goldhagen's *Hitler's Willing Executioners* revealed a dearth of historical research on the circumstances surrounding the positive, integrationist aspects of the Nazi regime. Apart from its flaws, it addressed the very important question of what it was that Germans wanted from the regime. A German tradition of "eliminationist anti-Semitism" may not be the answer, but Goldhagen's focus on the question is welcome.

1

"Help Build the Spaceship!"

In the early 1920s, Germany was a nation at a crossroads. Since its defeat in World War I in 1918, the country had been at war with itself. It convulsed under the strains of a lost war, an attempted revolution, and the fall of its long-established authoritarian political system to the new Weimar Republic. The nation was riven by violent clashes between communist revolutionaries and a shaky social-democratic government that was dependent on the support of influential conservatives who did not favor it in the first place. By 1923, the violence abated and the national economy stabilized under a major reform program, but the scars of its trauma remained in the form of angry, ultra-right-wing veterans' groups, unrepentant communists, and a society struggling to come to grips with the consequences of its World War I folly. In urban areas especially, the distractions of consumer culture assumed epic proportions as Germans found relief in smoky jazz clubs, subversive cabarets, and stunning new motion pictures. Yet a simmering, toxic resentment brewed beneath this rampant urban consumerism. Many right-wing conservatives rejected Weimar modernism, with its cultural decadence, its seeming political disorder, and its dismissal of tradition. They longed for a return of German order and power while rejecting democratic rule in all its forms. Anti-Semitism festered among those who believed that Jews were responsible for Germany's defeat and subsequent domestic chaos. According to their mythmaking, German Jews had stabbed them in the back in 1918 and sought Germany's permanent subjugation. The young republic had a brief reprieve, however, when a period of relative calm settled over Germany by 1924. In the midst of all this, a Romanian theorist of German descent wrote a short but profoundly important book that helped inject the concept of manned space travel squarely into German consciousness.

Hermann Oberth, the son of a German-speaking physician, was born on June 25, 1894 in Sibiu, Transylvania. After his service in the Austro-Hungarian Army during World War I, he studied physics at Cluj in Romania. When his year in Cluj was over, Oberth moved on to study in Munich, Göttingen, and Heidelberg, Germany. By his early twenties, the imaginative Oberth was thinking very seriously about the prospect of space travel. As early as 1917, the German Armaments Ministry rejected his idea of building a large, liquid-fueled rocket because officials thought the task impossible.[1] In 1923, however, when he was twenty-nine years old, he published his career-making book, *Die Rakete zu den Planetenräumen* (The Rocket into Interplanetary Space). It would go on to become, as one historian has put it, "the cornerstone of the Space Age."[2]

Oberth's groundbreaking book was the first in Germany to give serious academic and technological attention to manned space travel, which had been a popular science fiction topic for decades. *Die Rakete zu den Planetenräumen* offered a spirited, if turgid, argument in favor of the idea of manned spaceflight. A short volume of eighty-seven pages, it nevertheless covered nearly every important detail of spaceflight, including propulsion, guidance, life support, and reentry. Moving away from fanciful (and fictional) depictions, the book made interplanetary travel a solvable engineering problem that only lay a few years into the future. Even though Konstantin Tsiolkovsky, a little-known Russian schoolteacher who was interested in rocketry, arrived at similar conclusions before Oberth, and Robert Goddard, the reclusive American physicist, caused a major press flap with his passing ideas about sending an object to the Moon, the two were either relatively unknown or their methods were too complex to be accessible to the lay public.[3] Tsiolkovsky was in fact so obscure that Oberth had never even heard of him until 1924.[4] Though by no means a simple book, *Die Rakete* was far more available than Tsiolkovsky's writings, which were virtually unknown in the West, and far bolder than Goddard's cautious work. Even as thick as it was with complex mathematical equations, *Die Rakete* remained accessible enough to the interested layperson that its readers could envision the possibility of space travel.[5]

Oberth's book offered a number of futuristic ideas. First, he argued that the state of technology in the 1920s made it possible for man-made machines to climb "higher than the earth's atmosphere." Although more work lay ahead, man-made machines could actually achieve escape velocity and breach the Earth's atmosphere, carrying people into orbit. The rockets designed for this could also be capable of carrying human beings into space without harm, even in relative comfort. Finally, Oberth argued that within

---

[1] Hermann Oberth, "My Contributions to Astronautics," unpublished essay, presented at the XVIII International Astronautics Congress, First International History of Astronautics Symposium (Pre-1939 Memoirs of Astronautics), September 26, 1967, Belgrade, Yugoslavia; in National Air and Space Museum (NASM) File "Germany, 1920–1923," 8.

[2] Frank Winter, *Rockets into Space* (Cambridge, MA: Harvard University Press, 1990), 18–19.

[3] Tsiolkovsky's *Exploration of Cosmic Space by Means of Reaction Devices* appeared in Russia in 1911, and Goddard's famous paper "A Method of Reaching Extreme Altitudes" was published in 1920. See Winter, *Rockets into Space*, 10–11, 17–18.

[4] Oberth, "My Contributions," 17.

[5] Hermann Oberth, *Die Rakete zu den Planetenräumen* (Nuremberg: Uni-Verlag, 1960; original work published 1923).

a few decades, these spaceships were almost certain to be profitable "under certain conditions," which remained largely unspecified. These were indeed radical scientific and engineering concepts in 1923, and in the remainder of the book, Oberth set out the mathematical proofs for his remarkable ideas.[6]

For the most part, Oberth avoided bold statements of political loyalty in favor of strict adherence to what might be characterized as scientific neutrality. Pronouncements of nationalist goals or statements of political inclination were absent from *Die Rakete*. Oberth preferred to focus on the practical and theoretical problems of space travel rather than engage in the polemics that so many of his scientific and technical colleagues increasingly found themselves embroiled in. In the years to come, this "neutrality" would become a false front as the Nazi regime began to invest heavily in the development of science and technology at Peenemünde, but in the early 1920s, without the state's political and financial backing, Oberth was careful to remain largely nonpolitical concerning the theoretical possibilities of space travel and its associated technology.

Nevertheless, conservatives celebrated Oberth's ideas. The *Deutsche Allgemeine Zeitung* (DAZ), a right-wing newspaper, published a flattering review of *Die Rakete zu den Planetenräumen*. Although it noted that the technical means of space travel had not yet been realized, the DAZ agreed with Oberth's notion that a rocket could be sent up to a height of 100 kilometers and placed in orbit in the relatively near future. The DAZ noted that Oberth's idea was not exactly new, as Robert Goddard had reached a similar conclusion before him, but for the newspaper, what made Oberth's achievement noteworthy was his German heritage (despite the fact that he was a Romanian citizen). It noted the "German meticulousness" of Oberth's work and described as "happy news" the fact that a German engineer had been devoting so much time to the problem. The newspaper proclaimed, "For us, it is an uplifting feeling that in these years of the worst distress of the German nation, a German engineer has carried out valuable work toward a solution of this technical problem."[7] By connecting Oberth's work to his nationality, the DAZ offered a thin reed of legitimacy to Oberth's radical technological ideas in Weimar's influential conservative circles. It also helped to reconcile his ideas with many conservatives' fears that modern science and technology might somehow destroy the German soul.[8] By drawing the conclusion that the inspiration for the rocket lay in Oberth's German heritage,

---

[6] Oberth, *Die Rakete*, 7.

[7] "Die Raketen zu den Planeten," *Deutsche Allgemeine Zeitung*, December 2, 1923. See also Michael Neufeld, "Weimar Culture and Futuristic Technology: The Rocketry and Spaceflight Fad in Germany, 1923–1933," *Technology and Culture* 31 (October 1990), 744.

[8] See Jeffrey Herf, *Reactionary Modernism: Technology, Culture, and Politics in Weimar and the Third Reich* (Cambridge, UK: Cambridge University Press, 1984), for a full exploration of the conservatives' conflicted feelings about modern technology.

the DAZ made it clear that it was the German spirit that inspired technological advance and that could rescue the nation from its "worst distress."
The *Deutsche Allgemeine Zeitung's* search for signs of German renewal in
Oberth's work pointed to a nascent link between the interests of German
rocket enthusiasts and the nationalist right wing in Weimar Germany.

Nationalist interests aside, Oberth did openly acknowledge the possibility
of using a rocket as a weapon that could sow mass destruction. In a noteworthy passage in his 1929 book *Wege zur Raumschiffahrt* (Paths to Space
Travel), a more vigorous development of the ideas first raised in *Die Rakete
zu den Planetenräumen*, Oberth pointed out that the value of rockets was
not just in transportation but in weaponry as well. He raised the possibility
of using rockets to engage in chemical warfare by equipping the warheads
with poison gas. Oberth also suggested the idea of setting up a space station
and equipping it with a large mirror that could redirect the sun's energy,
changing local weather patterns and laying waste to entire cities (Oberth
actually offered the idea in 1923, but he had more fully developed it by
1929).[9] These acknowledgments, though short, are important. They are an
early civilian recognition of the rocket's possibilities as a tool of destruction,
rather than solely a scientific instrument imbued only with positive, constructive attributes. The early rocket pioneers were not necessarily devoted
militarists, but they were at least open to the possibility of using the fruits
of their labor for less than humanitarian purposes. In the years after World
War II, many German rocket engineers denied this, arguing that they were
never interested in building weapons, that they were forced by a brutal
dictatorial regime that brooked no opposition into producing missiles that
Germany could rain down on its enemies with impunity. On the contrary,
to Weimar rocket enthusiasts – nearly all of whom confessed to being influ-
enced by Oberth – the concept of building weapons was never completely
alien.

In any case, neither academia nor the lay public immediately embraced
Oberth's work. The University of Heidelberg rejected *Die Rakete* as a doctoral dissertation in 1922 because of its quirky subject matter, and the tightly
wound Oberth was fated to suffer the slings and arrows of other members
of academia for some time after the book was published a year later. When
first published in 1923, the book struggled.[10] Despite sluggish sales, *Die
Rakete zu den Planetenräumen* found its way into the hands of the right
people. It inspired a number of German enthusiasts to compose their own

[9] Hermann Oberth, *Wege zur Raumschiffahrt* (Bucharest: Kriterion, 1974; original work
published 1929), 199–200; Wernher von Braun, "Reminiscences of German Rocketry,"
Journal of the British Interplanetary Society 15 (May/June 1956), 145.
[10] Oberth, "My Contributions," 16. See also Hans Barth, *Hermann Oberth: Leben, Werk,
und Auswirkung auf die spätere Raumfahrtentwicklung* (Feucht: Uni-Verlag, 1985), 75–76,
93.

books on the possibilities of space travel. These included Max Valier's *Der Vorstoss in den Weltenraum: eine technische Möglichkeit* (The Thrust into Interplanetary Space: A Technical Possibility – 1924), Walter Hohmann's *Die Erreichbarkeit der Himmelskörper: Untersuchungen über das Raumfahrtproblem* (The Attainability of Celestial Bodies: Investigations into the Problem of Space Travel – 1925), and Hermann Noordung's *Das Problem der Befahrung des Weltraums: Der Raketen-Motor* (The Problem of Space Travel: The Rocket Motor – 1929).[11] The most noteworthy of these disciples was Max Valier.

Valier was born on February 9, 1895 in Bolzano in South Tyrol. He began his academic career by studying physics at Innsbruck from 1913 to 1915. During World War I, he served as a pilot in the Austro-Hungarian armed forces on both the Italian and Russian fronts, as well as in Romania. After the war, he studied astronomy, meteorology, and mathematics in Munich and Vienna. The dynamic Austrian wrote a number of books and articles on the occult as well as the idea of *Welteislehre*, or "glacial cosmogony," but found his calling after reading *Die Rakete*. He wrote to Oberth about a possible collaboration to build on the ideas first introduced in the Romanian's imaginative book.[12] Oberth eventually agreed to cooperate and sent Valier a number of calculations to work from. Valier's effort, the semipopular *Der Vorstoss in den Weltenraum*, was by no means an academically rigorous book. For that matter, it contained a number of glaring errors that spoke volumes about Valier's misunderstanding of Oberth's work. However, Valier was an irrepressible, energetic salesman with a gift for speaking and writing. His former colleague Hans Hörbiger, who first came up with the crackpot idea of glacial cosmogony, wrote somewhat disdainfully of Valier's talents that "He needs a topic to make his name a household word all over the world, to spread the impact of his writings and to fill his lecture halls, since he has to make a living for himself and his two families. And he is an excellent speaker who does not need to use any notes. But he also needs a gripping subject – and spaceflight makes converts of the most cautious adherents, while the mysticism of the WEL [glacial cosmogony] requires a public with greater technical background in order to generate some cash flow."[13] Hörbiger, who felt that Valier abandoned him for the more glamorous field of rocketry, was at least right about the Austrian's

---

[11] Max Valier, *Der Vorstoss in den Weltenraum: eine technische Möglichkeit* (Munich: Oldenbourg, 1924); Walter Hohmann, *Die Erreichbarkeit der Himmelskörper: Untersuchungen über das Raumfahrtproblem* (Munich: Oldenbourg, 1925); Hermann Noordung, *Das Problem der Befahrung des Weltraums: Der Raketen-Motor* (Berlin: Schmidt, 1929).

[12] *Welteislehre* (glacial cosmogony) was a theory first devised by Hans Hörbiger. It held, among other things, that the planets and moon were coated with ice. See also Ilse Essers, *Max Valier: Pioneer of Space Travel* (Washington, DC: NASA, 1976; Technical Translation TTF-664), 94–95. Barth, *Hermann Oberth*, 106.

[13] Hörbiger to Ley, June 12, 1927, in NASM File "Germany, 1920–1940, Correspondence."

talents. Valier's book sold briskly, going into a second printing in 1925.[14] The real importance of *Der Vorstoss*, however, was that it helped increase the sales of Oberth's book. In the end, it was the flashy Valier who proved to be the most adept at popularizing the staid, sometimes sullen Romanian's ideas.

Valier toured Austria and Germany in an effort to promote his and Oberth's work. He held dozens of lectures and tirelessly wrote illustrated articles on spaceflight for various magazines and newspapers. Many Germans proved to be open to these ideas, and his articles and lectures were quite well received.[15] Largely because of Valier's efforts, German popular culture embraced rocketry and space travel in the mid-1920s. The popularity of these concepts was meteoric, exploding quickly and fizzling a few years later, but only the Soviet Union surpassed Germany's level of popular fascination with rockets and spaceflight. In the United States and elsewhere in Europe, the idea received somewhat less notoriety.[16]

This popularity no doubt had much to do with Valier's energy, enthusiasm, and message. Ever the salesman, Valier trumpeted his own brilliance alongside the exciting potential of spaceflight itself, he and made no effort to disguise his ardent nationalism. Like the conservative editors of the DAZ, he linked the accomplishments of rocketry and spaceflight with the triumph of an innate German spirit. In a pamphlet he printed for a lecture tour he made in support of *Der Vorstoss*, Valier wrote that, "Despite its perfect scientific seriousness, [his lecture] also sensationally brings to all who hear it an undreamt-of enrichment of knowledge, an abundance of instruction, and enlightenment of the mysteries of the universe and their solutions through science and technology. Holding [the lecture] promotes the execution of this grand work of German spirit and daring in all parts of Germany."[17] Florid language aside, Valier's nationalist pride is clear. For him, the pursuit of space travel, with all its risks and rewards, was a task perfectly suited for a bold German nation. Such a task captured the individual inventor spirit that German scientists and engineers closely associated with their nationality.

In the politically supercharged environment of Weimar Germany, however, Valier's disquisitions on the value of specifically German technological innovations assumed a dangerous and partisan aspect. Many Germans were infuriated by their country's defeat and subjugation after World War I. Conservatives especially grasped at anything that they believed would revive

---

[14] Neufeld, "Weimar Culture," 730.

[15] Essers, *Max Valier*, 62, 123.

[16] For an introduction to the early Soviet space program, see Asif Siddiqi, "Deep Impact: Robert Goddard and the Soviet 'Space Fad' of the 1920s," *History and Technology* 20 (2004), 97–113 and "The Rocket's Red Glare: Technology, Conflict, and Terror in the Soviet Union," *Technology and Culture* 44 (July 2003), 470–501.

[17] Lecture Program, "Der Vorstoss in den Weltenraum," c. 1927, NASM File "Max Valier."

German national pride and illustrate to the world the superiority of German society and culture. Many saw technological advances as evidence of this superiority. Especially in postwar Germany, technological achievements upheld a sturdy sense of common national pride in an otherwise fractured environment of dissatisfaction and disillusionment. Technology, whether Germans liked it or not, would point the way toward a more prosperous future. For some, like Valier, the rocket was emblematic of the ability of technological advancement to function as a spur to national renewal.

These ideas were especially popular in the field of aviation, whose language Valier appropriated for his own work, and from which many important missile specialists emerged. Many advocates of flying inscribed the rapid expansion of public aviation and the burgeoning popularity of flight in the Weimar Republic with powerful nationalist meaning. They saw the growth of gliding and commercial flight, the establishment in 1925 of Germany's semipublic airline Luft Hansa, and the transoceanic voyages of the massive Zeppelin airships as a reflection of Germany's rebirth in the wake of catastrophic military defeat. German aviation pointed the way to a new, more robust nation that could meet the demands of ever-growing international competition among the European powers. Flight physically brought the world closer together. Germany, argued the leaders of the aviation establishment, should take advantage of the unprecedented proximity to others and the fundamental military and political benefits offered by aviation technology.[18]

Valier borrowed heavily from the aviation enthusiasts and cast his work in this aggressive language of increased national competition. In the English-language periodical *Aviation Mechanics*, he wrote of his desire to establish ongoing transatlantic rocket flights. After proclaiming the geopolitical importance of creating the world's fastest link between Berlin and New York, Valier wrote, "I want to state that it is not 'speed mania' which impels me to set the travel time [between Berlin and New York] so low; but it is a matter of technical and economic necessities [sic]." (Oberth also placed great value on the ability of rocket-equipped planes to shorten the flying time between two distant points.)[19] In justifying his desires in terms of economic and technological need, Valier acknowledged a prevalent feeling among Germans that aviation technology was an important way for Germany to parry its neighbors' competitive and hostile intentions. His writings situated him among those intellectuals for whom technological progress

---

[18] Peter Fritzsche, *A Nation of Flyers: German Aviation and the Popular Imagination* (Cambridge, MA: Harvard University Press, 1992), 132–184, and Guillaume de Syon, *Zeppelin! Germany and the Airship, 1900–1939* (Baltimore: Johns Hopkins University Press, 2002). See also Hermann Oberth, "Der Raketenantrieb bei Flugzeugen," *Flug* 10 (October, 1931).

[19] Max Valier, "Berlin to New York in One Hour," *Aviation Mechanics* 4 (November/December 1930).

was a necessary step in both the protection and advancement of the German nation.

By 1927, Valier's efforts led to the formation of the *Verein für Raumschiffahrt* (VfR – Society for Spaceship Travel). Willy Ley, a science fiction writer and space enthusiast, received a letter from Valier early that year. In it, Valier recommended to Ley that he organize a club in order to raise money for rocket experiments. He went on to suggest that Ley contact Johannes Winkler in Breslau (now Wroclaw, Poland); Winkler was a technically adept church administrator and fellow space enthusiast who would know how to go about setting up such a venture. Ley contacted Winkler, who agreed to Valier's scheme, and on July 5, the VfR held its first meeting.[20]

The purpose of the VfR, according to its charter members, including Valier and Winkler, was to create large spacecraft "which can be ultimately developed by their pilots and sent to the stars."[21] Above all, its members earnestly desired to experiment, but in reality the VfR spent much of its time raising money. In addition to membership dues, the VfR made much of its money by organizing recruitment drives to increase membership and by selling cheap souvenirs. Its leadership was quite adept at recruiting new members. Within a year of its founding, the society counted its members in the hundreds. By late 1929, that number reached over 1,000.[22] In 1927, less than a month after its establishment, nearly twenty percent of the VfR's members were engineers.[23] No information exists on the enrollment of VfR members beyond 1927, but the number probably increased as many well-known names in rocketry, including the Frenchman Robert Esnault-Pelterie and Oberth himself, enrolled in the VfR.[24] Winkler founded and edited the society's journal, *Die Rakete*, which was the first periodical exclusively devoted to rocketry. The journal was published regularly until December 1929, when it went defunct for lack of money. Writers contributed articles on the development of different types of rockets, propulsion systems, life-support measures, and various other aspects of spaceflight. The journal served as a forum in which VfR members could exchange views about such subjects and learn of other members whose interests coincided with their own. Through *Die Rakete*, VfR members were more able to keep abreast of

[20] Willy Ley, *Rockets, Missiles, and Space Travel* (New York: Viking Press, 1961), 136–137; Winter, *Prelude to the Space Age: The Rocket Societies: 1924–1940* (Washington, DC: Smithsonian Institution Press, 1983), 35.
[21] "Verein für Raumschiffahrt, E.V.," *Die Rakete*, 1 (July 1927), 82.
[22] Winter, *Prelude to the Space Age*, 36–37.
[23] "Verein," *Die Rakete* 1 (July 1927), 83.
[24] The VfR's periodical, *Die Rakete*, regularly published the names of its most well-known members. See also Willy Ley, *Rockets*, 118. Ley's recollections of the early rocket period are sometimes faulty, but they are nonetheless one of the most important sources of information on the VfR and *Raketenflugplatz* outside of Berlin. Ley, no friend of National Socialism, was one of the only rocket engineers to flee Nazi Germany. He escaped Germany in 1936.

theoretical and technological developments in spaceflight. Despite its short existence, *Die Rakete* became a respectable professional journal with some degree of international recognition in a very short time.

The specialized professional journal was essential for the society, but it did little to popularize rocketry beyond the already dedicated core of enthusiasts in the VfR. What truly caught the German public's imagination were the daring and fantastic experiments conducted by Valier and Fritz von Opel, the crown prince of the Opel automobile empire. Ignoring Oberth's more methodical methods, the two media-savvy experimenters conducted dramatic public tests with race cars equipped with black powder rockets in April and May 1928. Their first experiments took place in Rüsselsheim at the Opel headquarters on April 11 and 12, and the second, far more stirring test took place on May 23 in front of nearly 2,000 spectators on the Avus racetrack in Berlin.[25] Newspapers lent a great deal of coverage to both events, and many amateurs and even the military began to take notice.[26]

The fulsome publicity unleashed by these stunts resulted in an outpouring of new experiments conducted by Valier, Opel, and others. These were exciting, but often ridiculous, publicity-hawking events that were of little technical benefit. Large crowds gathered to watch rocket enthusiasts test their machines on train cars, gliders, an ice sled (operated by Valier himself), and even bicycles.[27] Newsreels, print media, and radio broadcasts helped to popularize these events.[28] This was the very peak of the rocket fad in the Weimar Republic. The rocket tests even helped to reinforce the desire of film director Fritz Lang, of *Metropolis* fame, to make a new film about space travel, which he entitled *Frau im Mond* (The Woman in the Moon).[29] Universal Film Corporation (UFA) contracted Oberth to launch a rocket on the date of the film's premiere, but Oberth was unsuccessful. The state of the technology was far too primitive, and the project was much too complicated to be completed quickly. It turned out to be a fiasco, driving Oberth to

---

[25] Essers, *Max Valier*, 140–156; Neufeld, "Weimar Culture," 733–734.

[26] *Volkische Beobachter* (Munich), Bavarian edition, April 15/16; *Vorwärts* (Berlin), morning edition, April 14; *Berliner Tageblatt*, evening edition, May 23, 1928; *Berliner Morgenpost*, May 24, 1928; *Die Umschau* 32 (June 1928), 487–488.

[27] Winter, "1928–1929 Forerunners of the Shuttle: The 'Von Opel' Flights," *Spaceflight* 21 (1979); Essers, *Max Valier*, 207, 209–210.

[28] Neufeld, "Weimar Culture," 736. Both the German domestic and international press reported on these tests. See, for example, "Raketen-Flugzeug Steigt," *Berliner Morgenpost*, October 1, 1929; "Raketen-Unfug und kein Ende!," *Flugsport*, January 9, 1929; "Raketen-start für Segelflugzeuge," *Luftfahrt* 8 (April 22, 1928); *New York Herald Tribune*, "Rockets Speed Sled in Test Near Munich," January 24, 1929; *New York Times*, "Valier Tries Rocket Sled," January 24, 1929; "Flying Bikes Fitted with Wings and Rockets," *Popular Mechanics* (June 1932).

[29] Ley, *Rockets*, 105–123. See also Michael Neufeld, *The Rocket and the Reich: Peenemünde and the Coming of the Ballistic Missile Era* (Cambridge, MA: Harvard University Press, 1995), 8–9.

the brink of a nervous breakdown. The test rocket was not ready for the film's opening and exploded at the launch site well after *Frau im Mond* premiered.[30]

Other, more scientific, experiments also followed. Some of these were conducted with the financial backing of large industrial firms. In 1929, Hugo Junkers, head of the Junkers Aircraft Company, lent his support to Winkler. Junkers had a particular interest in Winkler's work, seeing in it the potential to develop rocket-assisted takeoff devices (RATO, or *Starthilfe*) for large airplanes as well as out-and-out rocket propulsion for smaller ones. Shortly afterward, Winkler gained the distinction of designing Germany's first successfully flown liquid-fueled rocket. Valier also managed to secure the backing of Paul Heylandt's liquid-oxygen manufacturing firm *A.G. für Industriegasverwertung* (Industrial Gas Utilization Company) in Berlin and attempted to develop a rocket car that used liquid fuel instead of solid propellants.[31]

It was in these experiments with liquid-fueled rocket engines at the Heylandt Works that the flamboyant Valier met his death. Heylandt was a proponent of rocketry and enthusiastically agreed when Valier first approached him about using liquid oxygen for the purposes of rocket propulsion. At the time, liquid oxygen was primarily used for welding and in hospitals because it took up less storage space than gaseous oxygen. However, from the standpoint of rocket propulsion, liquid oxygen burned much more efficiently and powerfully than black powder. It therefore had much more potential as a propellant. On May 17, 1930, a group led by the determined Valier, assisted by Walter Riedel (who would become head of the design division at Peenemünde), and a young Arthur Rudolph (the future production chief at Peenemünde and Mittelwerk), was experimenting on a kerosene/liquid-oxygen engine. As Valier made a last close inspection of the idling engine, it exploded suddenly. Rudolph recalls

I was suddenly knocked over on my back. When I looked up the engine wasn't there anymore. I only saw a big stream of oxygen. I saw Valier reeling back and forth, and I saw Riedel running up to him and catching him under the arms to steady him, and I saw Valier's lips moving and then Riedel let go and ran towards the gate house to call for help. Valier walked a few steps and fell on his face. By that time I had gotten up, and I went to Valier and turned him over. He was bleeding profusely from the mouth. He had been hit in the chest by a piece of shrapnel. There was nothing I could do. Within a minute, he was dead.[32]

[30] Winter, *Prelude*, 37; Ley, *Rockets*, 124–130. Neufeld, "Weimar Culture," 737.
[31] Neufeld, *The Rocket and the Reich*, 10–11. See also Walter Riedel, "A Chapter in Rocket History," *Journal of the British Interplanetary Society* 13 (July 1954), 209.
[32] Printed in Thomas Franklin, *An American in Exile: The Story of Arthur Rudolph* (Huntsville, AL: Christopher Kaylor, 1987), 18–19.

The explosion that killed Valier led to a short public controversy that ended with a failed attempt in the Reichstag to ban rocket experiments altogether. Heylandt shut down the effort, but Riedel and Rudolph continued their work.[33]

Weimar's voracious consumer culture had already begun abandoning the rocket fad when the accident killed Valier. Without its most public champion, the decline of the so-called *Raketenrummel* (rocket craze) was assured. Despite a lavish premiere, *Frau im Mond* appears to have found only limited box office success.[34] Oberth left Germany for a well-deserved holiday in November 1929. His failure to build a usable rocket for the film left him exhausted. The VfR ceased publishing *Die Rakete* – which was operating at loss – in order to devote more of its meager resources to experimental activities. The organization thereby cut itself off from its members and lost financial support. The brief flash of popularity for spaceflight was disappearing as quickly as it had arrived. Into this bleak situation stepped Rudolf Nebel, yet another rocket enthusiast with a unique and irrepressible personality.

The unscrupulous Nebel first made his presence known to rocket enthusiasts a year earlier, in late 1928, when the technically inexperienced Oberth was searching for engineers to help him build the rocket for the premier of *Frau im Mond*. Nebel, equal parts con man and engineer, had some engineering experience and was a persuasive salesman. The World War I fighter pilot claimed to have started thinking about rockets as weapons in 1916, when he supposedly attached powder rockets to his biplane and used them in combat. After the war, he earned an engineering degree and went to work manufacturing ball bearings for the Swedish-German firm SKF-Norma.[35] Nebel's worldview reflected the growing archconservatism of many engineers and war veterans in the Weimar era. He joined the Stahlhelm, a right-wing veterans' organization, and lent his political support to the highly conservative German National People's Party (DNVP – *Deutsche Nationale Volkspartei*), which was almost as far to the right on the political spectrum as the upstart National Socialist Party that was coming together after the war in Bavaria.[36]

33 Arthur Rudolph Oral History Interview (OHI), NASM; Neufeld, *The Rocket and the Reich*, 11; Essers, *Max Valier*, 247–265; Franklin, *An American*, 19.

34 Neufeld, "Weimar Culture," 740; Paul M. Jensen, *The Cinema of Fritz Lang* (New York: Barnes, 1969), 79–92. UFA newsletters and film magazines in the Willy Ley Collection at NASM, Box 2700, Folder 164, and Box 2701, Folder 200. See also Ley, *Rockets*, 131 and Winter, *Prelude to the Space Age*, 37. Ley was a publicist for *Frau im Mond*.

35 Rudolf Nebel, *Raketenflug zum Mond – Von der Idee zur Wirklichkeit* (Düsseldorf: privately printed, 1970), 4–6. Much of Nebel's written work is of questionable veracity, though some of his wilder assertions have been proven correct by other sources.

36 Rudolf Nebel, *Die Narren von Tegel* (Düsseldorf: Droste Verlag, 1972), 16–17.

After working several jobs in Berlin, Nebel was hired by Oberth – without so much as a single interview to determine his qualifications – to assist in the *Frau im Mond* venture. That doomed project nonetheless revealed Nebel's dedication to spaceflight and his inscrutable methods. When it became clear that they would not be able to stage a successful launch before the premiere of the film, Nebel, not willing to expose the technology to a public failure, attempted a con instead. He tried to convince Oberth that they should film a rocket being lowered from a balcony, turn the picture upside down, and tell the public that it was a rocket ascending through the air. Oberth would have none of it. Nebel, however, was nothing if not dedicated. After that disappointment, he procured the engineering equipment purchased for them by UFA, and, with the leadership of the VfR, later regrouped in Berlin.[37]

By early 1930, Germany was sliding once more into political and economic chaos. The Black Monday Wall Street collapse in 1929 ushered in a massive economic depression that had lethal results for the fragile Weimar democracy. Unemployment skyrocketed, and elections held in July 1930 led to huge parliamentary victories for the Nazi and Communist Parties. This parliamentary polarization played itself out on the streets, where rival gangs of thugs from each party engaged in wild and bloody running brawls. The right-wing cabinet governed by emergency decree, and authoritarian rule increasingly replaced parliamentary democracy under octogenarian President Paul von Hindenburg, who grew more senile by the day. Disorder and political crises persisted until 1933, when Hitler assumed the Chancellorship and the Nazi Party was finally able to clamp down its domestic rivals.

In early 1930, when this political cataclysm began unfolding, Nebel spent a great deal of time searching for funding and a secluded area in which the VfR could conduct experiments. With the help of Willy Ley, he discovered funding sources from private donors and from the government.[38] Nebel also found an empty plot of land in Reinickendorf, a suburb of Berlin, in which the VfR could conduct its experiments. After a short period of negotiations with government authorities, the VfR gained access to the grounds in September. The site came to be known as *Raketenflugplatz Berlin* (Rocket Base Berlin).[39] It would go on to become the home of the most influential rocket group of the pre-Nazi period.

The Weimar Republic owned the unused land, which stretched across nearly two square miles. It was wholly unsuitable for manufacturing or settlement and the two roads that crossed it were little more than cow

[37] Ley, *Rockets*, 124–127; Winter, *Prelude to the Space Age*, 39; Barth, *Hermann Oberth*, 139–153. Hermann Oberth Oral History Interview, NASM.
[38] Rudolf Nebel, *Narren*, 72–75. Ley, *Rockets*, 136. Neufeld, *The Rocket and the Reich*, 16–22.
[39] Ley, *Rockets*, 136–137; Winter, *Prelude to the Space Age*, 41. Neufeld, *The Rocket and the Reich*, 14.

paths. Swampy lowlands and tree-covered hills dominated the landscape. There were no telephone connections, and the few buildings there were overgrown with weeds and brush. Inside the buildings, the workstations and living quarters were tightly cramped. The initial storage area doubled as a conference room, reception area, and office space.[40] Compared to the future missile base at Peenemünde, working and living conditions were primitive at best.

Despite these limitations, the *Raketenflugplatz* leadership, which Nebel had co-opted, was able to have their facility up and running within a relatively short time. The labor force at the *Raketenflugplatz* was free and plentiful, and Nebel's astonishing ability to procure goods and raw material at no expense meant that the group's limited financial resources could be dedicated elsewhere. The crushing economic circumstances of the Great Depression in Germany assured the enthusiasts of a large, inexpensive pool of skilled labor. Electricians, draftsmen, sheet metal workers, and engineers lived at the *Raketenflugplatz* and ate for free in exchange for work.[41] They converted one of the buildings on the grounds of the site into a dormitory in which employees slept. The workers received a daily ration from a nearby soup kitchen that Nebel had convinced to support the group. Indeed, Nebel's negotiating skills were excellent, and the *Raketenflugplatz* had to pay for little during its existence. The Siemens Corporation supplemented the food from the soup kitchen with cheap meals. Shell Oil provided free gasoline, and other firms supplied nuts, bolts, paint, sheet metal, liquid oxygen, and even a motorcycle.[42] Years later, Wernher von Braun offered a typical example of Nebel's skill at procurement. "Nebel," he wrote, "once talked a Director of Siemens Halske, A.G. out of a goodly quantity of welding wire by vividly picturing the immediacy of space travel. Our own use for such wire was extremely small, but Nebel offered it to a welding shop in exchange for the labor of a skilled worker, which we badly needed. ... Machine tools, raw material, and office equipment gradually accumulated as Nebel wove his spells around those who could spare them and who were vulnerable to space travel."[43] In this way, the *Raketenflugplatz* was able to scratch out a meager, yet fruitful existence.

The *Raketenflugplatz* touted the benefits of rocket development in a variety of ways. In the first place, its members promoted the rocket's potential nonmilitary, scientific, and economic applications. Engineers at Reinickendorf saw rockets being used for mail delivery, weather research, as

---

[40] Wernher von Braun, "Reminiscences of German Rocketry," *Journal of the British Interplanetary Society* 70 (May/June 1956), 127. See also NASM File "Germany, 1920–1940" and Ley, *Rockets*, 138.

[41] Von Braun, "Reminiscences," 127.

[42] Nebel, *Narren*, 86–87; Ley, *Rockets*, 138.

[43] Von Braun, "Reminiscences," 127.

passenger planes, and of course for space travel. Indeed, the lure of space exploration was a powerful one, and a primary motive for constructing the rocket in the first place. However, *Raketenflugplatz* members were also sensitive to criticism about the reality and purpose of flying into space. In 1932, one member of the team wrote, "After the most recent successes with rocket technology, the question of whether traveling by space ship and visiting neighboring heavenly bodies is realistic has come up again. For us rocket researchers, there is no doubt that space travel is possible."[44] This same writer, however, attached a deeper meaning to their goals. "Without doubt," he wrote, "space travel will be an expensive undertaking. But shouldn't it be possible to just once ante up for a cultural act of the first rank a fraction of the sum that one truly and uselessly ground away [*verpulvert*] during the World War?"[45] Journeys into space were not merely valueless exercises demonstrating human technological know-how. They were also major *cultural* achievements that could even play a unifying role in a time of great unrest. Using a phrase that became the de facto slogan of the *Raketenflugplatz*, one public appeal for funding cried out,

Help build the spaceship! This call goes out to everyone who wants to help with a new great act of German technology. As at the beginning of aviation, interplanetary travel is created first by unselfish promotion on the part of those who see great cultural progress in the problem of space travel. ... Only if we all unite will we be witnesses to the implementation of space travel, which has as its final goal the visiting of neighboring heavenly bodies.[46]

Such sentiment revealed frustration borne of the division caused by economic dislocation and depression, rampant unemployment, and chronic political chaos in the late years of the Weimar Republic. Even more, it held out the promise of a new, grand vision to heal these problems by uniting the nation behind rocket development and "visiting neighboring heavenly bodies." The concept of space travel was not just an abstraction to be realized by using brainpower. For the members of the VfR, the pursuit of successful spaceflight provided hope for a deeply divided nation. Success was laden with cultural and national meaning. Its pursuit provided a clarion call for political unity, and its attainment would be a profound cultural statement made by the Reinickendorf enthusiasts on behalf of a unified German nation.

    Not every motive, however, was humanitarian and pacifist. Partly because the right-wing Nebel, a supporter of the DNVP, was the public face of the organization and its de facto leader, the *Raketenflugplatz* also couched the goals of its work in a much more aggressive and militant language that echoed German right-wing discussions about the value of technology. The

[44] *Raketenflug*, 2 (February 1932).
[45] Ibid.
[46] Handbill, "Helft das Raumschiff bauen!" NASM File "Germany 1930–1940."

language they used attempted to arouse support for their research by harnessing German resentment of foreign restrictions imposed on the nation in the wake of World War I. Like Valier, they expressed their work on the rocket in terms that advocated Germany's resumption of world power status. One handbill printed by the group bemoaned the restrictions placed on Germany by its foreign rivals.

For decades, German scientists and technicians have worked on the problem of the rocket. Finally, tangible results are within reach. For the continuation and expansion of our findings, we are missing that which we have the least help with – money. Foreign nations have made monstrous efforts to tear the results of our studies away from us. Hindering these efforts must lie in the heart of every German. Everyone should give according to his means so that the fruit of our decades-long labor will not escape us. Through the solution of the rocket problem, Germany, at least in an economic and cultural sense, will strike a blow for the quick reconstitution of its international standing.[47]

Written most often by Nebel, these appeals portrayed the rocket engineers as victims of an unfair, nefarious international effort to undercut Germany's national potential by limiting German technological achievement and plundering their nation of its resources. They tapped into an increasing resentment of supposed foreign interference felt by many Germans. At the same time, these statements endowed the rocket, and indeed, its creators, with the ability to rid Germany of its subjugation through an unprecedented act of technological achievement.

Nebel was a skilled propagandist with a manipulative streak and penchant for stretching the truth. In an early fundraising appeal for the VfR, probably printed in 1930, Nebel decried the tide of money flowing out of German hands and into the West. Following on the arguments made by fellow engineer and Nazi ideologue Gottfried Feder's about "interest slavery," Nebel wrote,

The German nation pays 75 gold marks per second, 4500 gold Marks per minute to its enemies! This means slavery for all eternity. Our primary duty must be to cast off these bonds of slavery. For this, we need a new weapon! Under the motto, *"Help build the Spaceship!"* preparations for this goal were made and the Verein für Raumschiffahrt was founded. Join the Verein für Raumschiffahrt![48]

---

[47] Handbill, "Raketenflug Aufruf!" Herbert Shaeffer Collection, NASM, Smithsonian Institution Photo Number 77-6008.

[48] Undated Nebel Handbill, NS 19/1795, *Bundesarchiv-Lichterfelde*. Feder's early speeches in Munich made a deep impression on Hitler, who recognized in them both their propaganda value as well as a similarity with his own developing economic ideas. See Ian Kershaw, *Hitler, 1889–1936: Hubris* (New York: Norton, 1998), 123, 138. For Feder's economic thought, see Albrecht Tyrell, "Gottfried Feder and the NSDAP," in Peter Stachura, ed., *The Shaping of the Nazi State* (New York: Barnes & Noble Books, 1978), 49–87.

Absent from Nebel's appeal are any references to using the rocket for peaceful means. Instead, the rockets would deliver national salvation from the oppressive bonds of western "slavery." No longer would Germany have to suffer from the onerous reparations payments or the crushing economic and psychological burden of national poverty. For Nebel, the small groups of rocket enthusiasts that were slowly coalescing under the leadership of the VfR in Berlin and elsewhere were not simply collections of amateur hobbyists playing with children's toys. Rather, they were daring, forward-looking, technological wizards who would lead Germany back to world prominence through their development of the world's most futuristic and advanced technology.

Indeed, Nebel was not above considering the rocket's uses as a weapon that was capable of having a dramatic impact on the nature of modern warfare. In another early pamphlet, Nebel noted a number of ways in which the rocket would alter military realities in the twentieth century. According to the engineer, liquid-fueled rockets made possible "a qualitative improvement in armaments as well as [the fighting of] a war that can be conducted with 1000 engineers in the place of an army of millions."[49] At the time, this may have been taken as a taste of the more preposterous hyberbole that Nebel was known for, but in retrospect, he was also carrying strategic thinking about deploying these weapons to another level. With no small prescience, Nebel imagined future warfare and saw rockets, among other things, deployed as anti-aircraft weapons, bombarding enemy positions, propelling fighter airplanes through the skies, and, ominously, delivering poison gas to distant foreign targets. He even argued that rockets manned by pilots would be able to deliver powerful warheads to precise targets such as munitions depots, airfields, industrial areas, fortifications, and city quarters. Distance would mean nothing. Echoing the sentiments of his colleagues in aviation, he wrote that "Long-distance rockets with gyroscopic steering [*Kreiselsteuerung*] can strike any point on the Earth's surface that one wishes. It can, for example, travel from Berlin to Paris in five minutes, to London in six minutes, to Moscow in twelve minutes, to New York in thirty minutes, and to any other point on the earth's surface in fifty minutes."[50] The self-promoting Nebel also understood, in word if not in deed, that secrecy was a potent element in maximizing the use of rockets as weapons, writing that "Disguising [*Tarnung*] and financing it as a mail rocket ensures at the same time the permanent readiness for national defense."[51]

Nebel's arguments about the military use of the rocket clearly played up their utility in both civilian and military capacities. Though Nebel was not

[49] Nebel, "Raketen-Torpedos," *Raketenflug* 14 (1927).
[50] Ibid. In an article in the *Berliner Zeitung am Mittag*, Nebel also touted the rocket's uses in air defense; see Rudolf Nebel, "Raketen Schiessen die Grenzen," *Berliner Zeitung am Mittag*, June 10, 1931.
[51] Ibid.

above saying anything to anyone to advance his cause, it is at least clear that he was quick to make use of the cover that touting the rocket's civilian uses would give to its darker and more destructive potential. Though the rocket did have clear peacetime uses, its military deployment was at least as important, if not more so, for the future of the nation. It would enable Germany not only to ward off foreign threats but also to stand off and destroy the nation's enemies with impunity. It was unimportant if numerous international treaties proscribed the use of poison gas and the bombardment of civilian targets. If employed this way, the rocket would perform perfectly. Defense of the nation in an era both of unbridled international competition and unprecedented German military weakness demanded that the nation keep these possibilities in mind.

Again, these arguments helped reconcile the progressive, modernist elements that seemed inherent to rocket technology with the more conservative discourse and militaristic demands of large and powerful segments of German society. The appeal of this most modern technology dovetailed perfectly with conservative interests when cast not in terms of its peacetime utilization, but rather its wartime capabilities. As usual, however, Nebel dramatically overstated his case. By 1931, rocket designers could barely keep a small rocket in the air for more than a few seconds, and the promise of intercontinental ballistic missiles was a pipe dream that required far more resources than the meager material that private enthusiasts could drum up during the Weimar years. Ironically, Nebel would prove to be unable and unwilling to work within the bounds set by the Reichswehr, the one institution that most clearly echoed his sentiments about the military applications of the rocket. It was also the only organization capable of offering him the kind of financial and technological support necessary to see the project through to its successful conclusion.

In any case, Nebel's rhetoric was nothing new among rocket enthusiasts or the Weimar engineering community in general. Paul Heylandt, whose support for rocketry emerged independently of the *Raketenflugplatz* and the VfR, also made claims of being able to reach distant locations in minutes. In April 1931, he demonstrated his newest rocket car for a gathering of journalists, claiming that his rocket engines could reach anywhere in Europe in twelve minutes.[52] In general the engineering profession viewed the world in distinctively conservative terms. The technical universities were hothouses of a virulent new strain of right-wing conservatism that stressed the ability of technology to lead the way to a renewal of the German spirit and nation. Like traditional German conservatives, they also rejected capitalism and democracy as modern and "soulless." For some, anti-Semitism was embedded in this antimodernist sentiment. They already blamed Jews for the "stab in the back," leading to Germany's defeat in World War I. Who better to

---

[52] *New York Herald Tribune*, "Motor Flight in Stratosphere Shown in Berlin," April 12, 1931; *New York Times*, "Sees Lightning Speed for Liquid Gas Planes," April 19, 1931.

blame for flaccid democracy, exploitative capitalism, and even Godless Communism? Engineering ideologues argued that German technology offered a solution. It was a manifestation of the innate creativity that sprang from the German soul and allowed for the reproduction of an organic, uniquely German *Kultur*. Invention and the production of new technology was actually a rejection of Weimar modernity. Instead, it was a different kind of modernity, one that supported reasoned, rational thinking alongside the irrationality of innate spiritual connections among Germans. The invention of technology was something more like an expression and restoration of German greatness. The public rhetoric of the *Raketenflugplatz*, therefore, was a tributary of this larger ideological stream.[53]

It was during the *Raketenflugplatz* years that the young, brilliant engineer Wernher von Braun, one of the key figures in rocketry in the twentieth century, made his first foray into the field. Von Braun's fairly wealthy family descended from Prussian Junker ancestry. His father, Magnus von Braun, was a high-ranking civil servant in the fledgling Weimar government, but his purported association with the Kapp putschists – an extreme right-wing group that attempted a failed coup d'etat in 1920 – forced him out of office. The elder von Braun then went into banking and maintained his close ties with future President von Hindenburg and the old reactionary elites of the former Kaiserreich. In the midst of the political turbulence of 1932, Chancellor Franz von Papen made Wernher's father the Minister of Agriculture in a reactionary cabinet of traditional elites just before Hitler came to power. After Hitler was appointed Chancellor, von Braun did not become part of Hitler's cabinet, but believed, as many conservatives did, that Hitler's movement could be harnessed and controlled. With this parental background, Wernher von Braun was in all likelihood reflexively nationalistic, but not necessarily sympathetic to the Nazi cause.[54] In any case, the nationalist histrionics of the *Raketenflugplatz's* advertising campaign posed no problem for the young engineer. Von Braun was fascinated with the lure of space travel and joined the group for this reason. If anything, the idea of Germany riding to national glory on the thrust of a rocket only made it a little easier for him to join.

The younger von Braun was a rocketry enthusiast whose interest in the technology began after he read Oberth in 1926 and was piqued by the

---

[53] For the nationalism of engineers generally, see especially Karl-Heinz Ludwig, *Technik und Ingenieure im Dritten Reich* (Düsseldorf: Droste Verlag, 1974) and Herf, *Reactionary Modernism*.

[54] Neufeld, *The Rocket and the Reich*, 13; Rainer Eisfeld, *Mondsuchtig: Wernher von Braun und die Gebürt der Raumfahrt aus dem Geist Barberei* (Reinbek bei Hamburg: Rowohlt Taschenbuch Verlag, 2000), 41–42; Magnus von Braun, *Wege durch vier Zeitepochen: Vom ostpreussischen Gutsleben der Väter bis zur Weltraumforschung des Sohnes* (Limburg an der Lahn: Starke, 1964), 234, 263. See also Neufeld's new biography of the younger von Braun, *Von Braun: Dreamer of Space, Engineer of War* (New York: Knopf, 2007).

experiments of Valier and Opel, as well as Fritz Lang's *Frau im Mond*.[55] Von Braun was a handsome eighteen-year-old prodigy who was about to begin university studies at the Technical University of Berlin when he came to the *Raketenflugplatz* for the first time. He was no ordinary first-year student, though. After he met von Braun, Walter Dornberger, the Army captain who would be von Braun's closest ally at Peenemünde, was struck by his energy and theoretical knowledge at such a young age. Von Braun seemed to clearly understand the problems inherent to developing a liquid-fueled rocket, and his ability to systematically dissect these problems far exceeded his age and position at Reinickendorf. For Dornberger, "In this respect, he had been a refreshing change from most of the leading men at the place."[56]

The rocket enthusiasts that von Braun joined in 1930 were a tightly knit group who plied a dangerous trade. Except for Nebel himself, individuals at Reinickendorf rarely took individual credit for design changes or major advances. Willy Ley recalled that "We never paid any attention to the question of who had thought of what, knowing that it was a long way from our experiments to definite shapes, and knowing also that our glory was a collective glory."[57] The group of regulars at the *Raketenflugplatz* was small, so everyone knew each other relatively well and managed to get along with each other. Many of them ate dinner together and grew to be friends, spending time with each other away from their work. When they conducted particularly successful experiments, they followed them up with the time-honored tradition of a long night of celebratory drinking in a local pub.[58]

The *Raketenflugplatz* was an island in a stormy political time, but the Reinickendorfers did not ignore the politics swirling around them. According to Rolf Engel, a *Raketenflugplatz* engineer and one of von Braun's closer friends, political allegiances among the group were divided evenly between Communism and National Socialism, but the differences never affected the group's work. Recalling this period years later, Engel wrote that "The emotional connection to the technical problems of rocketry and space travel were so strong that political loyalties never broke them."[59] In short, the dynamics of the small community of engineers at Reinickendorf and their

55 Von Braun, "Reminiscences," 125. When he was fifteen, von Braun met Valier, who, ironically, upbraided the young Prussian for conducting experiments without proper safety precautions.

56 Walter Dornberger, *V-2* (New York: Viking Press, 1954), 27.

57 Ley, *Rockets*, 142.

58 Wernher von Braun, "Behind the Scenes of Rocket Development in Germany, 1928 through 1945," in Wernher von Braun Papers (date unclear, late 1940s), 6–7, Space and Rocket Center (SRCH), Huntsville, AL. See also Neufeld, *The Rocket and the Reich*, 14–15.

59 Heinz Horeis, *Rolf Engel – Raketenbauer der ersten Stunde* (Munich: Lehrstuhl für Raumfahrttechnik, 1992), 24. Written by Horeis with help from Engel, this semiautobiographical book recounting Engels's time as a rocket engineer is deeply self-serving and must be treated with great care.

FIGURE 1. Three rocket enthusiasts confer in the cramped office of the *Raketenflug-platz* in Reinickendorf in November 1932. The two assistants were on loan from a local company. [National Air and Space Museum, Smithsonian Institution (SI 77-4214)]

passion for rocket development forged an extraordinary personal dedication to the success of the overall endeavor. One visitor to the *Raketenflugplatz* wrote, "The impression you took away with you was the frenzied devotion of Nebel's men to their work . . . they belonged exclusively to a world dominated by one single wholehearted idea."[60] The bonds fashioned by the members of the *Raketenflugplatz* would become an important factor in keeping the most skilled men together during the transition from privately funded, shoestring-budget rocket experiments to the "big time" of government sponsored research at Peenemünde.

In practice, the *Raketenflugplatz* resembled a craftsmen's workshop more than an advanced rocket experiment station. The group's work bore few markers of a systematic scientific and technological approach. The improvements they made on the rockets tested at Reinickendorf were almost always ad hoc, and informal meetings between three to six people could result in

[60] Dimitri Marionoff, with Palma Wayne, *Einstein, An Intimate Study of a Great Man* (New York: Doubleday, 1944), 115. Marionoff was Einstein's son-in-law who introduced Nebel to the physicist in 1932. In the weeks after this first meeting, Nebel tried unsuccessfully to solicit financial support from Einstein, who rightly regarded Nebel's overtures with suspicion. Nebel, *Narren*, 114, 121–123.

major but spur-of-the-moment design changes. The established experimental strategy was trial and error by brute force, and tinkering was accepted as a given. Technology itself was a source of binding energy for the engineers. The instruments developed at the *Raketenflugplatz* were some of the first, albeit modest, steps toward the large, liquid-fueled rocket. The primary test instrument was their so-called *Minimum Rakete*, or *Mirak*. It was about fifteen feet long and pulled along by the engine in its nose. The experimenters gave virtually no thought to guidance or steering, so it was nearly impossible to achieve controlled flight. Stability was not foremost on the minds of the men in Reinickendorf. They directed the bulk of their efforts to creating a rocket that simply worked semiconsistently and could achieve enough thrust for liftoff. The secondary status given to stability created less than favorable safety conditions.[61]

The experiments conducted by the Reinickendorfers were quite dangerous. Given their trial-and-error approach, they took huge risks even in static engine tests. Improper materials and poor engine assembly were the primary dangers. Weak welds at the seams of the engine could blow apart under the intense pressure generated by the combustion in the engine. Shoddy assembly of even the smallest components lent the same danger. The scorching heat of combustion might also burn through the metal used to build the engine and cause an instant explosion, spraying metal fragments in every direction. This work required no small amount of courage. The best the experimenters could do was to trust that their colleagues carried out their design and assembly tasks to the best of their ability. The personalized nature of the work at least attenuated assembly problems by ensuring that the engineers and technicians had a vested interest in the successful experiments. Still, participants in the experiments knew that they assumed huge risks; explosions and failures inevitably occurred.

This could happen for any number of reasons. Improperly mixing the fuels, usually liquid oxygen and gasoline, led to violent explosions that destroyed entire test stands. After one notably terrifying explosion, Ley kept a piece of shrapnel that he found embedded in the handle of a shovel as a reminder of the dangers they faced. Inadequate cooling systems resulted in the walls of the engine's combustion chamber superheating, melting through, and exploding. This was a problem that plagued the V-2 designers through 1942. On one of these occasions, Ley recalled that they "ducked quickly and with great disregard for curiosity." Even when the engines worked, the danger was not over. The experimental rockets had no guidance system and were not designed with high-speed aerodynamics in mind. Flying rockets went astray, buzzing the engineers at very low altitude or crashing near their own test site.[62] The Reinickendorfers truly were taking great risks in their work.

---

[61] Nebel, *Narren*, 99–116; Winter, *Prelude to the Space Age*, 41–43; Ley, *Rockets*, 140–154.
[62] Ibid., 143. Neufeld, *The Rocket and the Reich*, 73–109.

Despite the danger, the experiments were obviously important from a technical point of view, but they were also key cultural events that were fundamental to the dynamic environment of the *Raketenflugplatz*. Thrown together as they were in ramshackle buildings, conducting risky experiments on untried technology, the rocket enthusiasts mastered new challenges, fulfilled personal ambitions, and experienced the drama of bringing an entirely new technology into existence. This bonding experience fostered a sense of community in a competitive world, and in its own way it helped ease the myriad of anxieties in the Weimar Republic's harsh political, social, and economic climate. Seen this way, the group's everyday, very personalized work and the process of testing and experimentation did not just lead to improvements in rocket technology. They also enhanced the active identification of the engineers with each other and with the project. The result was a small corps of engineers and technicians that were intensely dedicated to the complex and hazardous job at hand.[63]

Even so, their small number and lack of resources imposed difficult limitations on their work. At this early stage, the idea of just getting a rocket to fly consumed the *Raketenflugplatz* members, and they dedicated the bulk of their time to designing a workable propulsion system. They did not have the resources to tackle guidance and steering issues at the same time, so they set these problems aside until they achieved consistently satisfactory engine performance. With this approach, accidents because of unstable, unguided rockets were unavoidable, and one incident led to serious curtailing of their experiments. At the end of 1931, a rocket flew off course and crashed outside the grounds of the *Raketenflugplatz*. At least two rockets had crashed before this one, but the third damaged a barracks belonging to the local police force. No one was injured, but the damage was finally enough for the angry police to descend on the launch site. After a stretch of negotiations that lasted several days, the police imposed a number of restrictions on the tests. The Reinickendorfers could fill the test rockets with only a minimum (five kilograms) of fuel, and the new engines had to undergo three successful static tests before they could be launched. The police ordered the experimenters to inform them before every launch, and they permitted launch tests only Mondays through Fridays from 7:00 in the morning to 3:00 in the afternoon. Finally, the engineers were not allowed to launch rockets on windy days for fear the wind would blow the rocket off course.[64] These

---

[63] See S.F. Moore and Barbara Myerhoff, eds., *Secular Ritual* (Assen: Van Gorkum, 1977). Moore and Myerhoff point out that ritual analysis can be meaningfully used when studying events that are not necessarily sacred or religious. See also Hugh Gusterson, "The Rituals of Science: Comment of Abir-Am," *Social Epistemology* 6/4 (1992) and *Nuclear Rites: A Weapons Laboratory at the End of the Cold War* (Berkeley, CA: University of California Press, 1996).

[64] Ley, *Rockets*, 148–152. See also Grzescinsky to Nebel, October 17, 1931, in NASM File "Germany, 1930–1934."

regulations effectively capped the size of any rocket they attempted to build. They also seriously restricted the number of actual launch tests that could be run, drastically slowing research in guidance and control, the area in which it was most needed. For the men of the VfR, who dreamed of massive rockets capable of delivering large payloads across the continents, this was a most difficult arrangement.

The *Raketenflugplatz* was unquestionably an important institution in the earliest days of German rocketry. Nevertheless, in a professional scientific and engineering sense, their operating methods were completely inadequate, given the demands of the technology. Though the Reinickendorfers were a dedicated group that worked hard and experimented often, their approach was amateurish. They consistently failed to keep important data measurements such as pressure distribution, fuel flow, and exhaust speed, and they rarely recorded the results of their launch experiments. The result was an ad hoc, trial-and-error approach to a technology that demanded advanced theoretical and scientific testing and regularized, systematic data keeping. Resource procurement and dedication was always inconsistent, resulting once again in an inability to systematically work through the complex development issues. Moreover, even though the engineers and technicians got to practice their skills, they did not make any wages or a salary at the *Raketenflugplatz*, earning only room and board for their work. Though the individuals at the *Raketenflugplatz* made some technical progress and earned valuable experience while there, in the late 1920s and early 1930s, the most well-known, aggressively experimenting rocket outfit in Germany was limited by a lack of resources, restrained by police regulations, and held back by its own amateurish approach to experimentation. That it was led by Nebel, an ethically suspect con man who made as many enemies as friends, certainly could not have helped matters.

This was not necessarily the case everywhere. At the same time that Nebel was pulling together the VfR at the *Raketenflugplatz*, other groups conducted important research on liquid-fueled rockets elsewhere with much less fanfare and self-promotion. One of the few experimental teams that received limited corporate support was the small group led by Max Valier, which was financially supported by the industrialist Paul Heylandt. On the staff of this group and serving as assistants to Valier were two figures who would go on to become very important in the future development of the V-2: Arthur Rudolph and Walter "Papa" Riedel.

Born in 1902 in Königswusterhausen, just outside of Berlin, Riedel was the son of a locomotive engineer and a homemaker. From 1921 to 1928, he worked as a civil engineering technician for two different construction firms. In December 1928, the Heylandt Works hired Riedel to work as a research engineer. After Valier's death in 1930, Heylandt reassigned Riedel to other tasks within the firm. Riedel never lost his job during the Great Depression and was at least spared the worst of its upheavals through the early 1930s.

He demonstrated his rightist political inclinations in March 1933, voting for the Nazis in the elections that solidified Hitler's grip on power shortly after he became Chancellor. Riedel would go on to join the party in 1937, when party enrollment was reopened after a four-year hiatus.[65]

Riedel's coworker was Arthur Rudolph, born November 9, 1906 in Stepfershausen. Like von Braun, Rudolph was also swept up by the space-flight fad that washed across Germany in the 1920s. He was fascinated by Valier and Opel's stunts on the Avus, read many articles on rockets and spaceflight, and saw the film *Frau im Mond*.[66] In 1930, he graduated from the factory technical school in Berlin with a major in mechanical engineering.[67] Serendipitously, Heylandt hired him to work as a draftsman a few weeks after he graduated in the spring of 1930. In this capacity, Rudolph met Valier, with whom he worked as an assistant.[68] After Valier's death, Rudolph continued to work on rocket engines against Heylandt's express orders. He successfully redesigned the fuel-injection system of the engine model that had earlier malfunctioned and led to the accident that killed Valier in 1930.[69]

In 1931, Rudolph joined the Nazi Party and the SA (*Sturmabteilung*, or Stormtroopers), the Nazis' brown-shirted thugs known for their violent street brawls.[70] Though there is no evidence of his participation in any street fights, Rudolph did participate in rallies in which he performed the expected duties, carrying a banner and singing the Horst Wessel Song, the unofficial anthem of the Nazi Party.[71] According to postwar interviews, Rudolph joined the Nazis because, like many in the German middle and lower-middle class, he feared a Communist revolt. A year earlier, in 1930, the severe economic crisis of the Great Depression created an army of unemployed, and the Communist Party (KPD) capitalized on this development with large political gains that were surpassed only by the Nazis. According to Rudolph, one of his coworkers convinced him that only the Nazis were

---

[65] Walter Riedel Dossier, Box 371, Entry 1B, RG 319, Records of the Army Staff, National Archives and Records Administration (NARA).

[66] Arthur Rudolph OHI, NASM.

[67] OSI interrogation of Arthur Rudolph, October 13, 1982, printed in Thomas Franklin, *An American*, 189–190. Because the institution Rudolph attended was a small technical school, not a larger technical university, Rudolph graduated with a certificate in mechanical engineering and might be professionally described as a technician (*Techniker*), rather than one of the academically trained engineers (*Diplom-Ingenieure*) produced by the technical universities. In German parlance, however, both degrees qualified one to be called "Ingenieur." See Konrad Jarausch, *The Unfree Professions: German Lawyers, Teachers, and Engineers, 1900–1950* (New York: Oxford University Press, 1990) and Ludwig, *Technik und Ingenieure*, for a nuanced discussion of the differences.

[68] OSI interrogation of Arthur Rudolph, October 13, 1982 in Franklin, *An American*, 191.

[69] Arthur Rudolph OHI, NASM.

[70] Arthur Rudolph Dossier, Box 636, Entry 1B, RG 319, Records of the Army Staff, NARA.

[71] Rudolph OSI interrogation, Franklin, *An American*, 283.

capable of meeting the needs of the unemployed while beating back the communist threat.[72] Berlin, where Rudolph lived and worked, was a center of KPD activity in the 1920s and 1930s. In addition to being confronted with catastrophic unemployment, Rudolph was exposed daily to the rhetoric of both parties, and his aspiring middle-class sensibilities precluded him from lending his support to the KPD.[73] Widespread Nazi propaganda efforts to gain the support of technicians and engineers certainly played their part in garnering Rudolph's support as well.[74] Rudolph's enrollment in the party and SA a full two years before the Nazi accession to power indicates that, for whatever reason, he did indeed support specific planks in the party platform and was ideologically predisposed to at least some of the goals of the National Socialist project.

Rudolph was also not above working with the Nazi Party or the Army when it suited his interests to do so. In 1932, Heylandt was forced to fire Rudolph because the Depression was ravaging *Industriegasverwertung*. Rudolph and Alfons Pietsch, his foreman from the Heylandt Works who was also fired, were determined to continue their rocket work. In the spring, they went hat in hand to the local head of the Berlin SA for financial support. The SA expressed interest in sponsoring the two rocket specialists but had no money to offer them, and they were forced to look elsewhere. Rudolph and Pietsch then unsuccessfully attempted to secure the backing of the Kaiser Wilhelm Gesellschaft, which was the leading state-sponsored scientific foundation in Germany. They also applied for assistance to various industrial interests without success.[75]

Despite their failure to garner the support of the party, the scientific community, and big business, the two men kept up their efforts. In the spring of 1933, after the Nazis came to power, they applied to the Army for support, which they received in the form of a contract to build a new engine. Pietsch squandered the money and disappeared, leaving Rudolph to explain why they had no money left and only a half-completed rocket engine. Walter

---

[72] Arthur Rudolph OSI interrogation, Franklin, *An American*, 192–196. Arthur Rudolph OHI, NASM, 26.

[73] The best English-language examination of the KPD is Erich Weitz, *Creating German Communism: From Popular Protest to Socialist State* (Princeton, NJ: Princeton University Press, 1997). See also Klaus Michael Mallmann, "Milieu, Radikalismus und lokale Gesellschaft: Zur Sozialgechichtes des Kommunismus in der Weimarer Republik," *Geschichte und Gesellschaft* 21/1 (1995), who makes an excellent case for the importance of locality in shaping Communist demands and protest.

[74] For the shape and appeal of this propaganda, see Herf, *Reactionary Modernism*, Ludwig, *Technik und Ingenieure*, and Gerd Hortleder, *Das Gesellschaftsbild des Ingenieurs: Zum politischen Verhalten der technischen Intelligenz in Deutschland* (Frankfurt: Suhrkamp, 1970).

[75] Arthur Rudolph OHI, NASM, 15–16. OSI Interrogation of Arthur Rudolph, October 13, 1982, in Franklin, *An American*, 198–199. There is no record of Pietsch's political background.

Dornberger, the fast-rising artillery officer in the Army Ordnance branch who was so impressed by von Braun the previous year, allowed Rudolph an extra three hundred Reichsmarks to finish the work. Rudolph received neither salary nor expenses from the Army and was forced to live off of the paltry unemployment insurance offered by the State. Nevertheless, he did manage to finish the engine and successfully test it in front of his Army benefactors in August 1934. Impressed by Rudolph's work, Dornberger hired the hungry, impoverished engineer to come work on missile development at its Kummersdorf proving grounds. One of the stipulations of his employment was that he leave the SA, but he was permitted to remain a member of the Nazi Party.[76]

THE ARMY ORDNANCE OFFICE AND LIQUID-FUELED ROCKETRY

The Army's enlistment of Rudolph's talents was part of a larger effort by the Ordnance branch to develop missile technology. Lieutenant Colonel Karl Becker, the talented head of the ballistics and munitions section of Ordnance, spearheaded the effort. Becker, who held a doctorate in engineering from the Technical University of Berlin, first took an interest in rocketry in 1929.[77] His interest was kindled by the popularity of amateur rocketry during the second half of the 1920s as well as the Reichswehr's secret rearmament projects in the later years of the Weimar Republic.

Military rocket development in Germany took place in the larger context of German rearmament in the 1920s and 1930s. After World War I, the Treaty of Versailles set strict limits on the size of the German military. It abolished Germany's air force entirely and restricted the country to a modest army and even smaller navy. In the mid-1920s, the Weimar Republic experienced a brief period of foreign and domestic stability. Guided by Prime Minister Gustav Stresemann, Germany began a diplomatic effort aimed at repairing its international reputation, concluding the Treaty of Locarno in 1925, which recognized Germany's western borders as determined by the Treaty of Versailles, and entering the League of Nations in 1926. Shackled with their own problems and seeing these moves as a moderation of German militarism, British and French supervision over Germany's military establishment loosened considerably. Spying an opportunity, Reichswehr Minister Wilhelm Gröner, a former general in the Kaiser's army, surreptitiously

---

[76] To be sure that Rudolph kept up his part of the bargain, a nonplussed Dornberger sent his deputy, Leo Zanssen, to check on the progress of Rudolph's work every week. Arthur Rudolph OHI, NASM. Rudolph File, Berlin Document Center, RG 238, Captured German and Related Records, National Archives.

[77] Michael Neufeld, "The Reichswehr, the Rocket, and the Versailles Treaty: A Popular Myth Reexamined," *Journal of the British Interplanetary Society* 53 (2000), 163.

began a rearmament program in the mid-1920s, secretly embarking on a weapons development and training regimen in the Soviet Union and illegally stockpiling arms. The Weimar cabinets in the late 1920s ensured that the Reichswehr would have a steady financial base from which it could increase its strength, and it slowly but surely expanded its arsenal of heavy weapons, including airplanes, tanks, and artillery pieces. The Treaty of Versailles never explicitly identified rockets as a restricted weapon, but that probably mattered to Becker only tangentially. The Reichswehr occupied itself with the legality of its rearmament program only insofar as it needed to be done in secret. Its rocketry champions were more concerned with building a radical new weapon that gave them a decisive strategic advantage over their enemies.[78]

The Weimar rocket fad no doubt drew Becker's interest and curiosity, but he saw the technology as far more than a popular distraction dreamt up to amuse Germans by shooting suicidal bicyclists down a racetrack. In 1929, he commissioned a study on the state of rocket technology. He placed Captain Ernst Ritter von Horstig in charge of the study. The results of his survey were discouraging. Although they noted periodic successes by some amateurs, very little sustained work on liquid-fueled rockets had been done outside of Riedel and Rudolph's endeavors at the Heylandt Works and Oberth's failed attempts at a stunt rocket for *Frau im Mond*. (Becker had, in fact, quietly supported the latter project, discreetly donating 5,000 Reichsmarks from the Reichswehr's coffers before Nebel drew his ire for publicly discussing the Army's support. Later, Becker was furious with Nebel and refused to support him further.) Industry and technical universities had little interest in developing rocket propulsion, and the *Raketenflugplatz* had not begun to coalesce in any meaningful way when von Horstig revealed the results of his study to Becker. Even when the Reinickendorfers began serious experimentation, they did not keep detailed records of their work, an issue that Dornberger would later point to as a reason for friction between the Army and the amateur group.[79]

Becker's interest in the rocket's potential was too great to give up just because of the underdeveloped state of the art. In a December 1930 meeting with several important Reichswehr officers in charge of rearmament, including General Alfred von Vollard-Bockelberg, the chief of Army Ordnance, and Colonel Erich Karlewski, the head of the Ordnance Testing branch, Becker pitched the idea of developing a long-range military rocket. Becker

---

[78] Hans Gatzke, *Stresemann and the Rearmament of Germany* (New York: Norton, 1969); Wilhelm Deist, "Die Aufrüstung der Wehrmacht," in Militärgeschichtliches Forschungsamt, ed., *Das Deutsche Reich und der Zweite Weltkrieg* (Stuttgart: Deutsche Verlags-Anstalt, 1979), 379–392; Edward Bennet, *German Rearmament and the West, 1932–1933* (Princeton, NJ: Princeton University Press, 1979), 36–38; Neufeld, "The Reichswehr," 164.

[79] Winter, *Prelude to the Space Age*, 41–44, 51; Dornberger, *V-2*, 19–20.

surveyed its military applications, noting its potential as a longer-range and more devastating substitute for heavy artillery, still able to serve as a delivery system for poison gas. He also noted that its surprise employment would give Germany an immense tactical and strategic edge over its enemies. Becker was honest about its primitive development status, and he repeated von Horstig's sobering assessment of the current state of the technology. A great deal of work still lay in front of them.

Becker probably did not have to do much convincing. The Reichswehr officer corps was generally predisposed to progressive tactical and operational thinking and conservative political thinking, a situation that bode well for rocketry proponents. The Ordnance staff's interest in the potential of rocket technology was certainly more sober than the breathless enthusiasm of the amateur rocketeers, but it was no less militantly nationalist. Colonel Karlewski, the Testing branch chief, endorsed Becker's points. His own argument for the technology bore an eerie resemblance to the amateurs' nationalist rhetoric. He told the gathering that "Along with remote guidance, infrared and ultraviolet rays, etc., [the rocket] belongs to the areas from which one day the revolutionary new invention may emerge that Germany has been waiting for in order to achieve rapid liberation. We must stick to our oars in these questions in order to possibly overtake the other powers. If we do not do something in this regard, or do not do it quickly enough, someone else may one day surprise us with the new weapon."[80] For Karlewski, Germany's weakness lay in the fact that it was subject to foreign oppression and control, its proud military unbearably weak compared with that of its rivals. The rocket offered a way out of this bind. By successfully exploiting its military potential, Germany could resume its rightful place among the chief powers in the world. Karlewski also revealed the Army's persistent worry that another nation might beat them to the punch on rocket development. This concern about foreign military would become a familiar argument to justify V-2 development and production years later at Peenemünde. German military officials were convinced that rocket development in other nations, especially the United States, either equaled or surpassed their own. In any case, the clear message to Becker was to proceed, albeit cautiously.

In 1930 and 1931, German amateur rocket groups were still the key to the development of the liquid-fueled rocket as a weapon, despite the Army's endorsement of the technology. In 1931, Becker and his assistants devoted a great deal of their time and effort to searching for a qualified amateur group with whom to cooperate – while studiously avoiding the self-promoting Nebel and the *Raketenflugplatz*. Despite some intermittent contact with Heylandt, their search came to very little. Once more, Nebel stepped in, convincing Becker to revive their failed relationship by

---

[80] "Sitzungsbericht vom 17.12.1930 über die Raktenfrage," M.I. 14/820 (V), 1, 26–27, Imperial War Museum. Thanks to Michael Neufeld for showing me this important document.

promoting the *Raketenflugplatz's* success with liquid-fueled rockets since 1930.[81]

In April 1932, Becker, intrigued by Nebel's supposed success, wrote to Nebel and asked him to demonstrate their new rocket at the Army's proving ground at Kummersdorf outside of Berlin.[82] The test took place two months later, in June. Several important officials from Ordnance, including Becker, Walter Dornberger, and Dr. Erich Schumann, a physics professor at the University of Berlin who would go on to become a prominent figure in the formulation of Nazi science policy, attended the demonstration. In order to keep the Army's interest in rocketry secret, Becker ordered Nebel, Klaus Riedel, and von Braun to report with their rocket to Kummersdorf at 4:00 in the morning. In an attempt to keep the launch secret, but to the chagrin of a number of his colleagues, Nebel did not even go to the trouble of informing the *Raketenflugplatz* board of directors that they would be conducting this demonstration – a sign of growing dissension and frustration with Nebel among the experimenters at Reinickendorf.[83]

The launch demonstration was an utter failure. The rocket rose to a height of less than half a mile and crashed only a mile away. Almost immediately, Ordnance reiterated its dislike for Nebel. Its report on the launch stated that in addition to Nebel's clear inability to conduct work in secret, "the conclusion must be reached that, because he makes assertions against his better judgement, closer cooperation with Nebel is out of the question, even though he was able to produce a liquid-fueled rocket with an engine that worked well for a duration of several seconds."[84] Ordnance once again severed its relationship with Nebel in the middle of 1932 and shortly afterwards changed its focus from farming out rocket work to developing its own liquid-fueled rocket program.[85]

Besides the failed test, there was another important reason that Army Ordnance distrusted Nebel and decided to develop its own liquid-fueled rocket. Becker, who already thought Nebel a dubious character, despised the endless publicity seeking and the unsystematic, undocumented approach of the *Raketenflugplatz*. According to Dornberger, "We wanted to have done

[81] Nebel, *Narren*, 72–75; Neufeld, *The Rocket and the Reich*, 5–23.

[82] Nebel, *Narren*, 133–135; Wernher von Braun, "Behind the Scenes," 8.

[83] Wernher von Braun, "Behind the Scenes," 8–9; Nebel, *Narren*, 135–137; Hans Ebert and Hermann Rupieper, "Technische Wissenschaft und nationalsozialistische Rüstungspolitik: Die Wehrtechnische Fakultät der TH Berlin, 1933–1945," in Reinhard Rürup, ed., *Wissenschaft und Gesellschaft: Beiträge zur Geschichte der Technischen Universität Berlin, 1879–1979* (New York: Springer, 1979), 469–481. For Schumann's role in the formation of National Socialist science policy, see also Alan Beyerchen, *Scientists Under Hitler: Politics and the Physics Community in the Third Reich* (New Haven, CT: Yale University Press, 1977); Ley, *Rockets*, 155–156.

[84] Ley, *Rockets*, 155–156; Schneider Report, June 23, 1932, M.I. 14/801 (V), Imperial War Museum.

[85] Ordnance had been pursuing solid-fuel rocket development on its own since 1930. See Neufeld, *The Rocket and the Reich*, 16–17.

once and for all with theory, unproved claims, and boastful fantasy, and to arrive at conclusions based on a sound scientific foundation."[86] Nebel's penchant for exaggerated salesmanship and his group's nonexistent record keeping subverted both the Army's attempts at secrecy and any attempt to systematically assess the state of the technology and direction of development. Aside from the desire to keep Germany's rearmament program in general secret, the rocket's capacity for shock and surprise was essential to its deployment as a weapon. Ordnance wanted to be able to deliver the rocket unannounced, so as to terrify Germany's enemies into submission. If Nebel were to be involved in rocket development for the Army, his grandstanding would have made secrecy considerations impossible to maintain, thereby exposing Germany's rearmament and, in Ordnance's eyes, lessening the rocket's military effectiveness.[87] This episode marked the first time that secrecy began to play a determining role in German rocket engineers' professional development. Those engineers who were able to adjust to this new dynamic in their work would flourish within the confines of the Army rocket program. Those who could not adjust, like Nebel, were marginalized by the Army and virtually ignored by their colleagues.[88] The practice of secrecy would go on to become a major factor in the reproduction of the engineers' cultural lives at Peenemünde.

Nevertheless, Ordnance's difficult relationship with the *Raketenflugplatz* did have one positive result. Nebel had introduced von Braun to Becker and Dornberger, who were immediately impressed by the young engineer's intelligence and energy.[89] Several months later, the Army offered von Braun the chance to carry out his doctoral research on rocket development at Kummersdorf, which he accepted. Von Braun actually finished only part of his mechanical engineering program before becoming a doctoral candidate under the ubiquitous Schumann at the University of Berlin. His work began in earnest in December 1932, when he began research for his dissertation, "Constructive, Theoretical, and Experimental Contributions to the Problem of the Liquid Fueled Rocket," while working for the Army at Kummersdorf.[90] Von Braun was not yet an Army employee, but he received a decent stipend of 300 marks per month to work for the military.[91]

[86] Dornberger, *V-2*, 20.
[87] Walter Dornberger, "Denkschrift," c. 1943, Fort Eustis (FE) Files, FE 496, NASM.
[88] For Nebel's marginalization by the Army, see below as well as Michael Neufeld, "The Excluded: Hermann Oberth and Rudolf Nebel in the Third Reich," *Quest* 5/4 (1996).
[89] Dornberger, *V-2*, 26–27.
[90] Dornberger, *V-2*, 27; "Werdegang des Professors von Braun," FE 341, NASM. Wernher von Braun, "Konstruktive, theoretische, und experimentelle Beiträge zu dem Problem der Flüssigkeitsrakete," doctoral dissertation, University of Berlin, 1934, reprinted in *Raketentechnik und Raumfahrtforschung*, Sonderheft 1, 1960.
[91] Wernher von Braun Ordnance Contract, April 4, 1933, Wernher von Braun Papers, SRCH.

Nevertheless, as Michael Neufeld points out, "When von Braun began to work at Kummersdorf, Ordnance's own liquid-fuel rocket program can fairly be said to have begun."[92]

Walter Dornberger was von Braun's immediate supervisor and contact with Army Ordnance. Nicknamed Seppl by his close friends (including von Braun), Dornberger was born in the town of Giessen in Hesse on September 6, 1895. In 1926, the Army captain enrolled in the engineering program at the Technical University of Berlin.[93] Dornberger was part of the "study officer" program initiated by Becker, who, as an engineer himself, was concerned about the antitechnological assumptions of the old-line officer corps. This program allowed selected officers to gain valuable engineering training at the Technical University of Berlin.[94] In 1929, Dornberger assisted with von Horstig's survey of rocket development. He completed his Diploma-Engineer studies in 1930 and focused much of his work on developing solid-fuel rockets for battlefield use. A few years later, in 1934, Becker arranged to have him awarded with an honorary doctorate in engineering for his work.[95] Dornberger was perfect for the job at Kummersdorf. He was a smart, dynamic, and politically astute officer who would later provide the twenty-year-old von Braun with a healthy dose of friendly, fatherly advice during his career in the Third Reich. Possessed of a sharp technical mind of his own, he developed an excellent relationship with his young protégé, and the two would make a formidable pair of bureaucratic infighters – a skill that would become absolutely essential during the Nazi period.

Kummersdorf Proving Ground was located about seventeen miles south of Berlin. It had a test stand for powder rockets already in place, but Ordnance built two new work buildings and a new test stand for liquid-fueled engines. These facilities were a major improvement over the third-rate setup at Reinickendorf. The new test stand, completed in December 1932, was made up of three concrete walls that were twelve feet high and eighteen feet long. Large metal doors completed the enclosure, which was covered by a retractable roof. Built into one of the concrete walls was an observation room that housed the testing crew as well as instruments used to measure flow rates, pressure, temperature, thrust, and other critical components of the test process. Large tanks built onto the walls automatically pumped

---

[92] Neufeld, *The Rocket and the Reich*, 23. Von Braun's role at Kummersdorf is examined in greater detail in Chapter 2.

[93] Walter Dornberger IRR Dossier, Box 371, RG 319, Records of the Army Staff, NARA. Dornberger's penchant for wearing leather Sepplhosen when off duty earned him the nickname. Dieter Huzel, *From Peenemünde to Canaveral* (Englewood Cliffs, NJ: Prentice-Hall, 1962), 72.

[94] W. Phillips, "Karl Becker," Obituary, *Zeitschrift des Vereines deutscher Ingenieure* 84 (May 4, 1940), 293–294. Ebert and Rupieper, "Technische Wissenschaft," 469–480.

[95] Dornberger IRR Dossier, NARA.

liquid oxygen and alcohol directly into the engine, thereby disposing with the dangerous task of manually pouring liquid oxygen, and an automated measuring system calculated fuel consumption during tests.[96] In 1932–1933 von Braun and his tiny staff were limited to using one half of the test stand, but the facilities at Kummersdorf were a major improvement over those at the *Raketenflugplatz*.[97]

Hitler's appointment as Chancellor on January 30, 1933 was an important moment in the history of German rocketry. The Nazi-dominated government passed a series of laws that dismantled Weimar democracy and gave Hitler full dictatorial powers. The Nazis ruthlessly began to do away with rival parties and organizations.[98] The Army, which managed to maintain nominal independence from the party, took the opportunity over the next year to eliminate amateur development and open experimentation. Becker had long despised the amateurs' very public approach to their work and considered the secret development of the rocket to be paramount, but the liberal-style Weimar democracy made it impossible for Becker to act on these concerns. With constitutional controls removed by Hitler and the Nazis, Becker seized the chance. By the end of 1934, the Army had either co-opted the work of the amateur groups by hiring their leading experts or forced the collapse of nearly all of the groups themselves.[99]

In January 1934, Walter Riedel came from the Heylandt Works to join von Braun at Kummersdorf.[100] His hiring was a part of the Army's effort to suppress the amateur groups and consolidate liquid-fueled rocket development under its sponsorship. Riedel's addition provided von Braun with key development expertise. He also recognized that much of what they were doing was entirely new, and he put in place an extremely thorough system of quality control to limit accidents and mistakes. Von Braun recalled that "Hardly a rivet or washer in our experimental A-3, A-5, A-9, and particularly the A-4 [missiles that would all be developed either at Kummersdorf or later at Peenemünde] can have escaped his personal scrutiny."[101] When Peenemünde opened in 1937, Riedel would go on to head the Design Bureau for a time.

---

[96] Dornberger, *V-2*, 23–24.

[97] Von Braun, "Behind the Scenes," 63. Initially, von Braun had only one technician under him.

[98] Ian Kershaw, *The Nazi Dictatorship: Problems and Perspectives of Interpretation* (New York: Routledge, Chapman, and Hall, 1993), 65–81.

[99] The story of the Army's campaign to suppress the rocket groups is complex and somewhat obscure. Michael Neufeld ably sorts out the details. See Neufeld, "The Excluded" and *The Rocket and the Reich*, 23–32. See also "Rolf Engel vs. the German Army: A Nazi Career in Rocketry and Repression," *History and Technology* 13 (1996), 53–72.

[100] Riedel IRR Dossier, National Archives.

[101] Wernher von Braun, "Reminiscences," 131–132.

THE END OF THE VfR AND *RAKETENFLUGPLATZ*

Even before the Army's campaign to eliminate the amateur groups, Nebel's folly had already begun the collapse of the VfR and the *Raketenflugplatz*. In the summer of 1932, shortly after Nebel's disastrous test at Kummersdorf, Franz Mengering, an engineer with friends on the Magdeburg city council, came to the *Raketenflugplatz* touting the bizarre idea that the Earth actually existed inside a sphere. He wanted to test his idea by launching a rocket and seeing if it would crash against the outer edge of the sphere. Even though Nebel probably thought the idea incredibly foolish, this was a perfect chance for him to put his opportunistic fundraising skills to use. He succeeded in obtaining 35,000 marks from the city of Magdeburg for his "Magdeburg Pilot Rocket." After securing money, the VfR attempted to build a rocket that was capable of launching a human who could jump out of it with a parachute once it reached maximum altitude. This launch was to take place in Magdeburg during Pentecost in 1933. The attempt to build the rocket was an embarrassing failure (though Magdeburg's city fathers were not terribly put out – their city benefited from the extraordinary publicity around the event). The leadership of the *Raketenflugplatz* began to distance themselves from Nebel's activities. Nebel eventually launched a much smaller version of the pilot rocket, but the damage had been done.[102]

Nebel's questionable business methods were also beginning to catch up with him. As early as February 1930, unbeknownst to the VfR leadership, Nebel, as Treasurer of the organization, filed a bankruptcy petition for the society and allegedly began cooking the financial books.[103] Nebel was supposedly embezzling the VfR's money for his own purposes. Three years later, in September 1933, Hans-Wolf von Dickhuth-Harrach and Willy Ley, the respective President and Vice President of the VfR, discovered this scheme, accused Nebel of fraud, and attempted to expel him from the society. Citing the close ties between the *Raketenflugplatz* and the VfR, von Dickuth-Harrach also severed the ties between two organizations. Nebel in fact may have been the victim of trumped up accusations, and the local prosecutor refused to bring him up on any charges. The incident bears some of the hallmarks of a power play by VfR leadership eager to lessen his influence, and Nebel was able to battle against his expulsion for several months.

Von Dickhuth-Harrach explained his actions in a subsequent issue of *Raketentechnik*, the VfR's last regularly published newsletter. His explanation is noteworthy for the ardent message it sent about the VfR President's political outlook. He cast his decision to expel Nebel in terms of the larger "cleansing" of the economy then taking place as a result of the National

---

[102] Ley, *Rockets*, 157–160; Nebel, *Narren*, 125–128; Winter, *Prelude to the Space Age*, 44–46.
[103] Ley, *Rockets*, 157.

Socialist seizure of power. Von Dickhuth-Harrach wrote, "This highest ideal [a 'clean economy' – *saubere Wirtschaft*]," he wrote, "which was unfortunately almost completely lost in German intellectual circles during the years of Marxism, has become honorable again, thanks to the will of our Führer, Peoples' Chancellor Adolf Hitler. Hopefully it will soon be held in the same esteem as it was before the war; that is, each German feels in his flesh and blood that he cannot act anything less than honestly and openly." He went on by writing that the nation's worst enemies were not those who openly supported the Republic's political and economic initiatives, but rather those who disguised their own petty self-interest by acting "decently in speech and emphasizing their usefulness to the community."[104] According to von Dickhuth-Harrach, Nebel's behavior, with its duplicity, deceit, and lack of communal spirit, represented all of the worst characteristics of the Weimar Republic. Nebel's proclivity for self-promotion was emblematic of an era of scandal and self-aggrandizement. Von Dickhuth-Harrach, who became President of the VfR in 1931, had aligned the society with the National Socialist policy of eliminating the supposed corruption, venality, and moral depravity of the Weimar Republic. By appropriating Nazi rhetoric to justify his actions, he shifted the political terms of the work even farther to the right. It had been a nationalist statement of technological virtue, but it became a radically conservative achievement that embraced a regime that boycotted Jewish shops, burned "un-German" books and art, and sanctioned all manner of thuggery in support of its new government – all cloaked in the idea of sweeping aside the egoistic tendencies of the flaccid Weimar Republic and reviving the honorable idea of service to the nation.

Nebel also had other problems in 1934. In its campaign to eliminate competing groups, the Army denounced him to the Gestapo for talking openly about rocket experimentation and military development. The secret police arrested him in June 1934 during the Night of the Long Knives, the bloody purge of SA leadership. The Gestapo only detained him briefly; Nebel used his close connections with Franz Seldte, the leader of the ultra-Nationalist Stahlhelm organization of military veterans, to get himself released. In later years he periodically reemerged as a thorn in the side of the Army developers, but because of the Army's stranglehold on rocket development, he never again rose to prominence in the field.

Nebel's fall from the leadership of the amateur rocket circles is indicative of larger trends among German rocket experimenters during the early days of the Nazi regime. The expanding influence of Becker and the Army Ordnance Office was one major development. Ordnance favored proceeding

---

[104] Hans-Wolfe von Dickhuth-Harrach, "Saubere Wirtschaft," *Raketentechnik*, November 1, 1933, 1, in NASM File "Germany, 1920–1940, Correspondence." Von Braun was elected to the board of the VfR in September 1932 but dropped out of the organization in 1933. *Raketenflug* 7 (December 1932); von Braun, "Reminiscences," 130.

from a sober, rational, realistic assessment of the capabilities of rocket technology as they stood at the beginning of the 1930s. They sought out talented individuals, such as von Braun and Rudolph, who could carry on their work anonymously and with strict attention to scientific and technical detail, an approach that would produce systematic, measurable, repeatable results. Nebel, on the other hand, failed them utterly in this regard. The failure of the *Raketenflugplatz* to consistently measure and record the experimental results was symptomatic of Nebel's thoroughly amateur approach, and it also led to frequent technical problems, wild inconsistencies in experimental findings, and exaggerated assertions about the level of rocket technology. Short of seeing a rocket in flight, Ordnance had no way to know for sure about the exact state of the art. When Ordnance officers saw for themselves how undeveloped the technology was and how badly fabricated Nebel's assertions were, they determined to develop the technology themselves, co-opting the necessary personnel from the amateur groups and professionalizing the work by demanding a more systematic approach.

The pressure on the amateur groups by Army Ordnance to cease their activities proved irresistible. Nebel's arrest was one thing, but other VfR members also received intimidating visits from the Gestapo, possibly at the urging of Becker and the Army Ordnance Office. Another blow to the *Raketenflugplatz* came from something far more prosaic. In the middle of the Magdeburg pilot rocket episode, a city official arrived in Reinickendorf with a huge water bill for the *Raketenflugplatz*. Leaky faucets in some of the unused buildings accrued a large bill over the years of the rocketeers' residence on the site. Because the chronically destitute *Raketenflugplatz* had no money to pay the bill, the city cancelled its lease to the land. With the steady pressure from the Army and the loss of its experimental station, the *Raketenflugplatz* fell apart. Some of the engineers there went to work for Siemens, bringing with them some of the equipment and what few documents they had. Through von Braun, these men, including such luminaries as Klaus Riedel, Hans Hüter, and Kurt Hainisch, were eventually hired to work at Peenemünde.[105] The first chapter of their storied careers was closing, but they would soon discover a whole new world laid out before them.

---

[105] Ley, *Rockets*, 160–161; Winter, *Prelude to the Space Age*, 48; Riedel, Hainisch, and Hüter questionnaires, Folder "Boston," Box 703, RG 165, Assistant Chief of Staff, Enemy POW Interrogation File, 1943–1945, NARA.

2

"At Peenemünde, They Have Created a Paradise"

For the small group of civilian employees under von Braun, including "Papa" Riedel and Arthur Rudolph, going to work for the Army was a life-changing experience. Gone were the days of ad hoc development and short means. Purse strings were tight in the early period at the Kummersdorf Proving Ground, but the increased pace of rearmament ordered by Hitler, combined with the excellent backing provided by Becker and other influential military figures, meant that it was only a matter of time before the program's resources began to match its members' ambition. Increased military and technological demands mingled with the German engineering profession's traditional emphasis on loyal service to the state and with the developers' own personal and professional aspirations. The small, struggling group of engineers and space enthusiasts hired to work at Kummersdorf on behalf of the Nazi regime found in the Army an organization that paid decently and gave them the resources they needed to exercise their technological imaginations. In return, the engineers and technicians agreed to work methodically, scientifically and – just as important – quietly. The Army, governed by a near manic desire to maintain strategic surprise and assisted by Hitler's dismemberment of the Weimar constitution, closed down most amateur groups and hired the best developers for itself. Then it cloaked the work in a shroud of secrecy that made participation in these activities a sign of rank, privilege, and prestige. The developers' accommodation to the Nazi regime had begun.

KUMMERSDORF PROVING GROUND, ARMY ORDNANCE, AND THE ROOTS OF "BIG" GERMAN MISSILE RESEARCH, 1933–1937

From its first moment in power, Hitler's regime made rearming Germany's weakened military its top priority. For the Nazis, Germany's future depended solely on rebuilding its armed forces. All other government expenditures were secondary. As early as February 8, 1933, nine days after Hitler ascended to the Chancellor's post, he reported to his cabinet that "The next five years must be devoted to the restoration of the defense capacity of the German people," and he proclaimed that every state-funded work creation measure must be judged in terms of its value to this goal.[1] Accordingly, the government diverted hundreds of millions of Reichsmarks from other measures to pay for the illegal German rearmament.[2] In April, the Reich government initiated the "Second Armaments Program," which circumvented the standard budgeting process in order to provide extra money to the Army that was not included in the regular budget.[3] Hitler's approach to military spending flouted international accords and ignored practical limitations. His dedication to rearmament was the cornerstone of the military's loyalty to the Reich. Even though the Army's relationship with Hitler soured during the war, in the 1930s, its generals found in him a man whose military interests largely coincided with their own. The armed forces pounced on the opportunities for expansion and innovation that the dictator's aggressive armaments policy offered. They knew that they had Hitler's wholehearted support and continued to press for increasingly advanced and modern weaponry.[4] The historian Ian Kershaw has pointed out that this was not merely based on the desire to increase Germany's military strength. It was also one aspect of the armed forces' leadership's habit of "working toward the Führer," that is, consciously acting in accordance with what they perceived to be Hitler's own goals. None of this, however, happened in anything resembling a coordinated fashion. Hitler purposely set his competing bureaucracies at war with each other. The result was a chaotic, wholly unorganized rearmament program that resisted any attempts at rationalization.[5] By the late 1930s, Germany's rearmament process was a wild, uncoordinated, free-for-all. The Army, the Navy, and the brand-new Air Force (the Luftwaffe) sank billions of Reichsmarks into the development and production of new weaponry. Tanks, battleships, and bombers rolled off the assembly line in unprecedented numbers.

This accelerated pace of rearmament and the regime's uncoordinated, querulous administrative structures worked to the benefit of the missile's proponents. It allowed the ambitious Becker, who would be promoted to General in 1933 and assume the post of Ordnance Chief in 1938, and his fast-rising subordinate, Major (formerly Captain) Dornberger, to carve out a bureaucratic fiefdom that catered to the missile specialists' professional aspirations by putting development of a ballistic missile front and

---

[1] Quoted in Ian Kershaw, *Hitler: 1889–1936: Hubris* (New York: Norton, 1998), 444.

[2] Heinz Höhne, *Zeit der Illusionen: Hitler und die Anfänge des 3. Reiches 1933 bis 1936* (Düsseldorf: Econ-Verlag, 1991), 58.

[3] Michael Geyer, *Deutsche Rüstungspolitik, 1860–1980* (Frankfurt: Suhrkamp, 1984), 140.

[4] Kershaw, *Hitler: Hubris*, 437–444. See also Wilhelm Deist, *The Wehrmacht and German Rearmament* (Toronto: University of Toronto Press, 1981), 24–25; Klaus-Jurgen Müller, *Das Heer und Hitler: Armee und Nationalsozialistisches Regime, 1933–1940*, 2nd ed. (Stuttgart: Deutsche Verlags-Anstalt, 1988), 53–61.

[5] Ian Kershaw, *Hitler: 1936–1945 Nemesis* (New York: Norton, 2001), 9–18.

center in the German rearmament effort.[6] The determined developers at Kummersdorf profited from the skillful shepherding of Becker and Dornberger. The massive effort to rearm Germany as quickly as possible and without consideration for internal economics or external treaties helped to create the institutional environment for major technological innovation in the missile program by providing the necessary facilities, raw materials, and brainpower.

As von Braun could attest, the conditions at Kummersdorf were far better than at the *Raketenflugplatz*. Instead of unemployed engineers living in ramshackle quarters and squeezed elbow to elbow in cramped workstations, Kummersdorf offered the prospect of paid employment (employees at Kummersdorf earned between 2,400 and 8,000 Reichsmarks per year, depending on education and experience) while utilizing some of the best equipment that money could buy.[7] The launch equipment was far more advanced than any of the amateur groups' primitive setups. Safety considerations were much easier to maintain (though it was still dangerous – one of von Braun's colleagues was killed by an explosion in 1934) and angry policemen concerned about collateral damage to the surrounding areas did not restrict the work, as in Reinickendorf. Missile specialists in Kummersdorf could not have missed the improved quality of the working conditions provided by the Army. The fact that many of them were conservative nationalists only strengthened the links between the two until stronger institutional bonds could be forged.

Under the Army, secrecy increasingly defined itself as a major force in the culture of missile development. Aside from the strategic benefits of secret missile development, the exercise of secrecy proved to be a useful bureaucratic weapon that had the remarkable effect of empowering those who invoked it. It played a key role in obtaining necessary development resources and established missile developers as participants in activities that could only be the domain of a privileged few. Competition for money and resources was fierce in the pell-mell world of German rearmament, especially in the earliest years of the Third Reich. Despite Becker's concerted efforts on behalf of the missile program, administrators sometimes had to fall back on smoke and mirrors. They sometimes invented benignly misleading requisition requests, referring, for example, to a pencil sharpener as an "Appliance

---

[6] On the chaotic nature of the Nazi administrative hierarchy and the competition that developed within it, see Peter Hüttenberger, "Nationalsozialistische Polykratie," *Geschichte und Gesellschaft* 2 (1976), 417–442, perhaps the single-most influential essay in the historiography of Nazi Germany. See also Martin Broszat, *The Hitler State: The Foundation and Development of the Internal Structure of the Third Reich*, trans. by John W. Hiden (New York: Longman, 1981).

[7] For the salaries of civilian engineers at Kummersdorf, see the collection of professional dossiers in File "Boston," Box 702, RG 165, Records of the Army Chief of Staff, National Archives and Records Administration (NARA).

for milling wooden dowels up to ten millimeters in diameter." If this Nebel-esque masquerade failed, then, "We entrenched ourselves behind the magic word 'secret.' There, the budget bureau was powerless."[8] Dornberger offers a telling example.

Once, in the summer of 1933, we bought two boxes of Christmas tree sparklers. The idea was to use these sparklers inside the nozzle for igniting the first drops of oxygen and alcohol. A year passed. Then the Bureau of the Budget asked what Christmas sparklers were used for in the middle of summer. We replied tersely, 'For experiments.' But the Bureau of the Budget was not happy with this answer, and eight weeks later asked us what kind of experiments. We answered, 'Secret experiments.' Then they gave up.[9]

Dornberger's tactics drew a firm and early boundary between his very specialized missile developers, who had access to secret information, and others who did not. The demands of secret development provided the missile specialists with power to refute the prerogatives of those, like the budget bureau, who normally exercised a degree of power over the daily conduct of work at Kummersdorf.[10] It helped establish the boundaries within which a privileged few with the proper qualifications could operate relatively freely and unfettered by usual limitations. Ordnance strove to keep as few people as possible from knowing about their missile research, but in doing so, also established a dynamic in which those who did the research itself attained a certain amount of privilege that shaped what it meant to be a participant in this fantastic endeavor.

Supplied with better funding, equipped with improved facilities, and hidden behind Kummersdorf's fence, von Braun's small team surpassed the work of the amateur groups rather quickly. The year 1934 proved to be both personally and professionally rewarding for many there. In June, von Braun defended his dissertation, "Konstruktive, theoretische, und experimentelle Beiträge zur Problem der Flüssigkeitsrakete" ("Constructive, Theoretical, and Experimental Contributions to the Problem of the Liquid-Fueled Rocket"). The work reflected the major advances von Braun made in engine design and instrumentation virtually on his own in Kummersdorf in 1933.[11] The Technical University of Berlin awarded the brilliant twenty-two-year old a doctorate with high honors for his work, and his star was rapidly on the rise. By this time, the process by which he would endow this work with great personal significance was well under way. The long-time space enthusiast's research on rocket technology was groundbreaking, and he saw in his

---

[8] Walter Dornberger, *V-2* (New York: Viking Press, 1954), 37.

[9] Ibid., 38.

[10] See Chapter 1.

[11] Wernher von Braun, "Konstruktive, theoretische, und experimentelle Beiträge zur Problem der Flüssigkeitsrakete," doctoral dissertation, University of Berlin, 1934, reprinted in *Raketentechnik und Raumfahrtforschung*, Sonderheft 1, 1960.

work for the Army the fulfillment of many of his professional ambitions.[12] At the end of 1934 on the island of Borkum, von Braun's team staged successful test launches of two relatively small rockets, which were code-named A-2s but known affectionately by his group as Max and Moritz, after two ne'er-do-well German cartoon characters. Wernher von Braun and Walter "Papa" Riedel (joined at Borkum by Arthur Rudolph) worked on the A-2 project. They and their Ordnance sponsors were thrilled with their success.[13]

The Kummersdorf group's accomplishments with the A-2 test rockets proved to be an important turning point in the effort to create a larger, state-funded ballistic missile program. The great enthusiasm that the tests generated among Ordnance officials further loosened the Army's purse strings.[14] When Ordnance officers and von Braun presented their findings to their Army superiors in mid-January 1935, one officer's enthusiasm for the technology got the better of him. He argued that the Army should mass produce a larger version of the relatively primitive test rockets and use them to tactical advantage in artillery-style bombardments. An unenthusiastic von Braun threw cold water on the idea. The A-2s were, according to von Braun, inaccurate, unreliable, and might damage the case for larger weapons in the future.[15] Von Braun may have always had the idea of a manned space rocket in mind, but this was also an early indication of his willingness to conceive of his work in terms of its strategic military applications alongside his lifelong goal of spaceflight. Army officials accepted the young project manager's argument, and despite a moderate, but not unreasonable, degree of penny-pinching by the Army, the budget for liquid-fueled missiles grew continually throughout the early 1930s.

Larger budgets meant that the staff at Kummersdorf grew, the offices expanded, and the testing hardware improved. Von Braun's earlier spatial limitations were eliminated and the complexity of his facilities dramatically expanded. Driven by the need to test larger missiles (the next test model was dubbed A-3), Ordnance built a larger test stand for liquid-fueled engines that was surrounded by a blast wall and serviced by a locomotive that could tow large testing equipment and even complete rockets into firing position.[16] The staff dedicated to rocket development grew to seventy-eight people, and the

---

[12] Wernher von Braun, "Behind the Scenes of Rocket Development in Germany, 1928 through 1945," in Wernher von Braun Papers (date unclear, late 1940s), Space and Rocket Center (SRCH), Huntsville, AL.

[13] Michael Neufeld, *The Rocket and the Reich: Peenemünde and the Coming of the Ballistic Missile Era* (Cambridge, MA: Harvard University Press, 1995), 38. See also Arthur Rudolph Oral History Interview (OHI), National Air and Space Museum (NASM).

[14] Wernher von Braun, "I Reached for the Stars," Box 200, Folder 7, Wernher von Braun Papers, SRCH.

[15] Wernher von Braun, "Denkschrift," January 18, 1935, Fort Eustis (FE) Files, FE 727/a, NASM.

[16] Neufeld, *The Rocket and the Reich*, 36–54; Dornberger, *V-2*, 33.

research budget reached up to 80,000 Reichsmarks.[17] All of this was a far cry from the lean years at the *Raketenflugplatz*, but even so, increasing military demands on rocket technology and the growing ambitions of its supporters were beginning to make even the large proving ground at Kummersdorf too small for developing and testing large rockets.

Another aspect of the demand for improved rocket technology was driven by Ordnance's fear of foreign development competition. Ordnance officers increasingly argued that though Germany might have taken the lead in rocket development, other countries, especially the United States and Soviet Union, were showing signs of catching up. For the officers in charge of shepherding the German Army's liquid-fueled rocket program along, the presence of foreign development competition was a serious concern. Troublingly for the Army, they were receiving a number of reports of progress being made in this field, especially by Robert Goddard in the United States.[18] Unfortunately, there is little evidence relating to German intelligence on Soviet activities in the 1930s. Though none of this work even came close to approximating the scale or success of the German program, it was at least enough to give Ordnance justification for improved funding and expansion of the program. However spotty, this intelligence provided yet more impetus for officials in the armed forces to argue that continued missile development was of decisive importance to Germany's massive rearmament effort. In the future, Dornberger would seize on it to promote the highest wartime armaments priority level for missile development in an attempt to guarantee unlimited development and production resources. In the 1940s, his resort to citing foreign competition as justification for a project that consumed increasing resources would eventually put his development and production engineers under immense strain to produce. However, in the middle of the 1930s,

---

[17] Wernher von Braun, "Reminiscences of German Rocketry," *Journal of the British Inter-planetary Society* 15 (May/June 1956), 134. See also Volkhard Bode and Gerhard Kaiser, *Raketenspuren: Peenemünde 1936–1994: eine historische Reportage* (Berlin: Links Verlag, 1995), 43. For a history of the Kummersdorf Proving Grounds, see Wolfgang Fleischer, *Die Heeresversuchsstelle Kummersdorf: Maus, Tiger, Panther, Luchs, Raketen und andere Waffen der Wehrmacht bei der Erprobung* (Wölfersheim-Berstadt: Podzun-Pallas Verlag, 1995).

[18] In January 1936, the German Military Attaché in Washington sent a detailed report of Goddard's work on liquid-fueled rockets to Berlin, where Ordnance obtained a copy. The report contained information on the size, altitude capability, and speed of Goddard's instrument, which, though erroneous, gave cause for increased concern among Ordnance officials. In February, the General Staff forwarded a copy of the American Science Newsletter, which contained information about Goddard's ongoing work. Another report indicated to Ordnance the flight of a small rocket from New York City across the Hudson River to New Jersey on February 9, 1936. This particular launch never actually took place. See Boetticher, "Raketenversuche in den Vereinigten Staaten," January 7, 1936, T-78/434, RG 242, NARA; "Rocket Sent to 7500 Feet at 700 Miles an Hour, *Science Newsletter*, January 4, 1936, RH8/v.1945, *Bundesarchiv/Militärarchive Freiburg* (BA/MA), Auszug aus technische Nachrichten 27, RH8/v.1945, BA/MA.

Ordnance's concern about foreign development helped justify a push for a massive expansion of the program.

The idea for an expanded missile program first gained traction in early 1935, when the Reich Air Ministry (RLM) approached Ordnance about its missile work. Wolfram von Richtofen was the dashing head of the Technical Development Office, a World War I fighter ace, and cousin of Manfred von Richtofen, the infamous Red Baron. In February he took an interest in the exciting development work going on at Kummersdorf. The Air Ministry had been monitoring several different streams of forward-leaning development, including a small rocket project at the Junkers Aircraft Company and another one in Munich that eventually grew into the infamous V-1 "buzz bomb." In May, Captain Leo Zanssen, one of von Braun's superiors who would go on to become the military commander at Peenemünde in 1938, sent a memo to the Air Ministry that endorsed the idea of cooperation between the two organizations. Zanssen first noted that the use of rocket engines was perceived "primarily as a military weapon (a liquid fueled long range missile)," making it perfectly clear that Ordnance was not in any way interested in spaceflight – a point already known to von Braun and the rest of the civilian developers at Kummersdorf. Zanssen probably wanted this clear to everyone in the Air Ministry as well. He continued by writing that "A considerable development lead vis-à-vis foreign countries has been reached here, the relinquishment of which would be intolerable because the element of surprise is in the interest of national defense."[19]

Zanssen's memo also underlined the increasing importance that Ordnance attached to secrecy in missile development. The Luftwaffe, Germany's dynamic new Air Force, hoped to conduct a joint development venture with the Junkers Aircraft Company, but Ordnance was reluctant to join this effort because of secrecy considerations, which Zanssen did not believe private industry could ensure. In line with his superiors in Ordnance, Zanssen emphasized once again that this technology would have maximum impact if developed in secret and deployed by surprise. Few in Ordnance believed that industrial representatives understood this.[20] Ordnance was still not completely put off by the idea of cooperation, though. That following June, Ordnance convened a meeting with the RLM at Kummersdorf in order to address Ordnance's ongoing concerns and attempt to hammer out the terms of an interservice agreement. Von Braun, Rudolph, and von Horstig went as Ordnance's representatives, and Richtofen came with two representatives from Junkers.

In the meeting, von Braun presented a position paper that, as historian Michael Neufeld has noted, "Must be regarded as Peenemünde's birth certificate."[21] In it, von Braun outlined his position on cooperative development,

[19] Leo Zanssen to RLM, "Raketenflugzeug," May 22, 1935, FE 732, NASM.
[20] Ibid.
[21] Neufeld, *The Rocket and the Reich*, 46.

placing particular emphasis on the idea of the creation of a single facility dedicated solely to developing rocket engines for missiles and airplanes. His paper called attention to the advantages of cooperation between the Army and Luftwaffe, noting that "The difference between an engine for a free flying liquid fueled rocket and for a rocket plane does not come into question. Rather, it exists only in spatial arrangements. It is therefore advantageous that in the future, the development of the free flying liquid fueled rocket and the rocket engine for airplanes be carried out together in the same place."[22] For the young engineer, who also happened to be a flying enthusiast, inter-service cooperation to construct a facility solely for rocket development was the most efficacious path to continued improvement on a technology that he wholeheartedly embraced.

Von Braun's paper revealed the degree to which its author had adopted the Army's point of view as well as his own aspirations for rocket development. "For the implementation of this goal," he wrote, "it is desired that all new workers entering into this area of activity also remain [*bleiben*] in this 'experimental rocket center' [*Raketenversuchsanstalt*]. Section 1 feels that it is particularly important that it is agreed that the workers placed by the RLM for the development of new engines will later be taken over by Ordnance offices and/or the 'experimental rocket center.'"[23] Von Braun was intent on retaining as many specialists in one location as possible. He pushed this point for two reasons. In his time under Army employment, he had become thoroughly imbued with the Ordnance's goal of maintaining a monopoly on rocket development and what it viewed as proprietary information that emerged from the development process. Though Ordnance officers welcomed the RLM's financial and material contributions, at the administrative level, they nonetheless jealously guarded the secret developmental information and feared that the Air Ministry would make off with this knowledge and key personnel once they had attained what they sought from cooperation with the Army. Von Braun's position paper spoke to this fear of losing a monopoly on rocket development and sought to ensure this would not occur.

Von Braun's paper also points to his own vision of what professional rocket specialists should be and do. Once brought to the experimental rocket center, they would remain there, forming the nucleus of a like-minded group of technical specialists who would then work toward a common technological goal. His use of the verb *bleiben* indicated that von Braun did not simply mean for employees to live at the facility while they were employed there, only to depart for other projects at the whim of their superiors. Rather, these specialists should remain at the facility in order to focus their energy on continuing development of rocket technology. The new rocket center

---

[22] Wernher von Braun, Stellungnahme von Wa. Prw. 1 zur Entwicklung eines Raketenflugzeugantriebes in Verbindung mit RLM, FE 732, NASM.
[23] Ibid.

would serve as the physical locus of a new, cutting-edge technical profession. This arrangement would also allow the new institution to be the sole producer of new specialists by controlling the selection and training of newcomers from one central location.[24] Inclusion in this community would be hallmark of the increasing professionalization within the highly specialized world of rocket engineering. In his effort to shape this emerging specialization along the professional lines he saw fit, von Braun actively sought to determine the physical framework within which this new group of technical experts would carry out their tasks.

Finally, von Braun noted that although cooperation between the Army and Luftwaffe in the rocket venture made obvious sense, ties with private industry should not be fostered. He insisted that no documents produced by the military be made available to any private firms. In addition to the paramount importance of secrecy, von Braun contended that "There is the danger that profit-making opportunities would arise from development that the state has carried out at tremendous expense."[25] Rocket technology developed by the state, argued von Braun, should simply not be exploited by large industry.

It is tempting to view von Braun's statement on industrial exploitation of state-developed rocketry as a sign that the widespread Nazi anticapitalist rhetoric held some degree of appeal for him. Certainly, von Braun was imbued with the deeply conservative nationalist sentiment that was rife in the universities and that shot through his profession. As Neufeld has pointed out, however, the young engineer's mistrust of capitalism "drew less on National Socialist ideology than on centuries-old traditions of state ownership in Prussia and Germany."[26] Von Braun's worries centered on less political factors. His overriding concern was keeping the Army's monopoly on development with himself at the center of it. He also certainly believed that secrecy was important – strategic surprise was as key for him as it was for his military superiors. He did not want his weapon to lose its effectiveness if it was somehow exposed to industry. Though von Braun found great personal and professional satisfaction in the Nazi rearmament program and would join the Nazi party in 1937, he did so only after being requested to do so. The engineer was largely indifferent to the political maelstrom swirling around him. Rather than being an ideologue who invested heavily in Nazi

---

[24] See Lennart G. Svensson, "Knowledge as a Professional Resource: Case Studies of Architects and Psychologists at Work," and Charles McClelland, "Escape From Freedom? Reflections on German Professionalization, 1870–1933," in Rolf Torstendahl and Michael Burrage, eds., *The Formation of Professions: Knowledge, State, Strategy* (London: Sage, 1990). William J. Goode has also noted this phenomenon among the professions. See his article, "Community Within a Community: The Professions," *American Sociological Review* 22 (April 1957), 194–200.

[25] Von Braun, Stellungnahme, FE 732, NASM.

[26] Neufeld, *The Rocket and the Reich*, 46–47.

ideology, he was instead an opportunist who saw ample prospects under the regime to advance his own goals and concerns. The science writer Willy Ley wrote of von Braun, "Did we discuss politics? Hardly, our minds were always far out in space. But I remember a few chance remarks which might be condensed into saying that . . . the German Republic was no good and the Nazis ridiculous."[27] Ley's statement that he and von Braun never discussed politics is difficult to believe. The subject may not have been von Braun's favorite, but it was an unavoidable one in the tumultuous days of the early Nazi period. It was clearly on Ley's mind, given that he, after all, had the foresight and courage to leave Nazi Germany as early as 1935. But von Braun certainly paid only slight attention to most of what was happening around him and, having achieved a degree of official privilege, had no reason to leave Germany. He was an opportunist who used the resources that the regime put at his disposal to further his own cause.[28]

In any case, Ordnance officials subsequently followed von Braun's line of argument closely in their dealings with outside entities. They made it clear that they would only include Junkers in their plans if there was a way to ensure that the company would be able to adhere to the strict secrecy considerations that Ordnance thought necessary to implement: "[Ordnance] must insist that absolutely no drawings, documents, and so forth that are based our experiences in any way be made available to industry (not even the Junkerswerk), without special permission obtained from Section 1. The developments and research obtained here must remain in the hands of the developers here."[29] Again, though commercial exploitation may have been on the minds of the Ordnance representatives, their main concern was that their work be carried out in absolute secrecy. The only way to maintain such total seclusion was to pursue all development and production work from one central location rather than farm it out to various industrial firms.

Ordnance slightly scaled back its strict secrecy requirements later in 1935 and allowed a limited number of industry representatives access to rocket development. In the summer of that year, the Air Ministry brought Heinkel Aircraft into the rocket program. In September, Army officials agreed to this addition only after the few Junkers and Heinkel employees privy to the

---

[27] Quoted in Michael Neufeld, "Wernher von Braun, the SS, and Concentration Camp Labor: Questions of Moral, Political, and Criminal Responsibility," *German Studies Review* 25 (2002), 59.

[28] Neufeld, *The Rocket and the Reich*, 47. See also Rainier Eisfeld, *Mondsuchtig: Wernher von Braun und die Gebürt der Raumfahrt aus dem Geist Barberei* (Reinbek bei Hamburg: Rowohlt Taschenbuch Verlag, 2000), 70–74. Eisfeld argues that "opportunism might, in a specific situation, appear as necessary, even 'objectively' reasonable. In von Braun's case, this was equal, under given circumstances, not to advocating the Nazis' cause, but in fact to contributing to it."

[29] Protokoll über die am 27.6.35 in Kummersdorf stattgefundene Besprechung zwischen RLM, Wa Prw. 1 und Vertretern der Junkers-Flugzeugbau A.G., FE 732, NASM.

project signed a declaration protecting the secrecy of the work. The agreement read, in part, "The devices developed by the Army Ordnance Office for rockets should be used as engines for airplanes. In order to create functional designs, the absolutely secret documents must be made more accessible to the aircraft firms. Since this work must remain totally secret, the firms are obligated to make the documents handed to them accessible only to people given permission to see them by the Air Ministry." Ordnance representatives remained determined to keep the circle of initiates as tightly drawn as possible. The military limited the number of private industry specialists working on rocket technology to ten – four at Junkers and six at Heinkel. As a condition of being let in on the secret, Ordnance demanded that the industry employees carry out their experiments only in workshops that were totally off limits to other employees of the firms.[30] Despite their agreement, Ordnance officials still preferred that private industry not be involved at all. This ad hoc arrangement would come to an end when the Army and Luftwaffe parted ways in 1938, allowing Ordnance, with the exception of construction needs and parts supply, such as highly specialized gyroscopes, to bypass private industry and to concentrate nearly all of its developmental capability at Peenemünde.[31]

Technical successes, the fear of foreign competition, the overweening desire for secrecy, and the missile developers' own desire to build bigger machines led Becker and his assistants to begin thinking about the need for a newer, larger development facility. The prospect of cooperation with the Air Ministry made this idea even more appealing. The successful flight of Max and Moritz, the twin A-2s, made it clear even then that Kummersdorf was rapidly becoming too small for their work, and the firing range's location in the Berlin suburbs was not conducive to secrecy or safety.[32] Another Kummersdorf-like facility was not what Ordnance had in mind, though. The project leaders thought big. Safety and secrecy were of paramount importance, but Ordnance also wanted to keep all development-related activities centralized in one massive facility dedicated solely to developing a ballistic missile. Dornberger, who spearheaded this idea of "everything under one roof" (*Alles unter einem Dach*), wrote

We wanted to build, to build on a grand scale, and beautifully. … We wanted to investigate and develop on a single site everything that seemed essential to the effective employment of such a new and powerful weapon. We wanted to develop, not only the rocket itself, but also the necessary ground handling and testing equipment, and to study all its implications in the most diverse branches of technology and

---

[30] "Vereinbarung über Zusammenarbeit auf dem Gebiet der Rauchspur zwischen HWA, RLM, Junkers, Heinkel," September 2, 1935, FE 746-b, NASM.
[31] For the breakup of the Army–Air Force alliance, see Neufeld, *The Rocket and the Reich*, 54–63.
[32] Dornberger, *V-2*, 38.

science. We wanted to start with applied research and end up with a fully developed article ready for production in the factories. In short, we wished to put through on our own account a complete program. We needed a research and development site fully equipped with all the latest resources of science and technology.[33]

The project of missile development would be no small-scale program. A crucial consideration in the assembly of any new base was that its size and appearance match the importance its developers attributed to the weapon. Building "on a grand scale and beautifully" was essential. Dornberger's notion that all of the work, including development, assembly, and production, should be carried out under one roof fundamentally shaped the notions of how missile development should be carried out in Germany. Von Braun had advanced the concept of a single missile center where all development work took place and all missile experts were trained, but it was Dornberger who championed the idea and saw it through. The plan grew out of the Army's conservative culture of state-centered technical development and was reinforced by National Socialist anticapitalist rhetoric.[34] Both Becker and Dornberger believed that a single location that could more easily handle the variety of problems inherent in such a radically new technology was essential for the completion of the work. That they could more easily keep the work secret at such a place was never far from their minds either. For Ordnance and civilian leadership, the most advanced weapon required the largest, most advanced research, development, and production facility that could be assembled. Thus, even before the rocket specialists sketched out the A-4 as a concept and any equipment to be used in its development existed, the idea of an elaborate station for the development of ever-larger missiles had taken root. Ordnance was banking not only on the potential of the missile as a weapon but also on the skill of the people working in the program. Though it would prove a difficult period of development, the developers would not disappoint.

Cooperation with the Luftwaffe meant even more funding and support for missile development from both military branches. The financial commitment from the Army and Air Force was massive, and the developers' aims benefited from interservice rivalry to finance their work. In November 1935, the Luftwaffe promised five million marks for a new base. Becker was not about to let "that upstart Luftwaffe" show up the Army, and he made his own audacious promise of six million marks to the program.[35] As von Braun

---

[33] Ibid., 38–39.

[34] On the roots of the Army's anticapitalism, see Walter Goerlitz, *History of the German General Staff, 1657–1945* (New York: Praeger, 1962), 54–57. An excellent analysis of National Socialism's schizophrenic anticapitalism is Avraham Barkai, *Nazi Economics: Ideology, Theory, and Policy* (New Haven, CT: Yale University Press, 1990).

[35] Wernher von Braun, "Behind the Scenes," 18; "Reminiscences," 135. See also Neufeld, *The Rocket and the Reich*, 50.

put it in 1956, "In this manner our modest effort, whose yearly budget had never exceeded 80,000 marks, emerged into what Americans call the 'big time.'"[36] To the delight of the developers, millions of marks were flowing in support of the missile venture.

In January 1936, after several weeks of searching, Luftwaffe and Army officials settled on the area around Peenemünde, a tiny fishing village on the northwest tip of the island of Usedom on the Baltic coast, as the site for the single development facility proposed by von Braun earlier in the year. Von Braun's mother, who grew up in the nearby town of Anklam and whose father hunted on Usedom, recommended Peenemünde to her son. Although parts of Usedom were popular vacation spots, the village of Peenemünde was as remote as any in Germany. Electricity had only made its way there in 1928, and most of its 447 residents still burned oil for light.[37] From the developers' perspective, the quiet site was perfect for their massive new installation.

In 1936, events were taking a dramatic turn both inside and outside the missile program. In March, Hitler ordered German troops across the Rhine bridges and into the Rhineland, which had been demilitarized by the Treaty of Versailles. At the same time, the dictator tore up the Treaty of Locarno, another open repudiation of Germany's rivals, Britain and France. Neither nation was prepared to take action against the Reich, and both moves were wildly popular among Germans. For the missile specialists, events were just as exciting. In March, General Werner von Fritsch, the Army's Commander-in-Chief of the Army, visited Kummersdorf, where he viewed an impressive series of engine tests that made quite an impression. "How much do you want?" he candidly asked Dornberger.[38] By April, the Luftwaffe, which administered the building project at Peenemünde, approved final construction plans; groundbreaking on Usedom began in August. Over 10,000 workers under contract to civilian firms, Organization Todt, and the Reich Labor Service descended on the sleepy island to lay roads and train tracks, erect living quarters, and construct development workshops.[39] The missile developers, who were more than a little impressed by the Air Force's willingness to advance their cause, greeted the pace of this work and the decisive efforts of the Luftwaffe with great enthusiasm. Arthur Rudolph recalled years later that it was "entirely new, fantastic, unbureaucratic, fast moving."[40] "The guys were fantastic," he proclaimed.[41] Von Braun was thrilled, and he saw certain affinities between the missile developers and the Luftwaffe

---

[36] Von Braun, "I Reached for the Stars," Wernher von Braun Papers, SRCH.
[37] Bode and Kaiser, *Raketenspuren*, 24, 27–28; Neufeld, *The Rocket and the Reich*, 49.
[38] Dornberger, *V-2*, 38–39.
[39] Neufeld, *The Rocket and the Reich*, 49.
[40] Arthur Rudolph OHI, NASM.
[41] Thomas Franklin, *An American in Exile: The Story of Arthur Rudolph* (Huntsville, AL: Christopher Kaylor, 1987), 48.

officers, writing that they "were young, enterprising, and receptive, and did not suffer from the hidebound mentalities and masses of red tape which handicapped the Army and Navy."[42] "Here was action indeed!" gushed Dornberger.[43]

Ordnance officials still needed to determine what it was exactly that they wanted to create at Peenemünde. Indeed, the development of an *institution* for research and development took center stage even before Ordnance had any detailed conception of the twin objectives of a ballistic missile and rocket fighter. It was only after the site was chosen and funds dedicated to construction and development that von Braun, Dornberger, and Walter Riedel hammered out the technical outlines of their first major ballistic missile (everything up to then had been a test model). In the end, its technical outlines were fairly simple, even if the means by which they would complete their work were not. The three developers agreed that the missile would have a twenty- to twenty-five-ton thrust engine that was capable of a range of 250 kilometers. For operational purposes, it should be able to carry a one-ton warhead and fit through a standard railroad tunnel. There were still years of research and development in front of them, but von Braun, Dornberger, and Riedel had finally drafted the simple requirements for the A-4 missile, known to posterity as the V-2.[44]

THE RISE OF PEENEMÜNDE

The years 1935–1939 were exciting years for the members of the missile program, but they were disquieting ones for the rest of Europe. Domestically, the Nazi regime brought more of its malignant intent to bear on its so-called internal enemies, especially Jews, communists, and social democrats. The racist Nuremberg Laws of 1935 stripped Jews of their citizenship, forbade them from intermarriage with other Germans, and revoked their voting rights, among other noxious measures. Harassment of Jews increased, and signs reading "Jews unwelcome" popped up in cities and towns across the country. Those unwelcome Jews who wanted to leave Germany for an uncertain future in another country had to pay gargantuan exit taxes to the Reich and faced stringent immigration quotas abroad. Many Jews and communists languished in concentration camps like Dachau, Buchenwald, and Sachsenhausen. Their fates were no mystery to average Germans, many of whom benefited from the government's policy of expropriating Jewish property and passing it to "pure" Germans.[45] The year 1938 saw a spate

[42] Von Braun, "I Reached for the Stars," 58.

[43] Dornberger, *V-2*, 41.

[44] Walter Dornberger, "The German V-2," *Technology and Culture* 4 (Fall 1963), 398–399. See also Dornberger, *V-2*, 47–48 and Neufeld, *The Rocket and the Reich*, 51–52.

[45] Robert Gellately, *Backing Hitler: Consent and Coercion in Nazi Germany* (New York: Oxford University Press, 2001); Jeffrey Herf, *The Jewish Enemy: Nazi Propaganda During*

of violent and very public state-supported anti-Jewish hostility, including the destruction of the great synagogue in Munich and the terrifying *Kristallnacht* pogrom. Hitler's foreign policy also became increasingly belligerent. The Nazi government first absorbed Austria into the Third Reich in March 1938 and in September drove Europe to the brink of war with its demands for the Sudetenland in Czechoslovakia. The Munich Agreement that month exposed the hollow support of Britain and France for Czechoslovakia and ceded the Sudetenland to the Reich. Six months later, an emboldened Hitler, without fear of intervention, occupied Prague outright.

In the midst of all this, a stunning new rocket facility was rising from the earth at Peenemünde. Dornberger and Ordnance officials did not pinch pennies when it came to constructing their new base. In August 1936, workers began arriving at Peenemünde to begin constructing roads, rail lines, development workshops, an airfield, and living quarters for employees. The residents of the fishing village of Peenemünde were ordered to move, and the village was eventually destroyed.[46] The first 350 employees of the new base arrived in May 1937.[47] At the end of that year, while construction of the development workshops was still ongoing, Dornberger told von Braun and Rudolph that he also wanted to construct a production plant at the base. Both civilians, who were development experts with little experience in mass production, thought it was a terrible idea and protested loudly to Dornberger over it, but their superior went ahead anyway. In November 1938, just after *Kristallnacht*, Army Commander-in-Chief Walther von Brauchitsch gave the go-ahead to begin expansion of the facility to include the production plant. Construction of the factory began in 1939.[48] Factory planners estimated that the workforce required to man this plant would be approximately 5,000 people, but Usedom did not have the housing facilities for so many. Their solution was quite literally to build a new town for the employees, which came to be known as the "Settlement."[49] In March 1939, two weeks after German troops annexed the remnants of Czechoslovakia, Dornberger informed Becker, now Ordnance Chief, of the scale of the construction on the island. Among other things, they planned to install twenty kilometers of streets and roads, twenty-five kilometers of train tracks, a new harbor, six kilometers of four-foot-high dykes along the coast, 600 dwellings for employees, barracks for 4,000 construction workers, mess halls, a new administration building, and an apprentice workshop in the production

World War II and the Holocaust (Cambridge, MA: The Belknap Press of Harvard University, 2006).

[46] Bode and Kaiser, *Raketenspuren*, 27–28.
[47] Neufeld, *The Rocket and the Reich*, 55.
[48] Arthur Rudolph OHI, NASM. Wichtige Daten bei der Durchführung des Vorhabens Peenemünde, July 5, 1941, FE 342, NASM.
[49] Neufeld, *The Rocket and the Reich*, 114. Schubert Vortrag, June 7, 1939, BA/MA RH8/v.1206.

plant itself.[50] Planners also included walking paths, park benches, gardens, and a sports field in the Settlement.[51] The chief factory planner, Godomar Schubert, put the cost of construction at 180 million Reichsmarks.[52] Dornberger and Schubert wanted to build a modern, "model" industrial facility. All of the buildings, their technical equipment, and their accommodations for the employees were to be first rate.

While all of this planning and construction went ahead, the Third Reich plunged the world into war. On September 1, 1939, a gray, cloudy day in central and Eastern Europe, German troops, tanks, and airplanes stormed over the Polish frontier and swept eastward. Hitler had made a calculated gamble that Britain and France would abstain from fighting, but both pledged their assistance to Poland and declared war on Germany.[53] The outbreak of war would prove to be both a blessing and a curse for Peenemünde. At first, draft call-ups threatened to undermine the available workforce. Later, huge amounts of men and material would pour into the base, but political pressure to finish the A-4 project increased exponentially as the war continued.

The onset of war brought no rationalization or cooperation to the hectic and uncoordinated German rearmament effort. Competition for materials deepened, and the massive financial layout that Schubert forecasted caught the attention of powerful people in the Reich government. The extravagance of the project met with resistance from Armaments Minister Fritz Todt, who was making strenuous, if only partially effective, efforts to curb the massive consumption of raw materials at construction projects across Germany, especially for projects that showed no signs of immediate completion or success.[54] His efforts became especially sharp as the war lengthened. In 1941, Todt ordered, among other things, that all new buildings everywhere in Germany must be planned simply and sparingly, whereas aesthetic considerations were to play no role whatsoever in construction.[55] Todt was so aghast at the lavish construction in Peenemünde that he sent a special envoy, Eduard Schönleben, to the facility to rein in the ambitious construction project. Dornberger and Schubert, confident of their Army backing, brushed Schönleben aside. Schubert was incensed that the meddling Schönleben would have the temerity to try to restrict the construction at Peenemünde. In his chronicle of the Peenemünde production plant,

---

[50] Neufeld, *The Rocket and the Reich*, 113. Dornberger to Becker, March 31, 1939, FE 342, NASM.

[51] Entstehungsgeschichte der Fertigungsstelle Peenemünde, February 10, 1939, RH8/v.1206, BA/MA. Hereafter cited as Entstehungsgeschichte.

[52] Schubert to Speer, October 12, 1939, RH8/v.1206, BA/MA.

[53] Kershaw, *Hitler: Hubris*, 221–222.

[54] In April, Todt ordered a halt to all construction on projects that could not be completed by October. See Entstehungsgeschichte, April 6, 1940, FE 342, NASM.

[55] Fritz Todt, Richtlinien für behelfsmässige Kriegsbauweise, July 2, 1941, FE 831, NASM.

he wrote, "Under no circumstances can I accept Dr. Schönleben's views. Dr. Schönleben believes that what is possible at the front must also be possible at home." Dornberger and his colleagues reasoned that if employees at Peenemünde had the best living and working accommodations possible, they would perform better on the job. According to Schubert, Dornberger's position was simply that "The employees' happiness at work will suffer if the working conditions are too primitive."[56] For his part, an angry Todt wrote to General Friederich Fromm, Commander-in-Chief of the Home Army, to complain about Dornberger's efforts. "I am convinced," he wrote, "that the actual useful work toward the goal can be done quickly without increasing the laborers very much if we remember that we are living in a war and if the guidelines for makeshift construction are employed. In Peenemünde, they have created a paradise. The accommodations, the social provisions [*Sozialeinrichtungen*], clubs and apartments, the factory halls, the warehouses, all exhibit the highest degree of expense that one can possibly imagine."[57] Todt's complaints fell on deaf ears, however. Dornberger and Schubert pushed ahead with their own plans for Peenemünde. Just as Todt began to receive the powers he needed from Hitler to conduct a major overhaul of the war economy, he was killed in a plane crash while leaving East Prussia in February 1942.[58] Albert Speer replaced Todt, and the missile program would enjoy a great deal of support from the ambitious architect for most of the remainder of the war.[59]

Thus, throughout the 1930s and into the 1940s, civilian missile specialists working for the Army found themselves drawn ever closer to the regime that made their work possible. The armed forces welcomed their talents, financed their research, raised their professional status, and fed their creative energies by guaranteeing them the most technically advanced research facility in the world and dedicating millions of Reichsmarks to a project that a number of them had labored on in relative obscurity for years. This massive state commitment to missile technology also had a dramatic effect on the quality and pace of research. The rapidly expanding budget for missile development combined with high-level support for the work to override most of the military's remaining doubt about missile technology. The technical goals sketched out by Dornberger, von Braun, and Riedel for the A-4 missile were also clear cut, even if questions as to how to achieve these aims persisted. This clarity provided a collective technical focus that helped prevent the internecine strife between developers that was stunting Soviet missile

---

[56] Entstehungsgeschichte, October 1 to March 1941, FE 831, NASM.
[57] Todt to Fromm, July 30, 1941, FE 342, NASM.
[58] Franz Seidler, *Fritz Todt: Baumeister des Dritten Reiches* (Munich: Herbig, 1986), 367–369.
[59] On his early unequivocal support of the V-2 program, see Albert Speer, *Inside the Third Reich* (New York: Macmillan, 1970), 469–470. For a full account of Todt's battles with Army authorities over Peenemünde, see Michael Neufeld, "Hitler, the V-2, and the Battle for Priority, 1939–1943," *Journal of Military History* 57 (July 1993), 511–538.

development in the same period.[60] All of this meant an increased level of official support, professional independence, and personal satisfaction, even if it was carried out under the sponsorship of a regime that had plunged Germany into a cataclysmic war.

THE SECRET BEARERS

Secrecy was the central fact of life at Peenemünde when its new residents arrived there in 1937. The secrecy around the project had a decisive effect on how missile developers saw themselves and their work. It erected the framework within which the Peenemünde specialists came to understand their place and roles within German society. Indeed, the practice of secrecy was the very basis upon which the institution of Peenemünde remade individual identities. In both a positive and negative sense, the habitual and all-encompassing practices of secrecy remolded the specialists' identities, increasingly defining them as a community of elite weapons designers in the service of the Nazi state.[61]

Peenemünde was a unique community. Within this community, the practice of secrecy was the cornerstone of the process by which Ordnance and civilian administrators were able to bring together a large group of people with disparate political and social views, foster identification with the goals of the military installation, and encourage them to work cooperatively on the missile project. Anthropologists, sociologists, and ethicists who have studied secret societies have shown that secret practices are linked to specific ways of dealing with knowledge within a larger society and are a potent tool for destroying old identities in order to replace them with new ones. Initiation into secret groups is a key feature of social organization and identity. Once individuals are let into the circle of knowledge, secrecy acts on them to foster loyalty and community among those who practice them, while isolating those individuals who do not have access to the secrets being protected.[62] Sissela Bok argues that members of secret societies are united by "Secrecy itself: secrecy of purpose, belief, methods, often membership.

---

[60] In the 1930s, Soviet missile development was riven by internal personal and technical disagreements, which exploded into often life-threatening political disputes. Scholarly research has only recently begun to uncover these conflicts. See Asif Siddiqi's important article, "The Rockets' Red Glare: Technology, Conflict, and Terror in the Soviet Union," *Technology and Culture* 44 (July 2003), 470–501.

[61] It intellectually useful to view the research base at Peenemünde as a secret society. Georg Simmel has developed a somewhat stylized, though useful, typology of the internal dynamics of secret societies that has formed the basis for most of the sociological and anthropological work that followed it. Many of the characteristics he outlines compare well to conditions at the missile center. See Kurt Wolff, ed., *The Sociology of Georg Simmel* (Glencoe, IL: The Free Press, 1950).

[62] See, for example, Gilbert Herdt, *Secrecy and Cultural Reality: Utopian Ideologies of the New Guinea Men's House* (Ann Arbor, MI: University of Michigan Press, 2003), and Andrew

In this way . . . the secret societies promise the brotherhood and community feeling that many lack in their daily life. [They] give insiders [a] stark sense of separation from outsiders."[63] Part of the attraction of secret societies is that not only do members gain meaning in their own lives, but they also are able to participate in something beyond their own individual existence that they view as having an overwhelming importance for a larger cause. Secrecy, therefore, is often an adaptive, community-building process that can play a vital role in social life, enabling groups that hold communal secrets to achieve a particular set of objectives and decisively transforming the networks of relationships occupied by those who are subject to its practices.[64]

On the ground at Peenemünde, Ordnance created a huge, secret world that was almost totally isolated from the rest of Nazi-era Germany. From the standpoint of the demand for secrecy, the area was ideal. The physical separation of the facility was one important way in which employees were cut off from the outside world. Usedom was (and remains) a remote, heavily forested island located on the Baltic Sea approximately 100 miles due north of Berlin. It is separated from the coast by the Stettin Lagoon to the south, the Peene River to the west, and the Swine Channel to the east. The island was not directly connected to any major roadways and was accessible only across three bridges that military authorities guarded closely once hostilities began in 1939. The research center, several test stands, and many employee accommodations were located on the more isolated northern peninsula of the island. According to one contemporary, "A tight fence and stringent regulations" separated the Army establishment at Peenemünde East from the Luftwaffe facility.[65] Usedom's northern peninsula allowed test engineers to launch their experimental rockets on an eastward trajectory over the Baltic, thereby helping to maintain the secret nature of their work and ensuring that it did not crash over populated areas. The Army could also erect measurement stations along the coast to track launch tests. The largest settlements on Usedom were the small tourist destinations that were scattered along the coast southeast of Peenemünde. Before construction engineers arrived to transform the quiet peninsula, the closest train station was in Trassenheide, connected to the fishing village by a seven-mile footpath.[66]

Lattas, *Cultures of Secrecy: Reinventing Race in the Bush Kaliai Cargo Cults* (Madison, MI: University of Wisconsin Press, 1998).

[63] Sissela Bok, *Secrets: On the Ethics of Concealment and Revelation* (New York: Vintage Books, 1989), 46.

[64] See also Stanton K. Tefft, ed., *Secrecy: A Cross-Cultural Perspective* (New York: Human Sciences Press, 1980), 13–17.

[65] Peter Wegener, *The Peenemünde Wind Tunnels: A Memoir* (New Haven, CT: Yale University Press, 1996), 16.

[66] Peter August Rolfs, *Die Insel Usedom: Ein Heimatbuch und Reiseführer* (Husum: Husum Druck- und Verlaggesellschaft, 1991; original work published in 1933), 9.

The heavily forested island offered an abundance of natural camouflage, and construction planners attempted to remove as few trees as possible in order to conceal activities there. Despite the thousands of workers sent to the island, massive construction projects, and thunderous engine tests, one engineer recalled that "Peenemünde never lost its character as an isolated wilderness."[67] This isolation would prove a boon to the ongoing work on Usedom, but it also sealed off employees of the missile center from the rest of Germany, markedly limiting their contact with the outside world. This isolation made the facility a refuge in the violent war years, and it instilled in the Peenemünders the notion that the bloodshed and war wrought by the Nazi regime would remain at arm's length. When the destruction of the war burst upon them in the middle of 1943, it revealed the depth of their complacency and its terror shook them deeply.[68]

Physical isolation was only one way in which the Peenemünders maintained the secrecy of their project. The construction of the massive facility necessitated a huge expansion in the number of specialists who worked on the missile, and the demands of the war allowed administrators to bring in trained military and civilian personnel. Secrecy considerations, however, forced Ordnance to face the dilemma of luring skilled workers to Peenemünde without actually informing them of the kind of work they would do if they were hired. Prospective employees could not get wind of the ultrasecret work until they actually set foot on the base, and activities at Peenemünde could not be concealed without first properly educating the employees about the welter of rules regarding secrecy. The logistical problem of maintaining secrecy while interviewing and hiring new workers, thereby dramatically widening the circle of those "in the know," without informing them directly of the work going on at Peenemünde, was overcome by resorting to an ungainly, time-consuming process that itself turned to secrecy for successful completion. After obtaining the permission of the Army authorities, management at the base posted advertisements for skilled positions in major urban newspapers without actually making clear the location of the work, the employer, or the nature of the job to be done. The advertisements stated that interested individuals should send their applications to an anonymous address in Berlin.[69]

Once applications began arriving, they were screened for the requisite skills; those applicants who passed this screening received background questionnaires in the mail a few weeks later. Once managers received the questionnaires, they interviewed suitable applicants at an innocuous off-site location. They selected the best applicants on the basis of both their personal

---

[67] Dieter K. Huzel, *From Peenemünde to Canaveral* (Englewood Cliffs, NJ: Prentice-Hall, 1962), 50.

[68] I examine the British bombing raid of August 1943 and its effects in Chapter 4.

[69] Arthur Rudolph OHI, NASM.

character and technical knowledge. At this stage, the applicants, whose appetites were sometimes whetted with mysterious allusions to such things as advanced flight mechanics and guidance of flying bodies, still had no idea where the work was to be carried out, nor did political inclinations figure at all in decisions about whom to hire. Before he arrived at the facility, one engineer stated that he had no idea what went on there and that Peenemünde was for him "a Chinese word."[70] Managers of the responsible civilian and military offices discussed those applications that they considered the most promising (by now copied in quadruplicate) so that they could avoid any conflict over conscription. Meanwhile, Ordnance officials enlisted both the Army Counterintelligence Organization (the *Abwehr*) and the Gestapo to examine the selected applicants for possible links to supposedly dangerous domestic or foreign elements. After these investigations, the local police branches checked the applicants' background for any criminal behavior. Only after passing this rigorous application, background, and screening process were applicants promised a job. If they accepted, they were let into the secret of Peenemünde.[71]

Practical considerations sometimes overrode this formal hiring process when the work required the quick application of particular skills. Personal connections and recommendations proved to be exceedingly important, though secrecy considerations still dictated that caution be taken with information given to outsiders. For example, in 1935, a friend of von Braun's who worked at Kummersdorf introduced the development chief to engineer Bernhard Tessmann, who would go on to become an important figure in production planning at Peenemünde. On the basis of Tessmann's qualifications and on the recommendation of his friend, von Braun asked Tessmann to come to Kummersdorf, explaining only that there was "interesting work there and it [was] a good place for young engineers just starting out." He told Tessmann nothing about what kind of work was being done there, only that it was an entirely new field of research and development.[72] Von Braun also hired several of his former colleagues, Nebel obviously excepted, from the *Raketenflugplatz* days.[73] Many of the leading administrative heads were also

---

[70] Georg von Tiesenhausen, Interview with sociologist Donald E. Tarter, University of Alabama Huntsville (UAH), hereafter cited as "Huntsville Interviews, UAH." Tartar conducted over thirteen hours of videotaped interviews with former Peenemünders residing in Huntsville, a series he entitled "Our Future in Space: Messages from the Beginning." The result of this work, a comparative essay published as "Peenemünde and Los Alamos: Two Studies," *History of Technology* 14 (1992), attempts to compare the work environment at the German missile base with that at an American atomic bomb facility. The work falls prey to the widespread postwar myths about Peenemünde and utterly lacks any sophisticated understanding of life in the Third Reich.

[71] Richtlinien für die Werbung von Facharbeitern, RH8/v.1429, BA/MA. See also Dornberger, Abwehrauskunft über Dipl. Ing. Otto Muck, FE 366, NASM.

[72] Bernard Tessmann OHI, NASM.

[73] Von Braun, "I Reached for the Stars," Wernher von Braun Papers, Box 200, Folder 7, SRCH.

hired through their personal connections with individuals already in place at Peenemünde. Though a number of important people at Peenemünde were party members, there is no little evidence that political considerations had any influence in the hiring process.[74] Expertise, not party affiliation, guided the decision-making process.

One key hire who came to Peenemünde this way was Ernst Steinhoff. Born on February 11, 1908 in Treysa, near Kassel, Steinhoff received his Diploma-Engineer degree in 1933 from the Technical University of Darmstadt. During his studies, he became an avid gliding enthusiast, and after graduation he entered into employment at the German Research Institute for Glider Flight (*Deutsche Forschungsanstalt für Segelflug*), which was under the direct administrative control of the RLM. According to Dornberger, von Braun met his fellow gliding enthusiast at the school in 1939.[75] Dr. Hermann Steuding, a Darmstadt instructor who helped develop guidance theory for the technical work going on at Peenemünde, probably introduced them. On the strength of Steuding's recommendation, Steinhoff, a dedicated National Socialist who joined the Nazi party in 1937, began working at Peenemünde in July 1939 as chief of the Guidance section. In 1940, Steinhoff completed his doctorate in engineering at the Technical University of Darmstadt.[76]

Steinhoff was awed by the prospect of working at the base. On his first visit to the base, he witnessed an engine test and immediately went to Dornberger, begging to be brought aboard.[77] Steinhoff quickly demonstrated his worth, establishing useful contacts that yielded important long-term benefits. On the job, he earned a reputation as a demanding boss, but also for making sure that the people who worked hard for him received their due. Though he did not make many fundamental contributions to the basic design of the A-4, he was an excellent administrator with many useful contacts in the technical professions, many of whom eventually found themselves working at Peenemünde.[78]

Steinhoff was an ardent National Socialist, but he was not hired because of his party membership. Such a distinction meant little to the daily technical activities within the community at Peenemünde. At work, he rarely invoked National Socialist ideology. The idea of building a space-age weapon that would help Nazi Germany win its war appealed to Steinhoff, and that was enough for him. In short, he was defined not by his enthusiasm for Nazism but rather by his zeal for the technological work of missile development. Steinhoff's Nazi connections may have helped with his outside contacts to

---

[74] For a thorough discussion of the impact of Nazism at Peenemünde, see Chapter 3.

[75] Dornberger, *V-2*, 15.

[76] Steinhoff IRR Dossier, Box 400, Entry 143B, RG 319, NARA.

[77] Dornberger, *V-2*, 15–16.

[78] Steinhoff Basic Personnel Record, Box 703, RG 165, Folder "Boston," NARA. Steinhoff IRR Dossier, Entry 143B, Box 400, NARA; Dornberger, *V-2*, 15; Neufeld, *The Rocket and the Reich*, 101.

technical expertise, and it is possible that Steinhoff may have benefited in more subtle ways from his membership in the Nazi party, but his support for party principles was not the reason for his important position in missile research. Professional qualifications mattered most in the hiring practices and the day-to-day activities at Peenemünde.

All of the employees who arrived to work at the base discovered a layered security system that projected state power, kept interested eyes out, and closely regulated behavior inside the facility. Well-armed Army security units controlled the bridges leading to the island and checked the various travel papers of people who wished to gain entrance to Usedom. Their presence was especially prominent in the town of Wolgast, one of the primary crossing points from the mainland to Usedom.[79] Before hostilities broke out, tourists could still visit the island's beach towns, such as Zinnowitz, Karlshagen, and Zempin. After September 1939, however, only Usedom's residents had controlled access to the island. The area of Usedom from Karlshagen southeastward along the coast was not totally off limits and individuals could move about freely, but residents who lacked the proper pass that gained them access to the island were not allowed. The pass was valid for twelve months, and people who lost it were fined fifty Reichsmarks.[80] This area served as a buffer zone between the more sensitive grounds of the missile facility and the outside world. However, Usedom's residents were not allowed to venture on to the northern peninsula of the island (from Karlshagen northward), which was the site of the development workshops, test stands, production factory, and many accommodations. Only employees and guests of the facility could travel into this area. The entire peninsula was an area reserved for official military use (*Sperrgebiet*), and Ordnance did not allow anyone access to it who did not have the proper clearance.[81]

To have access to the base itself and to the secret information within it, employees were required to wear a unique aluminum badge that Army security issued to them upon arrival. The badges were different colors and shapes so that base security could easily recognize them. The badges allowed employees on to the base, but they also indicated to security officials where each worker was allowed to go while inside the facility. Peenemünde

---

[79] Kurt Bornträger Testimony, "Hitler's geheime Waffenschmiede Peenemünde," Dir. Jakob Kurzenhalt, Polar Film and Media, GmbH, 2001. In May 1943, when Peenemünde administrators began introducing concentration camp labor into production, SS Chief Heinrich Himmler instructed that the first SS guard posts be set up at the base gate at Karlshagen. This was one indication of the SS's growing role in the rocket program, to be discussed in Chapter 4. See also Neufeld, *The Rocket and the Reich*, 199.

[80] Horst Lukat Interrogation, RG 165, Records of the War Department General Staff, Entry 179, Enemy POW Interrogation File (MIS-Y), Box 647, NARA.

[81] Gustl Friedl Testimony, "Hitler's geheime Waffenschmiede." Friedl was one of von Braun's secretaries.

was divided into several different secure zones that corresponded to the specialized work in each area. These included separate zones for the Aerodynamics Institute, the guidance workshops, the production factory, and, after August 1940, the propulsion laboratories. Most workers were prohibited from freely entering areas in which they did not work. A technician in the guidance department, for example, could not just walk into the propulsion laboratories without a very good reason and the proper permission. Along with these badges, all employees had to carry identification papers with them at all times and present them upon request. Armed security personnel conducted regular identification checks on the factory train that ran from Zinnowitz to the development works. The guards also frequently rechecked badges and identification papers at the train stops.[82] In contemporary parlance, this was the first level of a "need to know" system that was designed to compartmentalize information. It limited how much any single person might understand about the project as a whole and complicated any potential efforts at espionage.

The use of the badge system made secrecy itself a sign of privilege. Badges entitled the individual bearer to physical access to the facility, making entrance to the base the reward of a select few.[83] The division of the base into security zones requiring one particular type of badge for entrance also strengthened the internal hierarchy of the Peenemünde community. Individual technicians and engineers working in their specialized, compartmentalized areas had very little way of learning the full scope of the work done elsewhere. Full knowledge of the work taking place at the base was limited to a select few. Only the highest-ranking military officers in the program, such as Dornberger, base commander Leo Zanssen, and their staffs, as well as civilian executives such as von Braun and Arthur Rudolph, had access to any area within the facility.[84]

After making their way past patrolling guards outside the base and through security posts, employees worked within a structure of layered security measures that were designed to enforce the demand for absolute secrecy. Foremost among these were oaths of secrecy. Sociologists have long understood that all secret societies seek to reinforce among their members the secrecy that forms the basis of the group. Oaths and threats of punishment are the central features in the effort to reproduce secrecy among initiates.[85] This was entirely the case at Peenemünde. A condition of employment was

---

[82] Huzel, *From Peenemünde to Canaveral*, 31.

[83] Anthropologist Richard Schaeffer discusses this phenomenon at length in his essay "The Management of Secrecy: The Ku Klux Klan's Successful Secret," in Tefft, ed., *Secrecy*, 161–174.

[84] Plaketten für Zugangsberechtigung zu den Sicherheitszonen von Peenemünde, 1938–1944, *Historisches-Technisches Informationszentrum Peenemünde*.

[85] Wolff, ed., *Simmel*, 349.

FIGURE 2. Employees at Peenemünde received security badges with unique sizes, shapes, and colors that both allowed them onto the base and indicated where they could go once inside. These two badges were worn between 1939 and 1941. [*Historisches-Technisches Informationszentrum Peenemünde*]

a signed declaration of secrecy and a sworn oath to remain quiet about the work at the base. New employees, whether civilian or military, pledged that they would uphold secrecy regulations and do their best to assist with personal and document security.[86] Finally, all personnel, military or civilian, visitors or employees, had to sign declarations of secrecy if they were present at any tests of missile technology.[87] "I have been informed," read one declaration, "by Herr Heinisch of the Army Research Station Peenemünde that I must keep silent to everyone about all knowledge of work and facilities at the Army Research Station Peenemünde and the Greifswalder Oie [a small island used for test launches] as well as about what I have seen personally or learned in conferences. It is communicated to me further that this oath of silence is a requirement as well as a prohibition issued from the Reich Government for the Guarantee of National Defense in the sense of section 92b of the Reich Penal Code. I have also been made aware that a transgression against this oath of silence is punishable according to the Law Against Bribery and Betrayal of State Secrets of 5/3/1917, as amended 2/12/1920, as well as the stipulations of section 88 of the Reich Penal Code."[88] These signed declarations formed the backbone of the effort to keep the activities at the base secret. The base commander's staff, which maintained files on all

[86] Dienstanweisungen Werk Ost, July 1, 1937, FE 348, NASM. Also see Guido de Maeseneer, *Peenemünde: The Extraordinary Story of Hitler's Secret Weapons V-1 and V-2* (Vancouver: AJ Publishing, 2001), 193. Maeseneer's book is one of the most recent works published by uncritical admirers of the former Peenemünde employees and glosses over a number of the more troubling questions about their activities. However, it is valuable in that most of it is based on conversations with the Peenemünders themselves and reveals much about the noncontroversial aspects of daily life at the base.

[87] I will examine the cultural dynamics of missile testing in Chapter 3.

[88] Verpflichtserklärung, October 1938, RH8/v.1215, BA/MA.

of the employees, retained the signed oaths and distributed them as needed to administrators throughout the facility.[89]

These requirements had a dual effect. Like the badging system, they sharply delineated the Peenemünders' universe of knowledge from the outside world. The work inside the fence at Peenemünde further became the specialized activity of the privileged few. This dynamic increased the employees' sense of segregation from the outside world, but it also increased their sense that they were favored with secret, and therefore important, information. The signed declarations also thrust state power squarely into the Peenemünders' world, spelling out the disciplinary measures in store for those who transgressed their vows. They made individual employees responsible for their own behavior and transformed the usual free exchange of information into a potentially criminal offense. In effect, they obliged the Peenemünders to hold the rules of their new community above those of their previous lives. These transformed obligations decisively impacted their interaction with the larger German society in which they lived.[90] On the practical level, Peenemünde employees could no longer discuss what they did for a living, where they worked, or their job frustrations and joys; they could not have any of the other usual conversations that constitute standard interaction with friends and family who lived outside the fence. This was highly covert work, a "black" operation of technology development. The signed declarations of secrecy helped make this point clear to everyone.

Efforts to promote secrecy and reminders of its importance were common. Base administrators put up posters around the facility reading "What you see, what you hear, when you leave, leave it here."[91] Another poster warned employees to "Be careful what you say – the enemy is listening!" Employees found their worlds regulated by rules of secrecy in other ways as well. Service regulations imposed tight limits on telephone conversations. Those who wanted to use a phone had to get permission from their supervisor, and conversations were to be short (perhaps to save money and lessen the strain on the island's telephone system more than anything else).[92] On a more informal level, nearly every employee found it safer and more security conscious to speak and write about the missile in simple euphemisms. At first, Ordnance officials referred to the missile as a "smoke trail instrument" (*Rauchspurgerät*). Over time, they just began calling the missile their

---

[89] Dienstanweisungen Werk Ost, July 1, 1937, FE 348, NASM. The basic outlines of these regulations remained in place throughout the war years.

[90] Anthropologists working on a variety of subjects have noted the transformational power of secrecy oaths. See Bok, *Secrets*, 21 and Michael S. Laguerre, "Bizango: A Voodoo Secret Society in Haiti," in Tefft, ed., *Secrecy*, 147–158. See also Hugh Gusterson, *Nuclear Rites: A Weapons Laboratory at the End of the Cold War* (Berkeley, CA: University of California Press), 68–100.

[91] Huzel, *From Peenemünde to Canaveral*, 31.

[92] Dienstanweisungen Werk Ost, July 1, 1937, FE 348, NASM.

"instrument" (*Gerät*). The subject of most of the correspondence at Peenemünde was the "Instrument A-4." The A-4 label given to the ballistic missile was itself purposely opaque. The "A" in these designations stood for "Aggregat" or "assembly."

These euphemisms and jargon became a part of everyday life at Peenemünde. Their use was so common that administrators at the base developed a formal list of code words and nicknames in order to regularize the terms used to refer to various parts and technologies. A new technological language emerged, understandable only by Peenemünde's secretive employees. The onboard radio receiver (*Funkkommandoempfänger*) was nicknamed "Honnef" and formally designated FT-Kdo-21b. The onboard telemetry transmitter (*Messwertsender*) received the code name "Messina," or Ms-1-92a. Even fire extinguishers received the oblique designation "Intra," and the launch platform came to be both formally and informally known as the "table" (*Tisch*). Thankfully for the employees, administrators made no attempt to give these items alphanumeric designations.[93] These terms not only obscured the objects' meaning and use to outsiders, they also represented a linguistic and bureaucratic barrier to membership in Peenemünde's increasingly exclusive club of technological elite. Without the proper initiation and training, technical specialists who otherwise had at least a moderately good theoretical knowledge of such types of technology would be hopelessly lost in the welter of coded terms used not only in written correspondence but also in oral communication.[94]

---

[93] Tarnbezeichnungsliste für Gerät A4, FE 330, NASM.

[94] This coded language is also the hallmark of the ever-increasing professionalization of rocket engineering. Professional, problem-solving groups use jargon to represent the ideas that emerge within this framework. The discourse created by the use of jargon provides a space of mutual understanding that is not commonly shared by others. This is a common trait in the professional certification of knowledge and the incumbent separation of individual professional groups from other segments of society. Individuals within professional groups travel over a common linguistic ground that both affirms their expertise and announces their social and experiential partition from the larger society in which they live. At Peenemünde, the missile specialists were no different. Even if some technical experts outside of their community understood the fundamental characteristics of some of the equipment they used, the outsider was not only unable to comprehend the use to which it was being put, but also completely incapable of penetrating the language used to refer to the technology in the first place. Only those specialists who had been initiated into the secret world at Peenemünde and given access to its forbidden knowledge were capable of reproducing the jargon by putting it to use. Secrecy, which in this case is the use of code words, enhanced the idea of a nascent profession in development at the facility. Though no one at Peenemünde referred to themselves as professional rocket engineers, they did experience a growing sense of professional elitism that was fostered by the utterly secret conceptual world that fundamentally shaped the way they viewed their work and the terms they used to discuss it. See Margatti Sarfatti Larson, "In the Matter of Experts and Professionals, or How Impossible is it to Leave Nothing Unsaid," in Torstendahl and Michael Burrage, eds., *The Formation*, 25–26.

Peenemünde administrators placed even more stringent regulations on the treatment of secret documents. Technical drawings, minutes of meetings, developmental correspondence, and administrative procedures were all more fungible, permanent, and often more specific than ephemeral conversations.[95] Indeed, security officials were well aware that these characteristics made documents the subject of increased interest and invited unwanted intrusion by prying eyes not only among foreign enemies but also among the Peenemünders themselves. Administrators at Peenemünde took great strides to limit the amount of information any single middle- or lower-level employee knew about activities at the facility. The more secrets an individual knew, the more damage would result were he or she to fall into enemy hands.

One of the most important ways that managers dealt with these concerns was to make every employee at the base absolutely cognizant of the need for good document security practices. The guidelines for document security were found in the service regulations handed out to each office and kept by the division heads. Ordnance first introduced these regulations at Peenemünde in July 1937. Management gave all new employees an overview of the regulations when they arrived at the facility, and administrators expected the Peenemünders to be intimately familiar with them. Every six months, on January 10 and July 10, employees signed a written attestation that they read the statutes and understood the regulations.[96] The service regulations outlined the center's organizational structure, the responsibilities of the division chiefs, the rules governing care and treatment of documents, employee responsibilities, and business trips. Though the heads of the administrative divisions amended them a number of times between 1937 and 1945, the broad general outlines of the conditions set forth in them changed very little, even if the more narrow details did alter over time. They offer a precise and thorough look into the daily practice of secrecy on the ground at the Army research center.

The regulations with regard to documents at Peenemünde were exhaustive. One test engineer remarked that "Office procedures and handling of classified correspondence were as cumbersome and strict as could possibly be."[97] A registrar in the administrative headquarters catalogued all incoming and outgoing letters according to their secrecy ratings into document registers. Administrators at Peenemünde generally employed three levels of secrecy for documents: top secret (*Geheime Kommandosache*), secret

---

[95] Simmel has noted that the written word is generally opposed to secrecy because it is more permanent than speech and "wholly unprotected against anybody's taking note of it." See Wolff, ed., *Simmel*, 352.

[96] Dienstanweisungen Werk Ost, July 1, 1937, FE 348, NASM.

[97] Huzel, *From Peenemünde to Canaveral*, 60.

(*Geheim*), and unclassified. Occasionally, a particularly sensitive document might bear the designation *Geheime Reichssache* or *Chefsache* (essentially, super top secret). Only the highest administrators at Peenemünde had access to these documents.

The document registrar distributed the resealed incoming mail by courier to the division that it pertained to. When it arrived in a particular division, only department supervisors and top administrative personnel were allowed access to documents rated top secret. Secret and unclassified documents were treated somewhat less stringently, but still with tight control. Only specifically and individually cleared employees could view these letters, and they had to do so under the supervision of their department supervisor. Workshop employees could only make copies of secret or top-secret documents with the permission and in the presence of the supervisor. Service regulations stipulated that files containing technical and developmental information, no matter what secrecy grade, were to be kept in the department supervisor's office and locked in a safe. The documents were then distributed from there, but regulations expressly forbade employees from removing them from their workshops. Documents containing information pertaining to the daily basic administration of the base were exempt from this rule. Only the department supervisor and his deputy were authorized to hold a key to the safe. If because of retirement, transfer, or even vacation, the department supervisor or his deputy were to be away from the workshops for an extended period of time, Peenemünde East administrators had the responsibility of making sure that he did not take any documents with him. Moreover, all of the personal papers of all employees in the workshops were the property of the base commander. If an employee left Peenemünde, the base commander and administrators of the development facility had the right to examine his personal papers to see whether or not they contained any classified information and, consequently, whether the personal papers should be retained at the base.[98]

Correspondence composed in the workshops was required to contain a list of the classified contents on the first page. Department supervisors sealed all outgoing letters in letter books and couriers delivered them to the Commander's Office. Most employees did not have courier privileges and were strictly forbidden from taking drawings, letters, and other documents out of the workshop. The base commander's staff examined all top-secret outgoing correspondence, and if the letters passed muster, they were resealed and sent off. Again, regulations permitted only specially designated couriers to carry outgoing messages from the base.[99]

Technical drawings created at Peenemünde posed an even greater security problem. Although typewritten documents contained information on the

[98] Dienstanweisungen Werk Ost, July 1, 1937, FE 348, NASM.
[99] Ibid.

function and design of parts and assemblies, they did not illustrate the parts layouts or how they fit together with each other. The technical drawings were the key to assembling a functional missile and therefore received the most stringent security precautions. Many of them were stamped with notices indicating the penalty for misusing the drawings. The hundreds of technical drawings for testing the A-3 missile, predecessor to the A-4, in late 1937 were stamped, "State Secret! This is a secret object in accordance with Section 88 of the Reich Penal Code (Version of 24 April, 1934). Misuse is punishable in accordance with the conditions of this law, provided that no other conditions of punishment come into question."[100]

A Drawing Administration division (*Zeichnungsverwaltung*) served as a repository and clearinghouse for the top-secret technical drawings. Regulations explicitly forbade employees from producing and storing design sketches in their workshops. If a sketch laid out the requirements for a part that needed to be developed, then it was forwarded to the Design Office, where illustrators produced the official design drawing. This step also helped to rationalize design changes. Employees were required to destroy their now unofficial hand-drawn sketches as soon as the Design Office created the official technical drawings. Officials in the Drawing Administration either forwarded the finished design drawing to Peenemünde's in-house developers, given only against a return receipt, or to a subsidiary private firm.[101] While the development and production shops inside the facility only required a special pass to receive these documents, drawings to be sent to subsidiary firms were packed in folders by the Drawing Administration personnel, bound with heavy tape, and sent to a document registrar's office for further packing and shipment.[102]

Peenemünders were conscientious about document security, but the explosive growth of the base after 1937, coupled with the steady technological advances, complicated document management and opened up gaps in the base's security measures. Peenemünde retained an element of craft-style production for parts in development, and constant tinkering made coordinating the management of design and production drawings extremely difficult. In reality, the management process for drawings did not function quite as smoothly as it looked on paper, especially when the regime prematurely demanded a start to mass production. The nearly continuous, ad hoc changes to design drawings complicated document security by undermining accountability for the documents. By 1940, the Drawing Administration was losing track of the documents it was set up to manage. Only in late 1942 did the

[100] See, for example, technical drawings enclosed in von Braun to Dornberger, September 20, 1937, FE 367, NASM.
[101] Dienstanweisungen Werk Ost, July 1, 1937, FE 348, NASM.
[102] Aktenvermerk über Überprüfung der Zeichnungsverwaltung in Peenemünde durch Oberstleutnant Krehnke am 29. u. 30.4.42, May 1, 1942, RH8/v.1215, BA/MA.

Drawing Administration and Design Office begin to assert more systematic control of the documents.[103]

The program's administrators also issued supplementary orders that complemented this host of regulations. Though Ordnance loosened its secrecy regulations somewhat in fall 1939 to permit greater contact between Peenemünde and the universities, officials still insisted that anyone with knowledge of the program follow strict secrecy guidelines. These orders became especially necessary with the massive expansion of the program that was inaugurated when the changeover to mass production brought hundreds, if not thousands, of people into the secret of Peenemünde. For the most part, these orders reflected new considerations because of the increased production of paperwork, greater number of business trips, and increased level of awareness of the missile project as the program ramped up for mass production in late 1942 to early 1943. Walter Dornberger published a set of orders around this time aimed at reinforcing the existing regulations and emphasizing the need for proper drawing and document management.[104] In July 1943, Heinz Kunze, Deputy Director of the A-4 Special Committee, the Armaments Ministry group detailed to coordinate raw materials delivery, development, production, and quality control of the missile, ordered a strict compartmentalization of information in all correspondence with firms outside of Germany proper. He directed that foreign companies under German control not even be informed of the existence of the A-4 program. All letters to them were to be categorized as top secret and references to A-4 development and production were to be made in only the most oblique terms.[105] Though orders such as this one only impacted the Peenemünders marginally, they are illustrative of the massive effort put forth to maintain a thoroughgoing sense of secrecy around the program, even as production created a situation in which more people inevitably became aware of the work. Even as the missile program experienced massive growth in the middle of the war years, the maintenance of secrecy around the work remained of paramount importance.

Dornberger reminded the Peenemünders of this in a major speech he delivered in 1943, just before mass production was scheduled to begin. "You must remember," he warned, "That every thoughtless word you speak about our work here, about our results, about our goals, can come to the ears of the enemy." He directed the Peenemünders to actively police themselves for secrecy violations, stating that "You yourselves are the best custodians of secrecy. ... Strike the gossipers on their big mouths [sic – "*Schlagt den*

---

[103] Ibid.

[104] Dornberger Circular "Geheimhaltung" (date unclear, likely late 1943), RH8/v.1254, BA/MA.

[105] Kunze to Leader of A-4 Special Committee Sub-Committee Leaders, "Geheimhaltung," RH8/v.1254, BA/MA.

Schwätzern aufs Maul"]. Get them on the hook and report them for punishment."[106] Loose lips, the General cautioned, could cost not only their own lives but also the lives of their families and coworkers. Anyone could be a spy, according to Dornberger, and Peenemünders should be wary of people who asked too many questions. The General then gave the assembled group an ominous warning: "Apart from the danger of the enemy, you also run the risk of being arrested and spending the rest of your days dressed in black and white as a prisoner. Be assured that in such a case, I will have no mercy."[107] Postwar memoirs and hagiographies of the "rocket team" totally ignore Dornberger's threat to imprison in a concentration camp, with no chance of release, those who broke secrecy regulations. It is perhaps the only direct surviving statement of such a warning from the program's senior military leadership. Dornberger had no objection to the use of force within the Nazi police state, especially when it served his ends. The threat exposed the Peenemünders even further to the violence, brutality, and capriciousness of the Nazi regime. With the Damoclean Sword of a place like Buchenwald or Dachau dangling over their heads, Peenemünders would understandably be careful not to break the rules.

The effort to keep activities as secret as possible became a part of the daily practical lives of people who lived and worked there. Employees integrated these regulations into their daily routines and took the restrictions that came along with them for granted. Adherence to the rules became a part of their common practical consciousness at home, on the way to work, at security checkpoints, and in the workshops. Peenemünders habitually looked over their shoulders and gave sensitive topics a wide berth. Otto Hirschler was a specialist in the guidance section who recalled that he was always circumspect in his conversations with coworkers in the train on the way to work. He and his colleagues found themselves glancing nervously around the train if the topic of work came up, and by silent consent, they refused to talk about sensitive subjects until they got to the lab.[108] Production manager Arthur Rudolph stated flatly that "It was *selbstverständlich*, it was understood, that you didn't [talk about the work]. You worked and didn't talk about it."[109] After engineer Herbert Lucht witnessed a test launch just after his arrival at Peenemünde in 1940 (probably of the A-5 missile, a smaller test version of what would become the V-2), his colleague told him "That is the most secret thing here in Peenemünde, and you can't say a word about it to your friends, at home, or at parties."[110] Werner Brähne

---

[106] Dornberger, Betriebsappell am 18.6.43, p. 4, FE 833, NASM.

[107] Ibid., 5.

[108] Otto Hirschler Testimony, Huntsville Interviews, UAH.

[109] Franklin, *An American*, 51.

[110] Herbert Lucht Statement, *Peenemünde: Schatten eines Mythos*, Dir. Matthias Schmidt, MJB Film- und Fernsehproduktion, 2001.

was a technical illustrator at Peenemünde who noted the curious fact that the possibility of receiving a death sentence for breaking the secrecy regulations did not seem to aggravate or upset the employees.[111] These attitudes speak to the strength of the regulations, but also are an indication of the degree in which they penetrated the practical consciousness of employees at the base. Observance of these, in anthropological terms, "rites" of secrecy at Peenemünde was nearly automatic.[112]

Secrecy, especially from the perspective of outsiders, also imparted a certain value to those practices that developed around it.[113] Before he became fully immersed in the Peenemünde community, test engineer Dieter Huzel recalled that "I admired from a distance those whom I believed had already achieved this higher order of existence – from the obscure language, the secrecy that seemed to shroud their actions, and from the occasional 'private notebook' tactics that some of them used."[114] The system in place that divided both the island and the missile facility into areas of greater or lesser prohibition placed an intrinsic value on the information and individuals who moved around in these areas. The practices of secrecy that Huzel pointed to, the coded, jargon-filled language, the curtain behind which they worked, and the very objects that were off limits, segregated the Peenemünde specialists and helped to foster the sense that they were a community of elites, one that worked at the very cutting edge of technological development.

Partially for this reason, secrecy regulations reinforced the hierarchy of authority at Peenemünde. Technological leaders controlled information by limiting access to a relatively small number of people. These individuals had to learn the proper uses of the secret information through a long process of group recruitment and training that was guided by their superiors. The authority of those in charge at Peenemünde, therefore, in part grew out of their larger knowledge of the technological activities on the island and their power to grant or deny access to secrets. Indeed, part of von Braun's leadership capability lay in his nearly omniscient knowledge of technical development that grew in part from his unfettered access to every secret in the facility. This knowledge cemented the strict administrative hierarchy at the base, a hierarchy that sociologists note is central to the effective functioning of all secret societies.[115] The authority of the leading administrators

---

[111] Werner Brähne, unpublished manuscript, "Die Mittelwerk GmbH. Eine Chronik über Firma und Werk," unpaginated, Gericht Rep. 299, Bd. 582, *Nordrhein-Westfälisches Hauptstaatsarchiv Düsseldorf, Zweigarchiv Schloss Kalkum* (HStaD-ZA Kalkum).

[112] I borrow the term from Gusterson, *Nuclear Rites.*

[113] Simmel noted that secret practices acquire value when others must do without them, and when those without them attribute special significance to those who practice them. He writes that "Inner property of the most heterogeneous kinds, thus, attains a characteristic value accent through the form of secrecy." See Wolff, ed., *Simmel*, 332.

[114] Huzel, *From Peenemünde to Canaveral*, 77.

[115] Wolff, ed., *Simmel*, 346–357.

at Peenemünde to decree regulations did not simply proceed from the positions accorded them by Ordnance. It also originated in their knowledge of developments at the base and was strengthened by the regulations published by the administrators themselves.

These practices formed the bedrock of all of the activities at Peenemünde. For new employees, the hiring process, background check, swearing of oaths, and signature of the service regulations were important moments that made the rules that bound them together in secrecy clear.[116] In an anthropological sense, these practices were initiation rituals that served to integrate individual newcomers into a larger, secret, and privileged group. In undergoing these rituals, neophytes were made aware of other members who worked at the installation, the formal and informal norms governing their professional existence, and of the stratification within the facility itself. Without these rites, which had the power to "rearrange and to transform allegiances, boundaries, and identities," secrecy could not be maintained.[117] For veteran Peenemünders, they reemphasized the fundamental importance of secrecy to their purposes and aided in tying them together as cohesive group.

The converse issue – limitations on public and professional discussion of the work – was not onerous for most of the specialists at the facility. Secrecy was an integral part of professional conduct among the Peenemünders. German professionalism, especially in the technical occupations, was defined as much by service to the state as by interaction with one's colleagues and membership in professional organizations. This concept had been drilled into them since their days in the technical universities and trade schools.[118] A willingness to work in secrecy was also a condition of their employment at Peenemünde. It was a given in this program, and those working at Peenemünde accepted it as a part of their duties. In some instances, the Army loosened its regulations when the need arose, but even these contacts were limited by an individual's need to know. For the Peenemünders, secrecy was a part of the professional code.

SECRECY, COERCION, AND CONSENT AT PEENEMÜNDE

Membership in a secret society like that at Peenemünde had certain coercive elements, both real and imagined, that affected individual perspectives. The physical security apparatus was impressive. The Army handled the bulk of the security work at the base throughout the war years. It provided guards

---

[116] Their working and living conditions, to be examined in the next chapter, also were important sources of binding energy.

[117] Bok, *Secrets*, 50.

[118] See, for example, Konrad Jarausch, *The Unfree Professions: German Lawyers, Teachers, and Engineers, 1900–1950* (New York: Oxford University Press, 1990), 3–24, 115–196, and Karl-Heinz Ludwig, *Technik und Ingenieure im Dritten Reich* (Düsseldorf: Droste Verlag, 1974), 103–159.

for the bridges connecting Usedom to the mainland, issued badges and identification papers, staffed the checkpoints scattered around the island, and
patrolled inside and outside the fence at Peenemünde. Later in the war, when
concentration camp prisoners arrived on the island, the SS supplemented the
Army's complement of guards around Usedom.[119] This was the large, visible security force on the base. A smaller, less visible security force was the
lynchpin of the system's effectiveness, however. This was provided by the
Gestapo, Hitler's dreaded secret police.

The Gestapo supplemented the Army's personnel security duties with its
own network of informants, who reported to the local office in Stettin, about
fifty miles southeast of Peenemünde.[120] Gestapo officials in Stettin took a
keen interest in the doings at Peenemünde. The earliest available record of
Gestapo operations in Peenemünde is dated November 1939. The massive
expansion of construction on Usedom because of the planned assembly plant
and accompanying worker accommodations made necessary the addition of
nearly another 1,500 laborers. In a striking contrast to the comfortable and
well-built accommodations for the technical experts, the facilities designed
to house the influx of construction workers were overwhelmed and strained
to the breaking point. The barracks built for them did not have enough beds,
nor were many of them heated against the Baltic winter. Mess halls built
for 1,000 men had to feed 3,000, forcing hungry workers to queue up for
over two hours. Many were lucky if the food was not all gone when they
reached the front of the line. These conditions led to a deepening discontent
among the workers and, worse, an ongoing and significant number of work
refusals.[121]

Army officials did not want potential worker discontent to delay construction at Peenemünde. In late November, the Army Counterintelligence Office
in Swinemünde contacted the Gestapo and requested their assistance.[122]
Almost six months later, officials from the office of the Reich Trusteeship
for Labor (*Reichstreuhänder der Arbeit*) in Pomerania, a government organization charged with acting as a liaison between labor and management in
large construction projects, also asked the Gestapo to help, and even recommended setting up a penal camp on Usedom to deal with the unhappy
workers.[123] Construction Directors Erwin Mahs and Heinrich Lübke (who
would become President of the Federal Republic of Germany in the late

---

[119] Niederschrift über die Besprechung am 7.8.44, August 8, 1944, RH8/v.1941, BA/MA.
[120] Stolze Bericht, "Bonner Bundepräsident Lübke," AV 7/85, Bd. 32, *Bundesbeauftragte für die Unterlagen des Staatsicherheitsdienstes der ehemaligen Deutschen Demokratischen Republik* (BStU).
[121] Untitled Gestapo report, author unknown, November 21, 1939, AV 7/85, Bd. 26, BStU.
[122] Unsigned telegram, Gestapo Office Stettin to Rühlmann, November 23, 1939, AV 7/85, Bd. 26, BStU.
[123] Reichstreuhänder der Arbeit to Gestapo Office Stettin (signature illegible), June 14, 1940, AV 7/85, Bd. 26, BStU.

1950s) from Baugruppe Schlempp, the Armaments Ministry organization that took over construction in May 1940, supported the idea of a penal camp, but they left the Gestapo in charge of organizing its supervision.[124] By September 1940, the Gestapo Office Stettin received full authority to monitor activities on Usedom. The official order granting this power indicated that they could rely on Mahs and Lübke for any help the Gestapo needed because "Both have proven themselves to be trustworthy."[125] Over the next several months and years, the Gestapo operated informants in Peenemünde that had access to the construction project as well as the research station's employees.[126] Army officials also sought the Gestapo's help in running background checks to ensure that prospective employees were not involved in any activities that the regime might deem untoward. Thus, the Gestapo did not have to surreptitiously infiltrate Usedom or engage in a bureaucratic battle with Army or construction authorities at Peenemünde in order to gain access to the base. Army Counterintelligence sought out the Gestapo's help so that Ordnance's research and production plans could proceed unhindered. Construction directors actively supported their efforts in order to keep the facility's frantic building activity moving forward. The result was active cooperation between Army Counterintelligence, Armaments Ministry representatives, and the dreaded secret police, not only to combat labor intransigence but also to ensure that secrecy was fully maintained on the island.

Army security officials and Gestapo informants on Usedom did a great deal to enforce the rules of secrecy. Watching over the scientists and engineers at Peenemünde and punishing security violators "as a matter of fact" provided an excellent opportunity for ambitious security officers.[127] Security forces on Usedom could only do so much, however. The employees themselves did the most to enforce strict secrecy and combat potential espionage or subversion. The internalization of the institutional regulations created a situation in which employees reined in their behavior and made sure to avoid breaking the rules. The presence of Army Counterintelligence and the Gestapo gave employees the sense that they were being watched and reminded them of the risks they ran if they broke the rules. Thus, security forces helped establish and reinforce the ubiquitous sense of surveillance around the project, but the employees also did their part to enforce the rules.

The feeling of police surveillance imposed a powerful form of discipline at Peenemünde. Its strength was in its randomness and invisibility. Peenemünders may have been able to spot the military police guarding

---

[124] Zusatzbericht zum Bericht vom 14.6.1940, June 19, 1940, AV 7/85, Bd. 26, BStU.
[125] Gestapo Order 5538/39, September 24, 1940, AV 7/85, Bd. 26, BStU.
[126] Stolze Bericht, "Bonner Bundespräsident Lübke," AV 7/85, Bd. 32, BStU.
[127] Huzel, *From Peenemünde to Canaveral*, 60.

the base, but the presence of the Gestapo was something else altogether. "One never knew when one was being watched," recalled an anonymous employee.[128] The Peenemünders understood that they could be observed, but did not know when, if it all, the gaze of the authorities fell on them. By default, employees assumed that they were always watched. Most accepted this as part of the bargain of working at Peenemünde, but this ubiquitous sense of surveillance meant that authorities could exercise constant control over employees even when they were not there.[129] The threat, if not the reality, of entanglement in the Gestapo's net was ever present. Ironically, the Gestapo probably did not have enough manpower to properly carry out its enforcement duties at Peenemünde. In 1937, it had no more than 7,000 employees to cover all of Germany. Even in August 1941, there were no more than 7,600 secret policemen in the entire country. Most of the arrests it made were the results of sometimes spurious tips from private citizens.[130] Its reputation for omnipotence was a myth.

An important factor in creating the feeling of police omnipotence was the preemployment background investigation that applicants submitted to during the hiring process. The Gestapo investigation transformed the employees from relatively anonymous people moving about in a large social milieu into individuals whose identity was closely known by the state and, therefore, subject to greater state control. In a totalitarian political system that possessed no scruples about invading the private lives of its citizens in the first place (and indeed obliterated the very notion of privacy), the sense of being under constant surveillance was intensified even further. Peenemünde employees were, in principle, under more scrutiny by the state than were citizens who were not involved in top-secret research. The Nazi regime stripped away the privacy rights of its citizens as a matter of course. The preemployment background investigations then drew the state's interest to individuals who applied for jobs at Peenemünde, enhancing the sense of being exposed at all times and encouraging individuals to closely regulate their own behavior. Peenemünders understood that the Gestapo was aware of who they

---

[128] "Bericht eines nicht genannten 'Peenemünder,' veröffentlicht in der Wochenzeitung 'Christ und Welt' im Juni 1950," Gericht Rep. 299, Bd. 158, HStaD-ZA Kalkum. This visibility is the key, according to theorist Michel Foucault. "Disciplinary power ... imposes on those whom it subjects a principle of compulsory visibility. In discipline, it is the subjects who have to be seen. Their visibility assures the hold of the power that is exercised over them. It is the fact of being constantly seen, of being able always to be seen, that maintains the individual in his subjection." See Michel Foucault, *Discipline and Punish: The Birth of the Prison*, trans. by Alan Sheridan (New York: Vintage Books, 1979), 187. Hugh Gusterson has demonstrated this to excellent effect in his book *Nuclear Rites*, 68–100.

[129] Werner Brähne, "Die Mittelwerk," Gericht Rep. 299, Bd. 582, HStaD-ZA Kalkum.

[130] Elisabeth Kohlhaas, "Die Mitarbeiter der regionalen Staatspolizeistellen: Quantitiative und qualitative Befunde zur Personalausstattung der Gestapo," in Gerhard Paul and Klaus-Michael Mallmann, *Die Gestapo – Mythos und Realität* (Darmstadt: Wissenschaftliche Buch Gesellschaft, 1995), 220–235.

were, and understandably they did what they could to avoid any unwanted attention.

Anonymous tips from private citizens at Peenemünde complemented the security effort. Fellow employees denounced one engineer for commenting critically on the war. The engineer spent a year in jail. Wilhelm Raithel got a sharp warning from a Nazi Party member and fellow employee to watch his tongue after he made unflattering remarks about the war.[131] There is also evidence, though quite spotty and poorly documented, that at least one, and perhaps as many as twenty, civilian employees were hanged by security forces between 1939 and 1945 for transgressions ranging from supposedly seditious comments to transgressions against secrecy regulations.[132] Even though these were exceptions to the rule in a facility with a population of nearly 6,000 at its peak, they made it clear to employees that they needed to be careful what they said and did.

On their own, employees narrowed the limits of allowable behavior and moderated any acts that could be considered inappropriate by state authorities. Wernher von Braun played an important role in this process of self-censorship. He vigorously guarded the secrets at Peenemünde and was unafraid to confront others about breaches in security. Von Braun regularly wrote letters and circulars to middle-level management and department heads reminding them of regulations governing secrecy and upbraiding them on the uncommon occasion when they did break the rules. As late as December 1944, when Germany stood on the brink of utter collapse and the war was irretrievably lost, Walther Riedel (known in the correspondence as Riedel III, no relation to von Braun's old friend Papa Riedel), the head of the Peenemünde Design Bureau, met with members of the Four Year Plan Institute for Transportation (*Vierjahrsplaninstitut für Kraftfahrzeuge*, or VfK) at Berlin Technical University concerning work on A-4 ground support equipment. The university detailed a secretary to facilitate the work in the room during the top-secret meeting. After several minutes and Riedel's repeated requests that she leave the room because of the secrecy of the discussion, an argument ensued with her superiors, and the angry woman departed in a huff. When von Braun received news of the episode, he wrote an acerbic letter to a Dr. Schmidt, the head of the group, in which he voiced his concerns about secrecy. His reaction to the incident reveals his own sense of propriety

---

[131] Wilhelm Raithel OHI, NASM.
[132] Manfred Kanetzky personal correspondence with author, April 27, 2004. Kanetzky is the archivist at the *Historisches-Technisches Informationszentrum Peenemünde* and has learned through conversations with former Peenemünders that at least one civilian was hanged, though documentary evidence that could prove this has not been found. He notes that others have cited up to twenty hangings. In each case, it is not clear if the Gestapo, Army, or civilian authorities charged the individuals and carried out the death sentences. Part of the problem with this information is that beheading, not hanging, was the preferred method of executing fellow Germans, at least until 1942.

as well as his overriding concern with keeping information about the A-4 to as few people as possible. "The improper tone and the general behavior of Frau Wolfe," he testily asserted, "exhibits a major lack of discipline. [We] are indignant over the above incident and the affront therein by one of your representatives. In the future, we will choose the meeting place for all further meetings with the VfK."[133] It is tempting to read a certain self-importance into this letter, but the subject line, "Secrecy" (*Geheimhaltung*), made von Braun's concerns clear. The control of secret information about the A-4 project simply could not be allowed to wane, even if Germany's fortunes in the war were. Even as his enthusiasm for the project began to diminish and the limits of the A-4's effectiveness became clearer by the day, von Braun proved himself to be more than willing to call onto the carpet those who breached the tight ring of secrecy around missile development, and his demand that the Peenemünders choose the location for future meetings with the VfK is indicative of his faith in the base employees' ability to tightly control access to secret information. Von Braun's own adherence to the secrecy regulations was fully automatic, a part of his identity as a missile developer. His internalization of Peenemünde's institutional regulations regarding secrecy was emblematic of many Peenemünders, and it resulted in a habitual, often proactive commitment to these rules.

Von Braun himself was not above the secrecy regulations either. In May 1943, he received an invitation to speak at a conference sponsored by the German Academy for Aeronautical Research (*Deutsche Akademie für Luftfahrtforschung*). The President of the academy, none other than Reichsmarschall Hermann Göring, stipulated that "the most secret things can and should be discussed at the conference" and promised "Nothing about it will be mentioned to third parties."[134] This was not enough for Dornberger, who was also keen on maintaining the Army's monopoly on missile technology and saw Göring as a rival. He did not give von Braun permission to participate and sent a curt response to the organizers: "OKH [the German Army High Command] refers to the Führer's order that the ongoing development in Peenemünde should be considered top secret, even super top secret [*Geheime Chefsache*]. Development may only be discussed when completion of the work is absolutely necessary. Since the Führer's order cannot be abrogated, Dr. von Braun cannot give a lecture on this topic."[135]

Secrecy, therefore, functioned in two ways. It segregated employees of the facility from the society around them while ensuring the loyalty of those who were let in on the secrets. On one hand, the practice of secrecy was

---

[133] Von Braun to Schmidt, "Geheimhaltung," December 2, 1944, RH8/v.1265, BA/MA.
[134] Deutsche Akademie der Luftfahrtforschung to von Braun, May 12, 1943, RH8/v.1960, BA/MA.
[135] Dornberger to Deutsche Akademie für Luftfahrtforschung, May 20, 1943, RH8/v.1960. Von Braun received his copy of this letter five days later.

a constructive process that built communal feeling and enabled individuals to achieve particular objectives. By restricting physical access, making documents the domain of a privileged few, and necessitating the use of jargon and code words, the practice of secrecy provided Peenemünders with a sense of privilege that compensated for the restrictions it placed on their professional world. Indeed, secrecy even enhanced the employees' notions of professionalism as it became a central feature of their everyday experience.

On the other hand, however, secrecy performed a negative function in that its constant practice quashed dissent or criticism by providing both a real and illusory sense of coercion around the work. Even if we are to take as truthful the postwar assertions of many engineers who argued that they had no control of the larger policy decisions when it came to determinations about labor deployment and treatment of the prisoners later in the war, surveillance became a mechanism for curtailing what might be construed in the Nazi context as politically deviant, inclining the Peenemünders to engage in very little, if any, opposition to the direction of the program. Thus, the overriding dearth of large-scale discord at Peenemünde was not simply a result of the specific technical vision of the project that guided employees down the same path.

An equally durable and meaningful means of evoking their collective focus was the internalization and automatic adherence to the rules of secrecy that guided the employees' behavior. Though individual employees disagreed, sometimes strenuously, their disputes were limited to more narrowly defined technical areas and never exploded into larger questions about the purposes of their work, the nature of the regime that sponsored it, or even the eventual use of slave labor to inaugurate mass production. This was a function of the secret society created at Peenemünde. Entrance to this society involved a thoroughgoing process of initiation and resocialization along the lines laid out by its members, and the Army and Gestapo ensured disciplinary compliance by providing a powerful, if, in reality, illusory, enforcement mechanism. This by no mean excuses the actions of engineers and technicians at Peenemünde. Employees there made individual decisions based on their own conceptions of right and wrong, but these choices were made in the context of an overarching dynamic of secrecy that acted to strengthen their identification with the project at hand while stifling public dissent. The result was a community of like-minded experts who automatically adhered to the dictates of the regime that made their work possible. The work itself that they did for the Third Reich would indeed be a challenge and a joy.

3

"It Was a Fantastic Life!"

Despite the restrictions that secrecy imposed on life at Peenemünde and the relative isolation of Usedom from the rest of Germany, employees at the missile facility found their lives to be personally rewarding and professionally stimulating. This buttressed the technical specialists' absolute dedication to Peenemünde's goals as an institution. In large part, this was a function of the lavish accommodations that the Army provided for them, the abundant opportunities for leisure and recreational activities (especially before the middle of 1943), and the exciting, well-paying work that they performed in a congenial, professional environment. Their happiness with life on Usedom encouraged employees to put an even more personal stake in the endeavor that brought them such good fortune. For them, the goals of the Peenemünde Army Research Station were intrinsically linked to both personal and professional satisfaction. Success in the A-4 project not only meant enhanced professional prestige, but also continued enjoyment of a comfortable life that provided plentiful social outlets and was relatively free, at least until August 1943, from the deprivations of war. Though they were not explicitly ideological in nature, life and work at Peenemünde were central to the social reproduction of support for the Nazi regime because of the subtle ways in which individuals at the facility came to identify their own goals and happiness with the mission of the institution, which was ostensibly to defend the Nazi regime from further harm.[1]

---

[1] Max Weber has shown that individuals in society relate to each other on the basis of sentiment. More recent studies have expanded on this point, arguing that sentiment is largely derived from group processes. Individual identities largely depend upon the groups within which people exist. Much of what people know, they learn from their social environment. They come to believe certain things to be true because the people around them repeatedly assert that they are. Once people hold these beliefs and discover that others within their given social networks share them, they take them as simple truths that have no need for further explanation or evaluation. See D. B. Clark, "The Concept of Community: A Reexamination," *Sociological Review, New Series* 21 (1973) 397–416. See also Michael A. Hogg and Dominic Abrams, *Social Identifications: A Social Psychology of Intergroup Relations and Group Processes* (London: Routledge, 1988); R. Scott Tindale, Catherine Munier, Michelle Wasserman, and Christine M. Smith, "Group Processes and the Holocaust," in Leonard S. Newman and Ralph Erber, eds., *Understanding Genocide: The Social Psychology of the Holocaust* (New York: Oxford University Press, 2002), 143–161; Serge Moscovici,

At Peenemünde, the engineers' sense of importance, professional achievement, career development, and prestige was largely a function of their feeling of belonging in a unique professional community. Peenemünders, isolated as they were from other outside sources of socialization, found their rewards largely within their own community. Their values emerged as a result of their individual place within the larger group dynamic. Peenemünders had a very high degree of solidarity and sense of self-significance, a feeling that grew out of their membership in this exclusive community. At bottom, this helped foster a sense among them that they felt themselves to part of an elite technical profession that performed a unique and profoundly important service for their nation. Their individual behavior was almost entirely informed by their conscious feeling of belonging in a community that espoused this very belief.

THE "PARADISE" OF PEENEMÜNDE

Despite Army Ordnance's protracted battle with higher regime authorities, especially Armaments Minister Fritz Todt, over the costs, amount of labor, and raw materials needed to construct the base, the Army's and Luftwaffe's largesse produced a first-rate technical facility with attractive and comfortable living accommodations for its employees. The island's undeniable natural beauty dazzled the employees. For nature enthusiasts, the site was extraordinary. An immaculate white sand beach stretched along the east coast of the island, running parallel to almost the entire base. Yards from the beach, a thick pine forest sprawled out across the island and provided an excellent habitat for deer, wild boar, and other animals.[2] The island was thick with waterfowl and game birds. Usedom was, for one engineer, "the most beautiful area, wooded area, you know, with lots of pine trees and leaf trees like oak trees and elm trees...a beautiful spot. ... We had beautiful birds in this area, all kinds of ducks."[3] Aerodynamicist Peter Wegener, who grew up in Berlin and as a child vacationed on Usedom, felt a warm nostalgia when he arrived at Peenemünde, recalling that "Most of my early summers were spent at one of the many resorts on Usedom. For a child from Berlin, the Baltic seashore – with its pure white sand, its dunes, and its hunting grounds for shells and amber – was the closest ocean holiday spot. ... The scenery, the smell, and the waters of Peenemunde were truly familiar."[4] The

"The Phenomenon of Social Representations," in Robert M. Farr and Serge Moscovici, eds., *Social Representations* (New York: Cambridge University Press, 1984), 3–69.

[2] Walter Dornberger, *V-2* (New York: Viking Press, 1954), 40.

[3] Bernhard Tessmann, Oral History Interview (OHI), National Air and Space Museum (NASM).

[4] Peter Wegener, *The Peenemünde Wind Tunnels: A Memoir* (New Haven, CT: Yale University Press, 1996), 17.

island's stirring natural beauty was (and remains) unquestioned, but it was not its only quality.

As Fritz Todt pointed out in 1941, the built environment at Peenemünde was stunning. Dornberger indeed built "on a grand scale and beautifully." Except for the production plant, most of the major construction was complete by 1939. The entire facility, from the tip of Usedom's northern peninsula to the tiny village of Karlshagen, stretched for nearly eight miles. The work was superb. Construction engineers who designed the facility built functional workshops, offices, and laboratories, but they arranged the buildings in an irregular manner and added many of their own unique flourishes to the structures in an effort to avoid drab monotony. The Settlement was the best example of the pleasant functionality that they strove for.

To enter the Settlement, workers and their families passed through the so-called Brandenburg Gate, a large stone building – complete with a large engraved swastika – with a tunnel that allowed cars to pass through. The gate also had bachelor apartments on either wing. Beyond the Brandenburg Gate to the east lay the grounds of the Settlement itself. It was based on the idea of the "Garden City," a concept designed to integrate aspects of the natural environment with the built environment. Right-wing architects in Germany, who saw this concept as a way to spiritually unite German families with their native soil, especially celebrated the idea. Two-story row houses and individual, detached family homes made up the Settlement, which was dotted by trees and personal garden plots. Planners laid the homes out in staggered rows to vary the Settlement's appearance, and they placed decorative architectural touches on the ends of the row houses to improve their appearance even further. Bakeries, cafes, a butcher shop, a grocery, and even a beauty salon and a bookstore opened there. Architects also built a school for the employees' children, a large sports field, and tennis courts. Reinhold Krüger, a technical apprentice at Peenemünde, was enormously impressed, recalling, "Above all, I was taken in by the new buildings in the clean, perfect town. For me, this was the epitome of German exactitude and cleanliness."[5] The entire Settlement lay just west of the beach, and was separated from it only by a stand of pine trees. "It was beautiful," recalled project engineer Werner Dahm. "Our house was right at the woods, and for my lunch time I could walk to the beach."[6] Most of Peenemünde's scientists, engineers, technicians, and military officers moved into the Settlement with their families when they came to the base.

Dahm's glowing assessment might have just as easily described "Peenemünde East," the site of the development workshops. This group of buildings was located just over a mile north of the Settlement and was

---

[5] Reinhold Krüger Statement, quoted in Volkhard Bode and Gerhard Kaiser, *Raketenspuren: Peenemünde 1936–1994: eine historische Reportage* (Berlin: Links Verlag, 1995), 38.
[6] Werner Dahm OHI, NASM.

connected to it by a modern electric railway, which by 1943 came complete with bright red cars modeled after the famous Berlin S-Bahn light rail system. Here, employees found their technical workshops, a dormitory for single men, the Aerodynamics Institute, and *Haus 4*, the large administration building that was home to the offices of Wernher von Braun and base commander Leo Zanssen. Designers also built two ornate clubs for officers and civilian workers. Rocket engine test stands dotted the coast of the peninsula to the north of the workshops, ending at Test Stand VII, the huge and complex launch site at the tip of the island. All of the structures were nestled into the island's dense forest, which was left intact during construction to improve camouflage. According to Wegener, "Administration buildings, laboratories, housing units, and test stands were widely separated according to a well-planned layout, and a bucolic atmosphere prevailed. The architecture was attractive, combining a resemblance to the older municipal buildings of the northern provinces of Germany with a touch of the twentieth century Bauhaus school."[7]

The massive missile assembly and storage hall was sandwiched between the employees' Settlement and the development workshops. Construction on it began in 1939. At 600,000 square feet, it was one of the largest freestanding industrial structures in Germany.[8] The building had a sawtooth roof that was designed to minimize air raid damage, but it also let in a great deal of natural light. The building was also extremely tall, designed to allow engineers and technicians to build the forty-six-foot-tall test rockets vertically, rather than horizontally.[9] Construction planners built the factory with the belief that it would hold rockets much larger than just the A-4. Eventually they meant the factory to produce the A-10, the first stage of a two-stage missile that, at nearly eighty-five feet, was almost twice the size of the A-4. That project remained on the drawing board throughout the war.[10] Obviously, they did not want to build a facility that was immediately obsolete, but this was also a reflection of their confidence that money would keep coming in, as would the political support allowing them to expand their work. In any case, this was the largest and, when combined with its electrical power needs and employment demands, the most expensive structure on the island. It completed a research, development, and production complex for

---

[7] Wegener, *The Peenemünde Wind Tunnels*, 17.

[8] Arthur Rudolph OHI, NASM. On the dawning realization of the size and complexity of such a structure, Rudolph, the building's chief planner, recalled, "I now felt as if I had been standing in the snow and making that little snowball and it began to roll and I could catch that ball any time I wanted to, but suddenly this snowball was an avalanche. A tremendous avalanche. And I got really scared."

[9] Bode and Kaiser, *Raketenspuren*, 36. Although administrators laid extensive plans for mass production in F-1, production was transferred to central Germany before they could be enacted.

[10] Arthur Rudolph OHI, NASM.

FIGURE 3. Accommodations at Peenemünde were lavish. This is one portion of the officers' mess hall at the base. [National Air and Space Museum, Smithsonian Institution (SI 77-12547)]

new weapons that, in terms of its size, complexity, and social considerations, was one of the most advanced in the world.

"IT WAS ABSOLUTELY WONDERFUL"

The island of Usedom had much to offer its new residents. Ruth Kraft, a data recorder in the Aerodynamics Institute, declared, "For those of us who came from central Germany or Saxony, the Baltic Sea was a wonderful experience."[11] Gerda Erdmann, whose husband was a lathe operator in the development workshops, stated years after the war that "Everything was wonderful. It's terrible that it's all broken down now."[12] One of the reasons that people grew so attached to their lives at Peenemünde was the undeniably good living conditions. The accommodations on Usedom were entirely pleasant and functional, and they had the advantage of being located on some of the best beachfront property in all of the Reich. Many single men lived in dormitories either in the development works or in the Settlement, but others – including von Braun – found rooms in guesthouses in Zinnowitz and elsewhere. Peter Wegener enjoyed his dormitory because he found so many interesting people there. Many of them became his good friends.[13] Those engineers with families were able to live in the well-equipped Settlement. Werner Rosinski, his wife, and infant child lived in one of the row houses there, but when they had a second child in 1940, they moved into a larger house that was only a three-minute walk to the beach. In the summer, he and his family ate breakfast and then took a stroll in the sand.[14] Rudolf Hermann's house was 200 meters from the beach. "Stepping out [of] the house," recalled the head of the Aerodynamics Institute, "Already you hear the noise of the sea."[15]

Outdoor recreational activities abounded in the summer, and the Peenemünders took full advantage of them. The quiet Peene River and the Achterwasser lagoon allowed for excellent sailing. Many of the more skilled and adventurous sailors at Peenemünde, von Braun among them, even enjoyed sailing to the Greifswalder Oie, a small island in the Baltic just north of Usedom. Several major launch tests took place there, but it also had a small inn, where, "despite all of the war rationing, one could

---

[11] Ruth Kraft Testimony, "Originaltonaussagen von Mitarbeiter der ehemaligen Heeresversuchsanstalt Peenemünde-Ost zum Thema Alltag in Peenemünde," *Historisches-Technisches Informationszentrum Peenemünde* (HTIZP).

[12] Gerda Erdmann Testimony, "Originaltonaussagen," HTIZP.

[13] Wegener, *The Peenemünde Wind Tunnels*, 19.

[14] Werner Rosinski Testimony, "Originaltonaussagen," HTIZP.

[15] "Memoirs of Rudolf Hermann," unpublished manuscript, p. 19, University of Alabama, Huntsville (UAH). This source is rather more like an oral history than a memoir. It is a transcript of an interview Hermann gave to Sandy Sherman in 1988 in Huntsville, Alabama.

FIGURE 4. A group of unidentified Peenemünders frolics at the beach, probably in 1943. This tightly knit community found great happiness in each others' company. [Author's collection]

always eat well."[16] Sunbathing and sports competitions during time off of work were also popular pastimes. Early every Wednesday morning in the summer, many people gathered for "morning sport" on the beach. They played handball, soccer, swam, or just went for a walk before gathering for breakfast, cleaning up, and catching the train to work. At the sporting field in the Settlement, employees also held friendly competitions and games between the various branches at the base. Many Peenemünders, including Dornberger and Zanssen, took their families to the interior of the island to pick the wild blueberries that grew there.[17] The base's high-ranking and more privileged officials, von Braun included, also enjoyed horseback riding in the woods with horses reserved for them in the riding stables.[18]

[16] Bob Ward, ed., *Wernher von Braun: Anekdotisch* (Esslingen: Bechtle Verlag, 1972), 41.
[17] Herbert Lucht Testimony, "Originaltonaussagen," HTIZP.
[18] Martin Middlebrook, *The Peenemünde Raid: The Night of 17–18 August, 1943* (London: Cassell, 1982), 20. Middlebrook's book is essentially a minute-by-minute account, drawn largely from oral history interviews he conducted, of the destructive RAF bombing raid on Peenemünde in August 1943.

In the winter, the main sources of entertainment were the theater and films (there were four cinemas on Usedom). Huzel's favorite was the cinema in Karlshagen, just south of the Settlement, which showed old films, "invariably of good quality," that never had anything to do with the war.[19] A local *Festzeitschrift* – essentially a community events calendar – kept people up to date about the social and cultural events happening on the base. *Feierabends* and *Kameradschaftsabends*, festive parties sponsored by the base's various administrative divisions and put on in the clubs, were consistently the most popular events of the year. Friends got together to eat, drink, tell jokes (sometimes at each others' expense), put on skits, play music, and sing songs.[20] Employees themselves often put on concerts and plays, which could be well attended by their colleagues. Rudolf Hermann, for example, was a member of perhaps history's most technically brilliant chamber music quartet, with von Braun, who played the cello, measurement specialist Gerhard Reisig, who played the viola, and aerodynamicist Heinrich Ramm, who played first violin. They sometimes performed together in public.[21]

The Peenemünders also established deep and durable social bonds with each other. Workers celebrated holidays and birthdays together, holding parties and exchanging handmade gifts.[22] They sometimes celebrated important events in the famous resorts in Zinnowitz. Huzel remembers these gatherings fondly.

> This town had been a swank seaside resort, and a number of restaurants were still operating. . . . The waiters wore white tie and tails; there were white tablecloths; and the food was pretty good for those times. . . . As wine was unavailable, it was acceptable for the customer to bring his own – which we usually managed to do. After dinner the waiter would spend twice the time with the ration coupons that he did with the bill. These were moments of pleasure stolen out of tragedy, and our humor was always high as we finally made our way back to House 1 [the bachelors' dormitory].[23]

Family life was pleasant as well. The demands of work and the war did not preclude many individuals from starting or expanding their families at Peenemünde. Most families had their own small but pleasant houses, and parents found plenty of neighbors who were willing to supervise their children when they both worked. Hermann's family gathered with his friends'

---

[19] Dieter Huzel, *From Peenemünde to Canaveral* (Englewood Cliff, NJ: Prentice-Hall, 1962), 130.

[20] See, for example, "Festfolge zum Kameradschaftsabend der Abteilungen TA/L, TA/Proj., TA/Che und TA/TB am Sonnabend, dem 15. März 1941, 16 Uhr in Schwabes Hotel Zinnowitz," HTIZP. The festivities included an opening speech by von Braun, a reading of Otto Schairer's poem "Deutschland dir mein Vaterland," and a performance of the song "Die echte deutsche Gründlichkeit."

[21] Hermann Memoirs, UAH, 19. Hermann played second violin.

[22] Herbert Lucht Testimony, "Originaltonaussagen," HTIZP.

[23] Huzel, *From Peenemünde to Canaveral*, 130.

families on the weekends to play games together at the beach.[24] Even though most of the people at Peenemünde were men, there were plenty of women on the island who worked as secretaries, clerks, typists, and measurement takers. Most of the women were single and lived in the *Kameradschaftsheim*, a fashionable hotel that was converted into a club and dormitory. Needless to say, this opened up many romantic opportunities for the young and single Peenemünders. Kiddy Luckman, a data recorder in the measurement group, often went to Zinnowitz with her friends in search of young and single men.[25] Wernher von Braun, that dashing, brilliant, gregarious young nobleman, skilled with the cello and the sailboat, had no trouble landing a steady line of girlfriends. Not surprisingly, some, like development foreman Horst Wiessner, even met their future husbands or wives at Peenemünde.[26]

Life on Usedom, then, was quite comfortable, even in the sharpened conditions of total war in Nazi Germany. Nearly everyone recognized this. Rationing demands made food scarce across the nation, but because of Peenemünde's location on the Baltic Sea, residents there were easily able to supplement their rations with fish and eel. Partially for this reason, Dieter Huzel was convinced the Peenemünders did not have it as bad as other Germans throughout the country.[27] Except for wine, alcohol was also plentiful. Chemists at Peenemünde distilled 75% ethyl alcohol into pure ethanol (alcohol), creating a potent moonshine that they added different flavors to.[28] Georg Tiesenhausen commented wryly (and, it might be said, alarmingly, in a gastronomic sense), "We had parties. Parties with rocket fuel."[29] Konrad Dannenberg held that the availability of these items in Peenemünde meant that they were not badly off there. "It was probably much worse all over the rest of Germany," he noted.[30] Nearly all Peenemünders were quite conscious that their lives were vastly better than the majority of their countrymen, and they embraced this fact. "It was a fantastic life!" remembered one former employee.[31] In the context of Nazi Germany, this secret life, largely free from the deprivations of the war and reinforced by an abundance of recreation and a tightly knit social community, sheltered the Peenemünders from the steadily mounting drumbeat of distress and destruction across the rest of the country.

[24] Hermann Memoirs, UAH, 19.
[25] Kiddy Luckmann Testimony, "Originaltonaussagen," HTIZP.
[26] For von Braun's personal life, see Michael Neufeld, *Von Braun: Dreamer of Space, Engineer of War* (New York: Knopf, 2007). See also Horst Wiessner Testimony, "Originaltonaussagen," HTIZP.
[27] Huzel, *From Peenemünde to Canaveral*, 80.
[28] Konrad Dannenberg OHI, NASM.
[29] Georg Tiesenhausen OHI, NASM.
[30] Konrad Dannenberg OHI, NASM.
[31] Werner Rosinski Testimony, "Originaltonaussagen," HTIZP.

„Hupp – Aber Liebling! – hupp – Was soll das bedeuten?
Ich war doch – hupp – nur zur Technischen Sitzung!"

FIGURE 5. Alcohol-fueled parties were important moments that built camaraderie at Peenemünde. In this cartoon, illustrated for the local *Festzeitschrift*, a tipsy Peenemünder comes home from the bar to find his bedding thrown on the floor by his wife. "But darling!" he protests between hiccups, "How now? I was only at a *technical* meeting!" [Fort Eustis Collection, FE 623, National Air and Space Museum]

All of these positive experiences and joyful events were moments of fundamental importance in the growth of the community of Peenemünders. Peenemünders lived, played, celebrated, and mourned together in ways that strengthened individual ties to the group. In a revealing comment about the strength of these bonds, technician Gerhard Rühr noted

This comradeship was present in Peenemünde and was not confined to professional or social groups. Whether one traveled by train or was in a club or in an air raid shelter or at a lathe, we all felt like one big family. Perhaps this spirit of togetherness was due to the fact that we had all, from the youngest apprentice to the general in command, come to this lonely island from all parts of Germany to witness the building of the A-4 rocket. . . . You might even meet von Braun at the dentist's.[32]

In this secret society, recreation, friendly get-togethers, and raucous evenings like the *Kameradschaftsabends* helped reaffirm the unity of this unique group. These were essential steps of the process in which the institution

[32] Quoted in Middlebrook, *The Peenemünde Raid*, 27–28.

of Peenemünde remolded the old, heterogeneous identities of its individual members into the closed, elite, and privileged community of Peenemünders. The reciprocal bonds that they established improved their performance on the job (or at least gave them a reason to do a good job – few wanted to get fired from Peenemünde) and spawned a highly developed sense of solidarity and loyalty among those specialists who came to live and work in this "paradise" on the Baltic.

"WE HERE ARE SUPER-ENGINEERS!"

Between 1937 and 1943, the development of missile technology in Germany made its most remarkable and important advances. This was particularly true in three areas. Perhaps the most difficult of these was in guidance and control, which had caused major problems in the Kummersdorf years. Beginning in 1939, Ernst Steinhoff's guidance department, assisted by university professors under contract to Peenemünde and by private industry, made several significant advances on electrical instruments and gyroscopes. In the area of liquid-fueled propulsion, the gifted but irascible Dr. Walter Thiel, after several difficult years, successfully spearheaded the design of the required twenty-five-ton thrust engine and pushed the engine's fuel efficiency nearly to its theoretical limits. Thiel did much of his work in Kummersdorf, but his group came to Peenemünde in 1940. Rudolf Hermann's aerodynamicists, against a chorus of ballisticians who argued that it could not be done, crafted the world's first fin-stabilized (or "arrow stable") supersonic body.[33] Without question, these impressive advancements were possible in part because of the major funding and support given to the work at Peenemünde.[34] The complicated technical problems of guidance, thrust, and supersonic aerodynamics could not have been solved without the proper dedication and delegation of authority and resources.

However, the availability of money and material only tells part of the story of the A-4's rapid development. The environment in which the intellectual resources – that is, the missile developers themselves – functioned, as well

---

[33] Donald MacKenzie, *Inventing Accuracy: A Historical Sociology of Nuclear Missile Guidance* (Cambridge, MA: MIT Press, 1990), 44–60. See also Gerhard Reisig, "Von den Peenemünder 'Aggregaten' zur amerikanische 'Mondrakete,'" *Astronautik* 4 (1986), 5–9, 44–47, 73–79; Michael Neufeld, *The Rocket and the Reich: Peenemünde and the Coming of the Ballistic Missile Era* (Cambridge, MA: Harvard University Press), 73–109.

[34] On some elements of Peenemünde's expansion, see the previous chapter. For a more detailed discussion, see Neufeld, *The Rocket and the Reich*, 41–143, and Neufeld, "Hitler, the V-2, and the Battle for Priority, 1939–1943," *Journal of Military History* 57 (July 1993), 511–538. Helmuth Trischler and Margit Szöllösi-Janze have used the term *Grossforschung* (Big Research) to describe the rise and dynamics of large-scale, heavily funded research projects that enlist the cooperation of university, industrial, and state resources in Germany in the twentieth century. See Margit Szöllösi-Janze and Helmuth Trischler, eds., *Grossforschung in Deutschland* (Frankfurt: Suhrkamp, 1990), 13.

FIGURE 6. A beaming Walter Dornberger, the Commanding General of the Army rocket program, poses with a buck killed on a hunt at Peenemünde. Dornberger, an avid hunter, relished the opportunities Peenemünde offered. [National Air and Space Museum, Smithsonian Institution (SI 12795)]

as the ways in which Army authorities motivated their work, were just as important. A project this complex and advanced required a committed group of experts to see it through. These talented specialists worked incredibly hard on behalf of a regime that obliterated enlightened notions of human rights, waged a genocidal war in Europe, and placed increasingly harsh demands upon its own population. What was it about their work that convinced them that this was the right thing to do?

Many postwar memoirs and histories of Peenemünde attempt to argue that the specialists on the base retreated into a sort of "inner migration" and became narrowly focused only on their work.[35] According to this line of reasoning, no political or military considerations played a part in their daily lives. To the extent that they reflected upon their work at all, they thought of it in terms of its ability to send humans to space, not its military applications. This argument has become a central part of the myth of Peenemünde that was built up by the former Peenemünders and their supporters in the years after the war. The essential fault with this interpretation is that it ignores much of the larger political and intellectual milieu in which the Peenemünders traveled. In reality, the military purposes of their work were clear and were embraced by nearly everyone. The terms in which regime authorities cast their work meant that employees at Peenemünde were confronted almost daily with the military, nationalist, and ideological implications of their work. In what ways did this confrontation occur? Did it have an important effect on the patterns of life and work at the base? Did it shape how the Peenemünders viewed themselves or their work? Rather than argue, as many defenders of the Peenemünde community have in the past, that their technological work was inherently apolitical, it is more useful to examine how their technological work was reconciled with the politico-military aims of the regime.

In the first place, more innocent considerations, such as space travel, were secondary or even unimportant for the great majority of Peenemünders, most of whom arrived during the war. Indeed, wind tunnel specialist Peter Wegener wrote that "During my time at the Baltic, I never heard a single remark about spaceflight. ... No one ever mentioned in my presence that the A-4 would be a stepping stone toward a moon flight. In my several meetings with von Braun, he never suggested this possibility, even in small social gatherings."[36] Wegener's remark is, admittedly, not common. Certainly, it is reasonable to expect that some people at Peenemünde did speak quietly about spaceflight, especially in the heady days before September 1939. Von Braun managed to bring in a few of the old *Raketenflugplatz* members who nurtured their dream, and von Braun himself certainly believed in it. Even so, in a facility of some 6,000 people, any thoughts of space travel were the

[35] See, for example, Wernher von Braun, "Behind the Scenes of German Rocket Development," Wernher von Braun Papers, Space and Rocket Center, Huntsville (SRCH). "Inner migration" is a term used by Alan D. Beyerchen in his important but now dated book, *Scientists Under Hitler: Politics and the Physics Community in the Third Reich* (New Haven, CT: Yale University Press, 1977).
[36] Wegener, *The Peenemünde Wind Tunnels*, 41–42.

exception, not the rule, and by 1942, such talk was politically dangerous. Their job was to create an operational ballistic missile in the shortest time possible in order to defend the regime and the nation that made their work possible. All other considerations disintegrated in the face of this one task. This was a truism at the base, and it was loaded with important geopolitical considerations. According to their military leaders, the Peenemünders' work allowed them to play a central role in Germany's very struggle for survival. They made this clear in no uncertain terms. Germany was a victim in this conflict, fighting a misunderstood war against the Bolshevik menace to the East that would have subjugated and enslaved Europe if their nation had not acted. As for the Peenemünders, they were the in the vanguard of the Germany's defense. Indeed, their work was of paramount importance for the security not only of the nation but also for all of Europe.

Dornberger made this essential point most clearly in an address that he gave in the middle of June 1943. Four months earlier, the Soviet Red Army had crushed the Sixth Army at Stalingrad and brought home to Germans the fact that they were indeed in the fight of their lives. In the main assembly hall of the production plant, Dornberger spoke to nearly 6,500 German employees and soldiers at the base shortly before they were scheduled to begin full-time production operations. The address contained a heady, self-serving mixture of militaristic nationalism, technological triumphalism, and Nazi ideology. It revealed how deeply Dornberger espoused many of the more aggressive features of National Socialism and the tone in which Army authorities cast the work at Peenemünde. The General began his long address by offering his assessment of why Germany was involved in the war in the first place. He explained away German belligerence by painting his country as the victim of Soviet plans to cast all of Europe under the Communist yoke. The Third Reich was only fighting for security in Europe, "So that our children will have better living conditions than we did, and so that no European state is ever in the position, out of envy or mistrust, of plotting a war of all against all." It was "Henceforth the task of the German armaments industry, which is organized for Total War, as well as the coordinated Europeans [sic – *gleichgeschalteten Europaischen*], to struggle against this foe."[37]

Turning to the tasks before the German engineering community, especially those at Peenemünde, Dornberger emphasized their importance in the current struggle. "They [Germany's enemies] want to come," he challenged. "Well, let them come. We will give them a proper reception. So that we can do so, so that we can pay them back for all of the damage they have done to Germany and the European nations, it is essential that the German armaments industry works continuously in order to put the best weapons in the hands of the best soldiers in the world. ... We wish to pay the English

[37] Dornberger, Betriebsappell am 18.6.43, p. 1, Fort Eustis (FE) Files, FE 833, NASM.

back for the terrible sorrow that they have caused to our country, especially our women and children, through their terror attacks." Dornberger also believed that German engineers and workers "Must through action prove every day that they, as outstanding representatives of the German nation [*Volk*], acknowledge our nation's claim to leadership in Europe."[38]

For Dornberger, the work of those at Peenemünde played "a decisive role" in the struggle. In this war "for the very existence of the German nation [*deutschen Volkes*]," all other concerns were secondary. Dornberger implored the Peenemünders to set aside all of their personal desires and concerns so that their work in the name of the nation could be quickly completed. "We must do our utmost in the unshakeable belief that we can bring our new technology into operation as soon as possible," he exhorted.[39]

Dornberger's call for sacrifice then turned to the conditions themselves at Peenemünde. He pointed out that the Army provided the Peenemünders with "the archetype of a National Socialist factory," complete with "the most modern technical and social considerations." He would, he announced, do everything in his power to ensure that conditions remained as pleasant and comfortable as possible, but he also warned that "I will ruthlessly come down on those who believe, to the detriment of the employees, that their own interests come ahead of the project."[40] Since the Army had provided for all of the cares that the Peenemünders could possible have, Dornberger felt that there was no excuse for them to concentrate on anything other than the tasks provided for them. As the "archetypal Nazi factory" (complete, he neglected to mention, with foreign slave labor), the production plant made high output possible by ensuring that workers there were well compensated financially, socially, and culturally.

The General, now in full form, then turned to what it would take to bring their project to a successful conclusion. Reminding them again that the nation was in the midst of total war, he explained that such times required every individual's last effort. To overcome the inevitable frustration, exhaustion, and human difficulties that would come with this work, he turned to what was fast becoming the Nazi regime's primary solution to the increasingly intractable problems facing the nation in the war. "The will," Dornberger lectured, "is always the best medicine. Total war demands total action from everyone."[41] Every last minute of work time must be dedicated to the work at hand. Employees at the facility "must be the bearers of the unshakeable belief in our success. You must fill the newly arriving employees with your spirit and inspire them with your acts. ... He who is careless

---

[38] Ibid., 2.
[39] Ibid., 3.
[40] Ibid., 4.
[41] Dornberger, Betriebsappell am 18.6.43, p. 7, FE 833, NASM.

works for the enemy."[42] After once more exhorting the Peenemünders to "free the German nation," Dornberger closed his speech with a crescendo of "Sieg Heils!" to "Our Führer and Supreme Commander."[43]

Dornberger's speech skillfully blended traditional German patriotism with Nazi ideological motifs and highlighted many of the unique factors that made missile development so successful at Peenemünde. Germany as a nation of victims was an old canard in the Nazi propaganda machine. Army leadership had also long since bought into the notion, exploited heavily by the Nazis, that Germany was the victim of ruthless foreign enemies who unceasingly sought to keep their nation under heel. As a career soldier and influential officer in the armed forces, Dornberger had worked tirelessly to overcome the limitations of the Treaty of Versailles and then restore Germany to what he perceived to be its rightful place as the most powerful nation in Europe. Moreover, as an engineer, Dornberger wholeheartedly believed that his profession must play an important role in this effort. Much like the Nazi propaganda regarding technology that became so prevalent in the 1920s and 1930s, Dornberger celebrated the engineers' unique contribution to strengthening German society.[44] The value of the engineer lay in his ability to provide the nation with the technological muscle it needed to fend off its enemies and bring about final victory over them. He exalted the engineers as the vanguard of this endeavor. Constructing the missile in a "model National Socialist factory" was to be the centerpiece of such an effort. By emphasizing the path-breaking nature of their work as well as its singular importance to the war effort, all while playing on the popular fear of the Soviet Union and the disdain for the Western Allies for bombing their cities into rubble, Dornberger had composed a powerful message that appealed to many Peenemünders. He did not once mention space travel.

Dornberger's subordinates reinforced the General's position. Earlier in 1943, Dornberger's Chief of Staff for liquid-fueled rocketry, Lieutenant Colonel Georg Thom, put together a self-serving but enthusiastic report on the missile program entitled "The A-4 Device in Total War" (*Das Gerät A-4 im totalen Krieg*). He gave it to high-ranking military authorities as well as to Peenemünde management. In his report, Thom emphasized that in total war, all means of achieving victory should be attempted and that all weapons, no matter what the cost, were important tools for achieving this goal. He argued that "In modern war, the necessary types of weapons can no longer be made dependent on their cost of construction and manufacture, on

---

[42] Ibid., 6.
[43] Ibid., 16.
[44] See especially Jeffrey Herf, *Reactionary Modernism: Technology, Culture, and Politics in Weimar and the Third Reich* (New York: Cambridge University Press, 1984), and Karl-Heinz Ludwig, *Technik und Ingenieure im Dritten Reich* (Düsseldorf: Droste Verlag, 1974.)

the more or less great number of front line and rear echelon troops that use them, or on correctly marshaling these weapons and soldiers – *they depend alone on the toughness and morale of the opponent and on the singular will to strike down the enemy through the employment of all means of war* [Kriegsmittel] *and all of the reserves of the nation* [Volkes]. ... *The most important factor is the will* [to embrace] *the totality of war!*" [emphasis in original]. Thom blamed Germany's loss in World War I not on any "shortage of inventor spirit," but on the unwillingness of military leaders to embrace all of the possibilities that this spirit could conceive. Moreover, Thom held that missile operations against England were important not only because they saved German lives and valuable raw material, but also because "In smashing the English nerve, the A-4 is far superior to the airplane." He concluded by writing that "In this decisive hour, Germany cannot be strong enough!"[45]

Thom's report placed Peenemünde squarely in the middle of the war effort. Armaments engineers supporting soldiers at the front must not spare any energy in their quest for victory. Total war was a ruthless, violent business that depended more than anything else on the will of the nations fighting it. In this regard, the missile was especially important because it was demonstration of German willingness to go to any lengths to achieve victory. Specialists at Peenemünde were developing a weapon would crush the British will because it could not be stopped (Thom's report included a hand-drawn illustration of a burning city quarter). Germany's willingness to use its technological expertise to go beyond its enemies' capabilities would mean victory in Germany's life-or-death struggle.

Thom's report also reemphasized the importance of the task before the Peenemünders. Clearly, they were building what their Army masters envisioned as an important, if not decisive, weapon. In its operation against civilian targets, it pointed the way toward victory. Sentiments like those of Dornberger and Thom established a particular milieu in which the Peenemünders operated. They helped cement what anthropologist Hugh Gusterson has, in his work on the Lawrence Livermore Nuclear Laboratory, called the "central axiom" of life at the facility.[46] At Peenemünde, this central axiom, the base's mission, was centered on the idea that the specialists were there to produce a weapon to defend their nation, even Western

---

[45] Georg Thom Report, "Das Gerät A-4 im totalen Krieg," RH8/v.1231, *Bundesarchiv/ Mitlitärarchiv* (BA/MA). Thom disingenuously argued that the loss of one bomber was equal to the loss of ten single-use missiles. Although there can obviously be no price on the lives of the bomber crew members, one can say that the resources, expense, and man-hours dedicated to missile operations, from development through mass production and deployment, were astronomically higher than those for bomber operations and of far less strategic value.

[46] Hugh Gusterson, *Nuclear Rites: A Weapons Laboratory at the End of the Cold War* (Berkeley, CA: University of California Press, 1996), 56.

civilization, against enemies who were bent on destroying it. This was at the heart of their reason for being on the base. All other concerns were unimportant. Used often, these powerful ideological messages were adopted by the Peenemünders and were accepted as a given. In other words, they became common sense.[47] Part of the experience of being a Peenemünder was coming to understand the significance that this mission played in establishing a framework for all other activities. It fostered the presupposition that working on the missile to defend the nation and regime was the sine qua non of their professional lives. Articulations of Peenemünde's mission helped to transform ideological drivel into simple truths for the Peenemünders.

Peenemünders were under no illusions about what it was they were doing. Peter Wegener and Dieter Huzel acknowledged that it was an important, if unspoken, assumption that building the A-4 was an essential contribution to the war. Wegener wrote that "Whatever personal opinions might have been held by individuals, the support of the war effort was uncritical: the technical work had to be done in the shortest possible time."[48] Huzel, whose memoir tends to whitewash the activities at Peenemünde much more than Wegener's does, explained with candor that the most important factor motivating their work "was the realization that the job was critical to the war effort, and that a long working day was merely a nuisance compared to the hardships endured by others."[49] Wind tunnel chief and local party leader Rudolf Hermann stated "Sure, already in Germany during the war, we were only concerned about getting a weapon ready for the war, nothing else."[50] In 1972, in an effort to justify his actions and to distance himself from what he called the "misuse" of his work, von Braun offered,

I deeply and sincerely regret the victims of the rockets; but there were victims on both sides. I repeatedly raised protests against the misuse of the rockets as tools of destruction. But war is war, and since my country found itself at war [*und da mein Land sich im Krieg befand*], I had the conviction that I did not have the right to bring moral criteria into the matter. My obligation was to help win the war, whether I had sympathy for the government or not. I had none.[51]

In his defense, von Braun probably did not have any sympathy for National Socialism. Nevertheless, his comment reveals how deeply these messages permeated the thinking even of those who were not true believers in Nazism. He knew perfectly well as early as 1933 that he was building missiles for the Army and made absolutely no protests about the use to which they were

---

[47] Clifford Geertz notes that "Common sense is not what the mind cleared of cant spontaneously apprehends; it is what the mind filled with presuppositions concludes." Clifford Geertz, *Local Knowledge* (New York: Basic Books, 1983), 84.

[48] Wegener, *The Peenemünde Wind Tunnels*, 41.

[49] Huzel, *From Peenemünde to Canaveral*, 84.

[50] Hermann Memoirs UAH, 18.

[51] Ward, *Anekdotisch*, 31.

being put. His argument about the victimization of Germany mirrored the rhetoric of victimization offered by the Nazis and their sympathizers (such as Dornberger).[52] According to von Braun, his country had "found itself" at war, rather than having actively and unilaterally embarked on war. Germans were equal victims of the war as their enemies.[53] His understandable, but reflexive, patriotism and nationalism only seconded the Nazi rhetoric, and he became fully imbued with Peenemünde's central mission of unquestioned service to the state.

The Peenemünders did not simply dismiss rhetoric such as Dornberger's and Thom's, which held out the promise of the continuation of their unique social and professional existence and drew on National Socialist ideological tenets to do so. Certainly, some questioned the assertions that the A-4 was as decisive a weapon as regime authorities argued, but they kept their doubts private.[54] The missile was a limited, inaccurate weapon. Many Peenemünders were aware of these limitations, but they nevertheless reflexively believed that the work was central to their nation's survival. It became axiomatic that missile development and production in the midst of total war played an essential part in helping Germany defeat its enemies, and therefore must be completed as quickly as possible.

There is also postwar evidence to suggest that the Peenemünders agreed with Dornberger's general rhetoric about the supposedly heroic position occupied by Nazi Germany in the world order. In June 1945, shortly after the German surrender, many of the leading Peenemünders, including von Braun, Dornberger, Riedel III, and many others, were in the custody of the U.S. Army. They were held at Garmisch, a resort town near the Austrian border, awaiting transfer to technical positions in the United States. Interested as the Army was in exploiting their expertise, they were also concerned about the group's political loyalties. Second Lieutenant Walter Jessel of the Army's Military Intelligence Service (MIS) conducted a security check of the personnel and wrote a report that in part assessed the Peenemünders' political and security liabilities. Jessel's report indicates that the specialists believed in the political messages that they received during the war. He wrote that among the Peenemünders in Garmisch, "There is almost nowhere

---

[52] German historian Andreas Hilgruber made the same argument in his book *Zweierlei Untergang (Two Kinds of Ruin)*, a work that earned him the opprobium of most of the historical community. See Andreas Hilgruber, *Zweierlei Untergang: die Zerschlagung des Deutcshen Reiches und das Ende des europaischen Judentums* (Berlin: Siedler, 1986).

[53] Von Braun made no mention of the Jews, but he presumably was including them as well. Recent research shows that Germans understood the fate that awaited the Jews at the hands of the Nazis. See Jeffrey Herf, *The Jewish Enemy: Nazi Propaganda During World War II and the Holocaust* (Cambridge, MA: The Belknap Press of Harvard University, 2006) and Götz Aly, *Hitlers Volksstaat: Raub, Rassenkrieg und nationaler Sozialismus* (Frankfurt: Fischer Verlag, 2005).

[54] Wegener, *The Peenemünde Wind Tunnels*, 43.

any realization that there was something basically wrong with Germany's war or the employment of V-weapons." In their conversations with Jessel, the Peenemünders refused to acknowledge German responsibility for starting the war and preferred instead to view their nation as a victim of foreign aggression. Nor did they have any compunction about using their work for destructive purposes.[55] Jessel's point was completely and conveniently forgotten by the Peenemünders after they came to the United States to begin their work. No memoirs, interviews, or histories written after the war by their enthusiastic and blinkered supporters raise the issue of war guilt, whereas nearly all of them point to some degree of regret that their rocket was employed as a missile. Closer to the truth is that even the less ideologically predisposed Peenemünders were imbued with the Nazi rhetoric of victimization and were unapologetic about the goals toward which they worked during the war.

Moreover, Jessel pointed out that nearly all of the missile specialists were convinced that war between the United States and Soviet Union was "around the corner." Jessel wrote that "They shake their heads in amusement and some contempt at our political ignorance and are impatient at our slowness in recognizing the true saviors of Western Civilization from Asia's hordes."[56] Undoubtedly, many Peenemünde specialists bought in to the aggressive anticommunist rhetoric peddled by Dornberger and prominent party ideologues during the war. Of course, anticommunism is not a crime, but to characterize the slaughter conducted by the Nazi regime as a service to Western civilization is, to the outsider, a misconstruction of the facts and an affront to the memories of the victims of Nazi aggression. Nevertheless, to Peenemünders, this was simply the articulation of the central axiom of their former institution. Dornberger's militant anticommunism and Thom's call to embrace total war inculcated and reinforced Peenemünde's central axiom, borne of Nazi rhetoric, that missile development and production was essential to the survival of the German nation. This powerful ideological message became an undisputed, unquestioned fact at Peenemünde.

The ever-present undercurrent of Nazi ideology was not unique to Peenemünde. Life in National Socialist Germany was permeated with ideological messages designed to imbue its citizens with the resolve, strength, and benefits of Nazi governance and the malevolence, corruption, and immorality of Jews, Slavs, Communists, and others. Josef Goebbels' propaganda machine, in its tireless effort to manufacture and maintain consensus among

---

[55] Osborne to Army Chief of Staff, G-2, USFET, Appendix A, Walter Jessel, Special Screening Report, October 29, 1945, RG 260, OMGUS/FIAT, Box 8, Folder 47.94, National Archives and Records Administration (NARA). My thanks to Mike Neufeld for bringing this document to my attention.

[56] Ibid. Of the Allies' refusal to acknowledge the "service" Germany supposedly performed for the western world, Jessel also wrote pointedly that this "Does not prevent them from playing with the idea of selling out to Asia's hordes if such recognition is not soon extended."

the German population, ensured that these ideas were inescapable.[57] The nearly constant bombardment of fanatical ideology laced with racist messages stunted millions of Germans' sense of right and wrong and withered their moral inhibitions.[58] Jessel's report indicates that important segments of Peenemünde's professional community were not immune to the fanaticism that poured from the Propaganda Ministry and others. The National Socialist and ideological components of life at Peenemünde cannot be ignored.

Political, social, and cultural life on Usedom was suffused with Nazi ideology. Photographs taken at the base show an active Hitler Youth group marching in a parade. A Nazi women's group took root on the island, and the local branch of the party regularly held well-attended meetings. Nazi Party membership was robust at Peenemünde. Many of the heads of administrative divisions and sections at Peenemünde, including von Braun, were party members.[59] A substantial number of Peenemünders in leadership positions showed some likely form of ideological commitment to party principles before or during 1933. Arthur Rudolph joined the Nazis in June 1931, well before their seizure of power. Rudolf Hermann joined the SA in 1933. Kurt Debus, the future head of operations at Test Stand VII (used for launch tests), also joined in 1933 and became a member of the SS in 1940.[60] Hermann was also the chief of the local branch of the party from 1941 until his departure from Peenemünde in 1943. Hans Lindenberg, one of Thiel's deputies in the

[57] On the scope and effectiveness of propaganda during the Nazi period, see Herf, *The Jewish Enemy*; David Welch, ed., *Nazi Propaganda: The Power and Limitations* (Totowa, NJ: Barnes & Noble, 1983), and Ian Kershaw, *The Hitler Myth: Image and Reality in the Third Reich*, 2nd ed. (New York: Oxford University Press, 2001). Martin Broszat also long ago offered a convincing explanation that "social motivations" bonded the German population to Hitler and the Nazis. See Martin Broszat, "Soziale Motivation und Führer-Bindung des Nationalsozialismus," *Vierteljahreshefte für Zeitgeschichte* 18 (1970), 392–409.

[58] Avraham Barkai, "The German *Volksgemeinschaft* from the Persecution of the Jews to the 'Final Solution'," in Michael Burleigh, ed., *Confronting the Nazi Past: New Debates in Modern German History* (New York: St. Martin's Press, 1996), 96.

[59] Though Army and regime officials reorganized Peenemünde's administrative structure several times, it generally retained the same core group of administrators involved in both development and production. For the various permutations of the facility's administrative structure, see Neufeld, *The Rocket and the Reich*, 285–288. See also administrative charts in files AV 7/85, Bd. 33, and RHE 28/83 USA, *Bundesbeauftragte für die Unterlagen des Staatssicherheitsdienstes der ehemaligen Deutschen Demokratischen Republik* (BStU).

[60] Arthur Rudolph Dossier, Box 636, RG 319, Records of the Army Staff, Counter Intelligence Corps (CIC), Records of the Investigatory Records Repository (IRR), NARA. Rudolf Herrman Dossier, Box 279, RG 319, IRR, NARA. Kurt Debus SS Officer Dossier, Roll A3343-RS-A5426, RG 242, NARA. Debus lied to Allied investigators after the war about his SS past, claiming that he was merely an SS candidate. See his Basic Personnel Record, Box 703, RG 165, Records of the Army Chief of Staff, G-2, Intelligence Division, Captured POW and Material Branch, Enemy POW Interrogation File, 1943–1945, Folder "Boston," NARA. From July 1962 to November 1974, he was the first director of the Kennedy Space Center, overseeing the Mercury, Gemini, and Apollo programs.

propulsion section, entered the party in 1934.[61] Six, von Braun among them, entered the Party in 1937 or later, but at least one of these, longtime rocket enthusiast Walter "Papa" Riedel, first voted for the Nazis in 1933.[62]

Though party membership is not always a good indicator of ideological commitment, it at least provides a rough guide to the extent to which Peenemünde scientists, engineers, and technicians were willing become fellow travelers with the party. Generally speaking, party membership among leading Peenemünders was higher than the average in Nazi Germany.[63] A sample of eighty-four background forms filled out by German specialists in the United States sheds some light on the numbers, which show a general affinity for National Socialism. Forty-three men, or 48%, indicated that they were formerly members of the Nazi Party. Of this number, eleven joined before or during 1933. Twenty-one entered the party between 1937 and 1939, and eleven joined after the outbreak of war, including three who joined in 1942 or later. Of the forty-one who were not party members, nine (22%) admitted their membership in organizations that had strong elements of Nazi ideology, such as the National Socialist Students' League or SA.[64]

The story of Anton Beier, an engineer employed at Peenemünde between 1938 and 1945, is typical. Born in 1906 in Upper Silesia, Beier earned an engineering degree at the Mittweida Technical College in Saxony and, in 1930, landed a job with the municipal utilities in the town of Ziegenhalls in Upper Silesia. In 1932, Beier lost his modest job to the Great Depression and spent approximately a year on the unemployment lists. However, 1933 was a pivotal year for him. In March, Beier embraced the Nazi movement, which was flush with victory after Hitler's recent appointment to the Chancellor's Office. In addition to joining the party, Beier also enrolled in the SS, where he would eventually rise to the rank of *Scharführer* (Staff Sergeant). His employment fortunes changed as well. He found a job with the Neustadt Customs Office in Upper Silesia, which paid him a modest 4,800 Reichsmarks per year. After four years in this office, he moved on to work for the Weigel Werke corporation, which specialized in the planning and installation of breweries. After a year with Weigel Werke, he applied for and received a job at Peenemünde, working as an engineer in charge of

---

[61] Hans Lindenberg Basic Personnel Record, Box 703, RG 165, Folder "Boston," NARA.

[62] Erich Ball, Gerhard Reisig, Ernst Steinhoff, and Bernhard Tessmann Basic Personnel Records, Box 703, RG 165, Folder "Boston," NARA. Walter Riedel Dossier, Box 371, RG 319, IRR, NARA.

[63] A lack of enough documentation makes any firm conclusions impossible. Historians must extrapolate from the small amount of documents that do exist.

[64] See the collection of background dossiers in Box 703, RG 165, Records of the Army Chief of Staff, G-2, Intelligence Division, Captured POW and Material Branch, Enemy POW Interrogation File, 1943–1945, Folder "Boston," NARA. This file does not include information on Wernher von Braun, who was both a Nazi party member (1937) and member of the SS (1940). Thanks to Mike Neufeld for pointing out this collection to me.

installing test stands. His work at the facility paid him a very respectable 9,600 Reichsmarks per year.[65]

Beier's case is typical for many in the technological professions throughout Germany in the early 1930s. He was the victim of the crushing economic conditions in the country during this period, losing his relatively low-paying position and spending a substantial amount of time among the ranks of the unemployed. At the same time he was receptive to the Nazi party's appeals, which, among other things, promised to raise the nation out of the swamp of the depression while glorifying the work of the German technical professionals. In the ten years between 1933 and 1943, his salary more than doubled, he found a prestigious job of major consequence, and he assumed a position of importance in the SS, his nation's most elite cadre of Nazis. Beier had handsomely benefited not only from Hitler's rapid rise to power but also from the Führer's aggressive rearmament efforts, in which the missile program assumed an important place.

A disproportionate number of party members held important positions in the management strata of the facility. This was due to a number of different factors that had less to do with ideological fervor than the larger demographic factors at work in Germany in the early twentieth century. Upper-level civilian managers at Peenemünde had many characteristics in common. Most, like Beier, were born in the same generation, between 1900 and 1913.[66] The young men born in these years experienced profound crises of war, revolution, and economic collapse. Their educations in the turbulent academic climate of the 1920s and 1930s tended to encourage their support for National Socialism. German universities, especially in the postwar years, bred a virulent political radicalism that excoriated democracy and saw the solution for Germany's problems in extreme right-wing politics. Many in this "generation of the unbound" quickly became disillusioned with both traditional and republican institutions of authority and found in National Socialism a vibrant third way.[67]

The majority of these men (von Braun was an obvious exception) came from middle-class or lower-middle-class backgrounds. In the economically lean years between 1918 and 1933, many of them undoubtedly felt themselves at the mercy of forces beyond their control. The emphasis placed by Nazi ideologues on the value of technology and the technological professions, hitherto disdained by the "old order" of conservative elites and "exploited" by the new order of capitalists and industrialists, encouraged

---

[65] Anton Beier Basic Personnel Record, Box 703, RG 165, Folder "Boston," NARA.

[66] Personnel dossiers indicate that 68 of the first 102 specialists from Peenemünde and Mittelwerk were born between 1900 and 1913. A further 25 were born in the years of World War I.

[67] Michael Wildt coins this term in his excellent study of the SS officers in the SS-RSHA, *Generation des Unbedingten: Das Führerkorps des Reichssicherheitshauptamtes* (Hamburg: Hamburg Edition, 2002).

young technical specialists to offer their support to a party that welcomed their particular talents and promised them a place of high esteem.[68] Moreover, the National Socialist party was on a dramatic rise with its decisive electoral victories in the early 1930s, precisely the point at which many of these individuals would begin to develop an adult political consciousness. The temptation to attach their wagons to a rising political movement that embraced technological work proved to be too much for many to resist. Von Braun was asked to join the party and did so out of concern for his job, but he certainly was aware of the possibility that extra favors might be curried through party membership. Finally, many owed their jobs and prosperity to the Nazi rearmament project.

These common socioeconomic and political factors were the primary forces in encouraging their membership in the party. Most Peenemünders came of age with the Nazi regime and its sectarian hostility to all other sources of political, social, and cultural ideas. Even those who were not uncompromising ideologues had no other way of conceiving their role except as to serve the state. Even if some individuals, such as von Braun, began to have concerns about the legitimacy of National Socialism late in the war, their doubts could barely find expression. But that was in the future. In the heady years between 1937 and 1943, few at Peenemünde doubted the legitimacy of National Socialism or their role in the Nazi state. They were raised in an environment in which individuals were taught to view the world around them exclusively in the terms cast by the Nazi regime.

The political climate established by National Socialism shaped the behavior of individual Peenemünders in disturbing ways. Many, like von Braun, were not particularly zealous believers, but they at least found in National Socialism an ideology that they could live with. Others, however, demonstrated more troubling behavior that was activated by the poisoned environment in which they lived. One of the tragedies of the Nazi regime is that it did not simply demonize those it considered enemies. It also encouraged average Germans to act out against its supposed adversaries. For example, Design Bureau Chief Walther Riedel (Riedel III, who took over Papa Riedel's job) was a proud, active party member who admitted to joining the Nazi

---

68 On universities and National Socialism, see, for example, Fritz Ringer, *The Decline of the German Mandarins: The German Academic Community, 1890–1933* (London: Wesleyan University Press, 1983; original work published 1969); Jonathan Harwood, "The Rise of the Party-Political Professor? Changing Self-Understandings Among German Academics, 1890–1933," in Doris Kaufmann, ed., *Geschichte der Kaiser-Wilhelm-Gesellschaft im National-sozialismus: Bestandsaufnahme und Perspektiven der Forshung* (Göttingen: Wallstein Verlag, 2000), 21–45; Steven Remy, *The Heidelberg Myth: The Nazification and Denazification of a German University* (Cambridge, MA: Harvard University Press, 2002). On technology and National Socialism, see Ludwig, *Technik und Ingenieure*; Herf, *Reactionary Modernism*; Kees Gispen, *Poems in Steel: National Socialism and the Politics of Inventing from Weimar to Bonn* (New York: Berghahn Books, 2002).

Party in 1937 of his own volition. Riedel III probably had anti-Semitic tendencies that Nazism encouraged. In 1939, before he came to Peenemünde, he moved into an apartment building in Berlin that also housed Jewish tenants. According to Walter Lucks, the building custodian, Riedel posted signs forbidding the Jewish families from using the elevators and from using the air raid shelters. The building manager recalled that "He showed an unpleasant behavior towards me, since I was in friendly contact with the Jewish tenants of our house. I know that Herr Riedel caused considerable difficulties to Herr Lucks, the former janitor of the house, because the latter also supported and was on friendly terms with the Jewish tenants." These statements, given after the war in 1948, are lamely contested in the testimony of Georg Häberer, one of Riedel's colleagues from a previous job and himself an active member of the Nazi party. Häberer told investigators that, as far as he knew, Riedel "Showed an entirely neutral attitude towards the Aryan question. I never heard that he made any anti-Jewish utterances," – a nondenial denial if there ever was one. Another former colleague of Riedel's stated that the engineer was "non-political," and, in contravention of Riedel's own statement, that he joined the party because it "had become necessary with regard to his professional advancement."[69] Weighing these contradictory statements together, it is perhaps fair to say that Riedel III was not a fanatical anti-Semite, but his feelings toward the Jews were such that he was receptive to the more vicious form of Nazi anti-Semitism. Whatever their source, he felt no compunction under the Nazis to treat the Jews as equal citizens.[70] Riedel's attitudes were encouraged by the regime's loosening of the constraints on German citizens vis-à-vis their Jewish neighbors.[71]

The case of Herbert and Ilse Axster also offers an instructive example of how the ubiquitous presence of Nazi xenophobia loosened constraints on people who lived on Usedom. Herbert Axster worked as Dornberger's Chief of Staff at Peenemünde starting in 1943. He was a respected lawyer before getting drafted and seeing combat duty on the Eastern Front. The circumstances behind his entering the Nazi party are unclear, but his record indicates that he officially joined in March 1940. Though various Germans who

---

[69] FBI Report "Walther Hans Heinrich Riedel," Walter Lucks, Ella Bagatsch, Georg Häberer, and Bernhard Förster Statements, Box 101, RG 65, Records of the Federal Bureau of Investigation, Entry A1-136AB, Records Declassified Under the Nazi War Crimes Disclosure Acts, Classification 105, NARA. Riedel also confessed to investigators his "complete ignorance" of the existence of concentration camps.

[70] Riedel disavowed Nazi ideology after the war. In 1949, he told U.S. agents that "He could now see things much clearer, and no longer believed in Nazi theories or principles... "; see Box 101, RG 65, Entry A1-136AB, Classification 105, NARA.

[71] For a description of these effects, see, for example, Yehuda Bauer, *Jews for Sale? Nazi Jewish Negotiations, 1933–1945* (New Haven, CT: Yale University Press, 1994), and Marion Kaplan, *Between Dignity and Despair: Jewish Life in Nazi Germany* (New York: Oxford University Press, 1998).

knew him alternately described Axster as an ardent Nazi or a "follower," he was not particularly active in the party. In contrast, his wife, Ilse, was the head of the local Nazi women's group, the *NS-Frauenschaft*, and she had a notorious reputation as a fanatical Nazi. The Axsters owned an estate in Kölpinsee and a farm in Loddin, both on Usedom and a short drive southeast of Peenemünde. Forced foreign laborers worked at both sites and residents in Loddin were well aware that neither Ilse nor her husband spent any energy caring for the prisoners on their estates. In particular, they accused Ilse Axster of badly mistreating them, either personally or by ordering the local police to beat the workers. Her husband, on the other hand, paid no particular attention to the laborers, but on one occasion the local residents accused him of beating a French farmhand for setting rabbit traps on his estate. According to a U.S. intelligence report, "The people in the village were of the opinion that the Frenchman had [set the traps] because he was hungry and did not get enough food from Axster." On another occasion, both Axster and his wife ignored the medical needs of a Polish foreign worker who had been injured on their estate until a village resident offered to take the laborer to the hospital in Swinemünde. According to the villager, his meddling enraged the Axsters. They warned him to mind his own business and to not pay too much attention to the concerns of the foreign laborers.[72]

Axster was a wealthy attorney whose Nazi record was not terribly distinguished. Nevertheless, he felt no legal or moral constraint on his behavior toward "his" foreign workers. Nazi-inspired racism and xenophobia, so much a part of the general texture of life during the war, expanded the limits of allowable behavior toward foreigners. Acts such as Axster's mistreatment of the French farmhand became at least more conceivable, and with that realization, the actual possibility that they could take place dramatically increased. The isolation of Usedom did not inhibit the introduction of Nazi ideology, and many residents there, including the Peenemünders, were susceptible to its appeals. Many at Peenemünde, even if they were not dyed-in-the-wool fanatics, viewed the world around them on these terms.

Nazi propaganda generally found at least sympathetic ears among most Germans, even those who did not join the party. The historian Norbert Frei has shown that Hitler's regime produced "powerful socially binding forces" that served as the glue for national unity. The Nazi idea of the *Volksgemeinschaft*, the so-called national community, was the foundation for this idea. It was based on the notion that, under National Socialism, class barriers would collapse and all Germans would feel themselves part of a larger German community that was based on their innate "Germanness." Certainly, average Germans, whose pride the Nazis restored, felt that their fortunes had improved in Hitler's first six years in power. In the prewar

[72] FBI Report "Herbert Felix Axster," Box 45, RG 65, Entry A1-136AB, Classification 105, NARA.

years, most Germans believed that their nation had undergone a national resurrection under the Nazis, and that they finally had their chance for a better life for themselves and future generations. During the war years, the Nazis enjoyed a great deal of success in their efforts to build a feeling of social equality among Germany's (Aryan) citizens. Despite their lack of enthusiasm for the war, they were willing to sacrifice on behalf of this feeling of national community and the party that forged it. Any remaining doubts melted in the face of early victories in Poland, Scandinavia, and Western Europe. By the end of 1942, when Germany began to reap the catastrophe it had sown, citizens still clung to the notion of the *Volksgemeinschaft*. Virtually unimpeded Allied air raids and the distant but grim news from the Eastern Front produced a resigned siege mentality that was based on the solidarity of shared suffering. People looked to each other for support, and this strengthened the idea of the national community.[73] This was an ironic and unexpected twist in the Nazi idea of the *Volksgemeinschaft*.

Under these circumstances, the constant barrage of National Socialist ideology could only help to fortify the Peenemünders' will to fight on or to make ever more sacrifices. Between 1937 and 1943, engineers at Peenemünde would have been stubbornly dull witted and impervious to the obvious to miss the meanings of these messages in their lives, an intellectual condition for which there is absolutely no evidence. The Peenemünders embraced the goals of the institution for which they worked, knowing perfectly well the purpose and aims of the regime. Though much of this was for reasons that were explicitly nonideological, the steady diet of National Socialist rhetoric was the kernel upon which Peenemünde's central axiom was articulated. Though this identification would be strained in the last eighteen months of the war, the relationship between the regime and missile program proved to be extremely durable.

Nevertheless, the Peenemünders' access to a comfortable life and the pull of ideological imperatives still does not fully explain the absolute dedication of Peenemünde's employees to the institution's goals. Another factor motivating their dedication had to do with the daily professional duties and everyday work dynamic at Peenemünde. Working at Peenemünde was by no means an easy assignment. This was mainly due to the intense pressure

73 Norbert Frei, "Peoples' Community and War: Hitler's Popular Support," in Hans Mommsen, ed., *The Third Reich Between Vision and Reality: New Perspectives on German History, 1918–1945* (New York: Berg, 2001), 59–75. One of Frei's many useful contributions in this essay is to point out that it is a mistake for historians to underestimate the pull of ideology, no matter how crackpot it may be, among average Germans. In addition, Avraham Barkai argues convincingly that "Traditional hostility towards Jews was deeply rooted in almost every section of the population." This deep-seated anti-Semitism, such a critical part of the *Volksgemeinschaft*, became even more widely accepted by the population at large once it had been elevated by the Nazi Party to a state ideology. See Barkai, "The German *Volksgemeinschaft*," 85.

placed by military authorities on the Peenemünders to produce a viable weapon. As early as September 1939, Dornberger secured an order from Walther von Brauchitsch, the Army's Commander-in-Chief, ensuring that the A-4 project would be guaranteed access to the resources it needed as the new war progressed. However, in return, Dornberger agreed that the missile would be ready two years earlier than originally scheduled, in 1941 instead of 1943.[74] Among the military hierarchy, this immediately raised unrealistic expectations of the weapon's state of readiness, and any delays would cast doubt on the Peenemünders' ability to create a functional weapon by a specific deadline. At the same time, it forced Dornberger to make even greater demands on the Reich's already strained labor and raw materials so that the missile's large production plant at Peenemünde would be ready to go into action when development was complete.[75] In any case, Brauchitsch did not give Dornberger carte blanche. He intended to keep a close watch on the program and demanded quarterly reports from Dornberger on the progress of development.[76] Later that year, Dornberger raised even greater expectations of the development work at Peenemünde. He forecasted, in addition to an operational A-4 by 1941, another missile, to be ready by summer 1941, with an extended range of 500 kilometers, and yet another missile with a four-ton payload and a range of 800 kilometers, which would be complete by the end of 1943.[77]

Much of the development pressure, then, came from Dornberger himself. He underestimated the complex problems involved in missile development, and, for political reasons, miscalculated the advances that his developers at Peenemünde could make in a short period. By promising so much, Dornberger put his engineers under extreme pressure to craft an operational ballistic missile in a nearly impossible period of time. He and his subordinates imposed these deadlines in an effort to maintain high-level official support for Peenemünde's activities. Dornberger's ambitious marketing of the missile's potential secured huge resources for the project, but it also brought massive political pressure on the developers to complete their work. From the earliest days of the war, they faced difficult demands and were forced to labor under impractical deadlines that, despite their best efforts, they could never meet.

For example, Leo Zanssen, base commander at Peenemünde, wrote in June 1940 that developers would have the first experimental rocket on the test

[74] Entstehungsgeschichte der Fertigungsstelle Peenemünde, September 6, 1939, RH8/v.1206, BA/MA; Von Brauchitsch to Dornberger, September 5, 1939, FE 342, NASM.

[75] On the battles over the priority of missile production, see Michael Neufeld "Hitler, The V-2, and the Battle," 511–538. Neufeld shows that the priority battles over the missile program were fought largely over the production plant and that development was only minimally affected.

[76] Von Brauchitsch to Dornberger, September 5, 1939, FE 342, NASM.

[77] Dornberger, Vortragsnotiz, December 14, 1939, FE 349, NASM.

stand by August 1941 – around the time that Dornberger had promised to
deliver the 500-kilometer-range A-4 – and a test run of twenty missiles would
be ready a year later. Zanssen reported that by the end of 1942, the production plant at Peenemünde would be turning out 500 missiles per year.[78] As
the deadlines approached and it became more apparent that this schedule
could not be met, Dornberger ordered his section chiefs to make monthly
reports to him so that he would have a better picture of the problems caused
by delays in each section.[79] Delays kept piling up over the next year, and in
October 1941, he wrote to the Army General Staff with a revised schedule
indicating that development would be complete by the fall of 1942 (over a
year after his originally promised delivery date) and that preparations were
under way to manufacture as many as 150,000 missiles, if Hitler would only
give the order.[80] Dornberger had to promise them something, but this was a
patently ridiculous number, given Germany's economic, military, and industrial capabilities. He himself probably did not really believe the figure, but
even so, his statements in support of such an absurd number put enormous
pressure on the developers at Peenemünde. Two months later, Dornberger
increased the pressure even more by announcing that in order to maintain
the support of the Army's higher authorities, the first experimental missile
had to be launched by the end of February 1942.[81] After the Peenemünders
failed to successfully launch a missile in that month, Dornberger wrote to
Speer that the first test launch would not be attempted until four months
later, in early June 1942.[82]

During and after the war, Dornberger blamed these delays on the varying
levels of wartime priority and, therefore, fluctuations in the availability of
raw materials and resources given to the work at Peenemünde. The convo-
luted, ever-shifting government priority rating system for wartime industrial
projects alternately ranked Peenemünde as anything from a superpriority
project to, during the Battle of Britain, a third-level priority.[83] The long
delays were not the result of priority battles, however. The sheer complexity of the technology was the most important factor in the Peenemünders'
inability to meet the shortened deadlines. Designing an engine that could
atomize and mix the propellants, feed them into the combustion chamber
at extremely high pressure, and produce the required exhaust velocity and
thrust, all while cooling the combustion chamber so that it did not explode,
was only one of the more daunting challenges facing the designers. Guidance,
control, and supersonic aerodynamics also presented their own seemingly

[78] Zanssen report, June 20, 1940, FE 357, NASM.
[79] Dornberger to Zanssen, December 16, 1940, FE 349, NASM.
[80] Dornberger to Koch, October 11, 1941, FE 341, NASM.
[81] Dornberger Aktennotiz, December 23, 1941, FE 728/E, NASM.
[82] Dornberger to Speer, February 3, 1942, FE 342, NASM.
[83] Neufeld, "Hitler, the V-2, and the Battle," 511–538.

insurmountable obstacles. The first static test model of the A-4 arrived on the test stand in October 1940. A raft of problems kept it there through the middle of 1941, even though Dornberger had promised delivery of the A-4 by January. Two other test models were dramatic failures, exploding on the test stand in late October and early November 1941. In early 1942, another delicate test model slipped out of its corset and crashed to the ground after being tanked with liquid oxygen (the extremely cold temperature of the liquid oxygen caused the fuselage to shrink). Two test launches, one in June and one in August, also failed. Other problems then delayed launch activities until October of that year.[84] This was an entirely new field, and mistakes caused by inexperience were inevitable. Missile development proved to be a much more difficult and demanding technology to bring into being than anyone among the military or civilian specialists anticipated.

By September 1942, technical glitches and development errors meant that the rocket had yet to fly successfully. The absolute failure to meet any of the deadlines assigned by Dornberger only increased the regime's pessimism about the missile. At the end of September, a despairing Dornberger wrote to Peenemünde that, delays aside, Hitler believed that the missile would never be accurate enough to effectively deliver a warhead over a long range. Armaments Minister Albert Speer, General Friederich Fromm (the Head of Army Armaments and Commander of the Reserve Army), and Field Marshall Erhard Milch, Göring's deputy in the Air Ministry, also all doubted the success of Dornberger's project. Dornberger admitted that with the current military situation, especially the massive consumption of material on the increasingly worrisome Eastern Front, it was understandable that they bridled at committing so many of their resources to a project whose prospects for success were entirely unknown. He wrote to the Peenemünders that "The fight can be conducted with many great prospects of success if the first successful launch experiment is behind us and the results of this test came quickly one after the other. I now have the impression *that we only have a few months' time* [emphasis in original] to produce proof of the success of our development, its suitability for factory production, and its usefulness at the front." Dornberger praised the efforts of the developers, but he cautioned that regime authorities were not interested in their difficulties. "They are only interested," he continued, "in when we will get how many pieces into operation. Only then can they direct the support that a project will give to the life and death struggle of a nation." He once more exhorted the Peenemünders to pour all of their energy into launching twenty test rockets by the end of December and to extend their working hours, taking no days off, in order to finally get their A-4s to fly.[85]

---

[84] Neufeld, *The Rocket and the Reich*, 155–156, 158.

[85] Dornberger to Peenemünde, October 29, 1942, FE 342, NASM. Dornberger's behavior regarding the development schedule fits a pattern common to the military-industrial complex

Developers at Peenemünde felt the strain of Dornberger's demands. Georg Tiesenhausen noted that "We worked under colossal strain at Peenemünde."[86] Propulsion specialist Konrad Dannenberg recalled that "There was always a lot of pressure. . . . I certainly felt the pressure."[87] Even if Dornberger had not agreed to consistently set overly optimistic deadlines, the Peenemünders likely would have found the work to be carried out under intense circumstances anyway. Their working hours were always long during the war; up to twelve hours per day. Nevertheless, the Peenemünders responded to Dornberger's call, often extending their shifts to around-the-clock work before important tests.[88] Despite the natural difficulties inherent in nurturing a new technology through its growing pains and the increased problems created by Dornberger's politically motivated development schedule, the specialists at Peenemünde were absolutely willing to make sacrifices in order to achieve the goals laid out for them by military and political leaders. Given these demanding goals, the pressure to produce successful results was massive. Their ability to work under this pressure spoke volumes about their professionalism and individual dedication to the goals of the missile program.

The seemingly unending technical difficulties and the mounting pressures that came with them drove some of the engineers to the brink of despair. Walter Thiel, the mercurial, fastidious, and absolutely brilliant propulsion group chief, wrote to von Braun in early 1943, several months before production would begin, that he was completely on the brink of breakdown. Thiel was dyspeptic over the difficulties of making the fuel pumps function reliably and simplifying the fuel-injector design. He left Peenemünde for a much needed vacation in March.[89] Guidance and aerodynamics continued to produce their own problems. Despite these tribulations and many others, the Peenemünders soldiered on. Dieter Huzel put it most succinctly, writing that "If there were technical difficulties that strained the so-called state of the art, there were also times that tried the mettle of the men at Peenemünde, for

in many nations. His predictions of technical performance, at best overly optimistic and at worst absurd, betrayed his strong desire, in this case politically motivated, to deliver the promised performance from the missile in an extremely short span of time, despite the fact that no project of this sort had ever been attempted. This phenomenon has been termed "self-efficacy" by psychologists. They argue that modifications and advancements across a variety of endeavors can be motivated by the *belief* that such changes are possible, even in cases in which there is no evidence to indicate that this shift is achievable. Furthermore, self-efficacy is a strong determinant in whether an entity attempts a given task, the degree of persistence when the group encounters difficulties, and the ultimate success of the effort. See Anthony Bandura, "Self-Efficacy: Toward a Unifying Theory of Behavior Change," *Psychological Review* 84 (1977), 191–215. Bandura is the first to conceptualize this theory.

[86] Georg Tiesenhausen OHI, NASM.
[87] Konrad Dannenberg OHI, NASM.
[88] Wegener, *The Peenemünde Wind Tunnels*, 20. See also Georg Tiesenhausen OHI, NASM and Huzel, *From Peenemünde to Canaveral*, 79.
[89] Thiel to Braun, March 16, 1943, FE 692/F, NASM.

above all this was a place of human beings. There were days when even the toughest minds seemed to run out of resources, only to bounce back full of new ideas, drive, and enthusiasm – often after a long and sleepless night."[90] Undoubtedly, the tasks taken on by the Peenemünders were some of the most difficult and complex of their careers. The pressure of the war only complicated matters. Nevertheless, the missile specialists at Peenemünde succeeded in bringing the world's first ballistic missile, a technology that existed only in the minds of science fiction writers and amateur enthusiasts, from the drawing board and into mass production in only seven years, a feat that took no small amount of determination and resilience. Huzel's remark points to their fortitude and reveals a strong professional identification with the development of this new technology.

The novel tasks performed in the workshops and production facilities were a part of the appeal of working at the base. Though it had come a long way from the days of short means and primitive experiments at the *Raketenflugplatz*, rocket technology as practiced at Peenemünde was still in its infancy. Von Braun's deputy, Eberhard Rees, who also helped set up the production plant at Peenemünde, said years later that "Rocketry at that time was quite new, and it was for engineers very, very interesting. Peenemünde was for most engineers a most interesting place."[91] Karl Heimburg agreed with Rees, remembering "Even for those who had no contact at all with the rocket fad of the 1920s, work at Peenemünde was incredibly exciting because it was so new, so radical."[92] The cutting-edge nature of the work helped drive the employees' enthusiasm and provided them with the energy they needed to continue in the face of military pressure and technical failure. Measurement specialist Gerhard Reisig stated with only slight exaggeration that "It was always exciting. ... I can't remember a single day at Peenemünde that was not exciting or at least interesting, because something was always up."[93] Rudolf Hermann came to Peenemünde "because I saw all the possibilities at Peenemünde with the rocket development, big problems to solve."[94] In tackling these problems, many, like Hermann, doubtless also saw an excellent chance for career advancement at the base that surpassed that offered by the university. Clearly, for many Peenemünders, the cutting-edge nature of their work spurred their excitement and was reason for many of them to continue, despite the strain of short deadlines and the growing pains inherent to a radically new technology.

The first-rate technical facilities themselves offered another incentive to work at Peenemünde. Dieter Huzel worked as an assistant to the chief engineer at Test Stand VII, where actual test launches were conducted. Test Stand

90 Huzel, *From Peenemünde to Canaveral*, 80.
91 Eberhard Rees OHI, NASM.
92 Karl Heimburg OHI, NASM.
93 Gerhard Reisig OHI, NASM.
94 Hermann Memoirs, UAH, 16.

VII was something like a space enthusiasts' theme park. It was equipped with a huge, modern service building, a large conveyor crane to transport test missiles, and an ovular earthen berm studded with advanced measurement equipment. At its center was the launch table, where engineers and technicians made last-second adjustments to the missiles before attempting the launch test. Huzel was thrilled to be there. "Finally!" he wrote. "Here was the break I had been seeking. ... Such an assignment would bring me right into the heart of the experimental rocket development, in the largest and most complete facility in the plant."[95] For Huzel, work at the test stand, from which the first man-made vehicle to reach space was launched, was a highlight of his career.

Rudolf Hermann's case is also instructive. He began his association with von Braun in early January 1936, when the young aristocrat went to the Technical University of Aachen to discuss Herrmann's advanced aerodynamic research at the small supersonic wind tunnel built by the university. The square-shaped tunnel was four inches on a side, and its maximum velocity was Mach 3.3, just over three times the speed of sound. Hermann used small models to design the shape of the A-3 test rocket, but when it became apparent the tunnel at Aachen was insufficient to meet the demands of the much more ambitious A-4, von Braun convinced Ordnance that Peenemünde needed its own wind tunnel. He offered Hermann the chance to run what would eventually be the world's largest and fastest facility, measuring sixteen inches per side with a maximum velocity of Mach 4.4. Casting aside the chance for a prestigious full professorship at the Technical University of Braunschweig, which the university offered the Docent at nearly the same time, Hermann jumped at the opportunity and came to Peenemünde on April 1, 1937. The excitement of the work and the unequaled technical resources made the offer too much to resist.[96] By the middle of 1939, Hermann's staff at the institute reached sixty, and by 1943, he had 200 employees at his disposal.[97] The talented aerodynamicist had made what by any measure was a significant professional step forward.

Many of the scientists, technicians, and engineers at Peenemünde either received civilian draft exemptions to work there or were already members of the Army. At the beginning of 1940, almost 1,700 of Peenemünde's total staff of over 2,000 employees had civilian draft exemptions, and this number dramatically increased as the Ordnance recruited and hired more scientists, engineers, and technicians over the next three years.[98] This mutually beneficial arrangement lasted until late in the war, when overwhelming personnel shortages forced the Army to seek more soldiers wherever it could find them,

[95] Huzel, *From Peenemünde to Canaveral*, 65.
[96] Hermann Memoirs, UAH, 16–17; Neufeld, *The Rocket and the Reich*, 86–87.
[97] Neufeld, *The Rocket and the Reich*, 88.
[98] Peenemünde employment survey, January 1, 1940, FE 357, NASM.

FIGURE 7. An A-4 test missile launches from Test Stand VII in 1943. [National Air and Space Museum, Smithsonian Institution (SI SI-90-69)]

while local party authorities attempted to conscript more and more individuals into the *Volkssturm* militia units. Even so, many Peenemünders survived such harrowing close calls. Guidance specialist Walter Hauesserman earned his Diploma-Engineer degree in 1938 from the Technical University of Darmstadt, but he was drafted in September 1939, just after the war broke out. Ernst Steinhoff, who had many contacts at TU Darmstadt, managed to get Hauesserman removed from the Army and sent to Peenemünde in December. There, Hauesserman received his civilian draft exemption so he could work uninterrupted in Steinhoff's guidance department. Between 1939 and 1943, Hauesserman split his time between Peenemünde and Darmstadt, but his brushes with military service were not over. In the middle of 1943, Hauesserman received orders to report to his old unit to join the fighting on the Eastern Front. By this point, the engineer had made several important contributions to the A-4's guidance system. According to Hauesserman, von Braun successfully intervened with Army authorities because it was imperative that Hauesserman be allowed to continue his work. Because of von Braun's intervention, Hauesserman kept his civilian draft exemption and went on to perform valuable guidance work on the A-4, the Wasserfall anti-aircraft missile, and advanced torpedoes for the Navy.[99] His technical expertise, therefore, offered him the chance to fulfill important professional goals while avoiding some of the worst horrors of war at the front.

A different group of specialists at Peenemünde did not receive draft exemptions but were actually pulled out of front-line military duty to work at Peenemünde. They were members of an Army unit, ordered into creation in late 1941 by von Brauchitsch, with the innocuous title of *Versuchskommando Nord* (VkN – Experimental Command North). The VkN, under the command of a Major Heigl, a career officer with no technical experience, was first made up of about 620 men, but quickly expanded to nine companies of about 300 men each, nearly all of whom had formal engineering or technical backgrounds.[100] The Army classified these soldiers as front-line troops on temporary duty in Peenemünde, which officially kept them off limits from civilian authorities who might wish to requisition them for any number of projects. Soldiers of the VkN worked in both the development workshops and the production plant.[101] Payment for these men was excellent, by Army standards. The officers received standard Army-scale salary, which was marginally less than they would receive as civilians. The enlisted men, all of whom were scientific and technical specialists, still earned their Army salary when the VkN first came into existence. Within a few months, however, the regime adjusted their pay so that they earned about the same amount of money as civilians at the base, which was a major increase over the standard Army salary.[102]

---

99 Walter Hauesserman OHI, NASM.
100 Organisatorische Massnahmen seit Führererlass (date unclear, likely early 1942), FE 692/C, NASM; Unidentified prisoner of war statement, NASM File "V-2 (A-4) Missile (Germany, WWII) Intelligence Interrogations." See also Guido de Maeseneer, *Peenemünde: The Extraordinary Story of Hitler's Secret Weapons V-1 and V-2* (Vancouver, Canada: AJ Publishing, 2001), 102–103.
101 Organisatorische Massnahmen seit Führererlass, FE 692/C, NASM.
102 Huzel, *From Peenemünde to Canaveral*, 34. Huzel himself was a member of the VkN, drawn from the Eastern Front in the winter of 1942.

VkN members found even more important reasons to look with happiness on the good fortune of being assigned to Peenemünde. Of course, the most important was that they no longer worried every day about getting killed in combat, especially those who came from the Eastern Front. Peter Wegener, who found himself assigned to Peenemünde in the spring of 1943 after serving on the Eastern Front, wrote that he was "continually mindful of the great advantage of not being involved in further fighting in Russia, fighting that turned increasingly into disaster for the German troops. I shared my father's frequently repeated view, based on his World War I experience, that in war, any place where nobody shoots at you is fine. I had no responsibilities for others or daily worries about survival."[103] Huzel, who also served in the Russian campaign, paints an even more vivid picture of the contrast between the front and Peenemünde. "The trying business of constant alert," he wrote, "the automatic feeling of guilt at the mere sight of a trim uniform, the old frustration of motion for motion's sake, were fast fading. ... Outside the summer air was fresh and clean, the afternoon sun bright and warm, and the war a long, dim way off."[104]

The VkN's daily routines were surprisingly casual, and the trappings of military life were almost nonexistent. When Wegener arrived at Peenemünde late in the evening and reported for duty, declaring in his best military voice his name and assignment, he was greeted by a man in pajamas who told him that he could have waited until the morning, to find quarters for the night, and to come back the next day.[105] Though many soldiers ate in the Army mess hall, individuals were not always expected to eat meals with their comrades. Rather, they were able to take meals wherever they preferred, either in the cozy restaurants of Zinnowitz (for which they still needed ration coupons) or one of the Army cantinas. During the morning roll call, the straggling, sleepy, half-dressed soldiers stumbled out of bed, formed terrible lines, and chided their sergeant for calling the roll too slowly. Huzel noted with delight, "From a strictly military point of view, this was a mess. Personally, it was a pleasure."[106] Beginning in early 1944, Army regulations even allowed certain VkN soldiers to wear civilian clothes. This was largely done for security reasons, as more VkN specialists made long and secretive trips to the various assembly plants across the Reich.[107] Military rank melted away in the face of professional qualifications. Observers noted the incongruity of a corporal who also happened to hold an advanced degree in engineering giving orders to his technically less-qualified superior officers

[103] Wegener, *The Peenemünde Wind Tunnels*, 19.
[104] Huzel, *From Peenemünde to Canaveral*, 39.
[105] Wegener, *The Peenemünde Wind Tunnels*, 13.
[106] Huzel, *From Peenemünde to Canaveral*, 36, 39–40.
[107] Storch to Heigl, January 15, 1944, FE 732, NASM. See also Storch to Heigl October 17, 1944 and Heigl to Storch, October 23, 1944, both in RH8/v.1941, BA/MA, and Storch to Heigl, January 3, 1945, FE 732, NASM.

when they were on the shop floor.[108] Indeed, military considerations were entirely secondary to technical ones. As it did for civilians, life at Peenemünde proved to be idyllic for soldiers who had only recently endured the savagery of the war, only to find themselves dropped into the middle of a virtual technological and scientific paradise.

VkN specialists adjusted quickly to their work. One report noted, "The new employees of the Northern Experimental Command have generally proven their value and clearly find happiness in their work."[109] A central component to this newfound satisfaction was the fact that the Army was finally putting their professional talents to use, rather than forcing them drive a truck or lug ammunition at the front. Huzel remembered his frustration, writing, "My duties on the Russian front made no use whatsoever of my degree and years of experience in engineering. I was a *Landser*, an ordinary foot soldier, and my real capabilities, along with those of thousands of other good technical people drafted in a similar manner, were lost to the now-desperate German war effort. ... Overnight, Ph.D.s were liberated from KP duty, masters of science were recalled from orderly service, mathematicians were hauled out of bakeries, and precision mechanics ceased to be truck drivers."[110] Once away from the front and in place at the Aerodynamics Institute, former infantryman Wegener wrote with satisfaction that "I was learning a great deal of fascinating science and engineering and was slowly adapting to intellectual challenges."[111] Thus, in a number of ways, the soldiers of the VkN endowed their work with a great deal of personal significance. Not only did it save their lives, but it also removed them from the most frustrating elements of military life, paid them very well, and catered to their professional aspirations by setting them to work on some of the most cutting-edge technology on the planet. In a world in which the alternative to their work was carrying a rifle on the Eastern Front, these considerations went a long way toward ensuring their unequivocal dedication to their work.

For military and civilian employees alike, absolute cooperation and teamwork in the missile endeavor was essential to their success. Indeed, despite the tight regulations governing secrecy, administrators at Peenemünde encouraged a great deal of collaboration between workshops. In a circular sent to Peenemünde in June 1942 that clarified the division of labor between various development and assembly branches, Dornberger emphasized that "*The clear, full understanding and cooperation of all divisions is*

---

[108] Helmut Hoelzer OHI, NASM.
[109] Organisatorische Massnahmen seit Führererlass, FE 692/C, NASM. The same report, however noted both that unskilled employees were still in desperately short supply and that there were ongoing failures to overcome this bottleneck. Peenemünde engineers eventually chose slave labor as the solution; see Chapter 4.
[110] Huzel, *From Peenemünde to Canaveral*, 23, 27.
[111] Wegener, *The Peenemünde Wind Tunnels*, 19.

the indispensable precondition for the success of the entire project [emphasis in original]."[112] To ensure this collaboration, administrative divisions and the workshops that made them up mutually supported each other, actively communicating questions, problems, concerns, and experimental results in order to most effectively utilize the little time available to them. Service regulations directed specific division and workshop managers to freely and punctually communicate information requested by their partners in other areas. Managers in charge of static engine tests and full-scale launch tests worked closely with the Aerodynamics Institute and the Measurement group. The Ballistics Office freely cooperated with development engineers in the Technical Office and Aerodynamics group.[113] The supremely complex nature of missile development meant that cooperation between specific divisions and specializations was absolutely essential to the project's success. By inserting provisions regarding cooperation between Peenemünde's specific technical divisions into the service regulations, the facility's administrators formalized a cooperative environment and made technical collaboration a hallmark of missile development.

This was the result of a set of thoughtful, conscious decisions made by von Braun and others regarding the best way to rapidly develop missile technology. Von Braun's ideas fundamentally shaped the emerging profession of rocket engineering in the middle of the twentieth century. For him, the absolute complexity of rocket and missile technology demanded that cooperation between diverse specialists be the permanent watchword. Writing after the war in an American periodical, he emphasized that "The missile field, extending as far as it does into technical areas as far apart as fuel chemistry and ultra-high frequency radio, stress analysis and supersonic aerodynamics, materials research and gyroscopes, pure mathematics and shop management, cannot possibly be encompassed by a single brain. As in baseball, good players are needed, but it is the quality of the teamwork among these players that decides whether they are big league or bush league."[114] Though his own case might be considered an exception, for von Braun, there could be no single individual capable of understanding all of the intricacies of such a difficult project. The mark of a professional rocket specialist was that he understood this. He continued, "Whether they are scientists, engineers, or mechanics, they must be given an opportunity to learn to appreciate the capabilities and accomplishments of their fellow team members. In guided missile development this is particularly important because there simply cannot be an argument as to what professional group

[112] Dornberger, "Entwicklung und Fertigung des Gerätes A 4," June 6, 1942, RH8/v. 1959, BA/MA.
[113] Dienstanweisung für das Prüffeld, Dienstordnung für das Ballistische Büro, Dienstordnung für die Abteilung Messgruppe, 1937, all in FE 348, NASM.
[114] Von Braun, "Teamwork," 38–39, Box 200, Folder 7, Wernher von Braun Papers, SRCH.

is more important."[115] Cooperation, then, was fundamental to such a difficult endeavor. These ideas are self-evident in retrospect, but in the 1940s, when the technology was so young, these were the basic building blocks of the profession. A good rocket specialist was only partially defined by his technical skill, whatever that may be. His – it was and remains very much a male profession – willingness and ability to work cooperatively with other technical and scientific experts was equally important.

Von Braun's ideas, although put to paper after the war, fundamentally shaped interpersonal relationships at Peenemünde during the Nazi period. Unfailingly, Peenemünders professed after the war that the emphases on cooperation and collaboration were important components of the success they experienced there. Helmut Zoike warmly remembered that "The main thing of the whole story was the teamwork that people had there." Zoike went on to credit von Braun with setting an excellent example of hard work, teamwork, and leadership.[116] Gerhard Reisig, the chief of the Measurement section, hailed von Braun's ability to set a cooperative tone.[117] The emphasis on teamwork enhanced the individual employees' active identification with each other and their work while offering them the chance to participate in a collaborative venture of surpassing importance. The result was the generally smooth day-to-day functioning of research and development as well as the establishment of strong bonds of community inside the work place that reinforced those already in place outside of it.

Personal and professional relationships on the shop floor at Peenemünde closely reflected von Braun's ideas. The atmosphere in the workshops was almost always friendly and cordial, with employees often referring to each other by their first names, no small feat in a deeply title-conscious society. Though there inevitably were moments of friction between individuals, work at Peenemünde was for the most part characterized by harmony between employees and office groups. Ernst Kütbach, an employee in the Measurement section, characterized the workshops as having "A highly tolerant feeling of camaraderie [*Kameradschaft*]."[118] Herbert Lucht remembered that "We were all equals, engineers, doctor engineers, and so forth. And that was always, in my opinion, good for us – this camaraderie."[119] When Peter Wegener arrived at Peenemünde, he found his supervisor at the Aerodynamics Institute, the highly respected Dr. Hermann Kurzweg from the University of Leipzig, to be "an exceptionally pleasant person. . . . In retrospect, I find it remarkable that this varied group, disregarding the individuals' particular

[115] Ibid., 41.
[116] Helmut Zoike Statement, *Peenemünde: Schatten eines Mythos*, Dir. Matthias Schmidt, MJB Film- und Fernsehproduktion, 2001.
[117] Gerhard Reisig OHI, NASM.
[118] Ernst Kütbach Statement, *Originaltonaussagen*, HTIZP.
[119] Herbert Lucht Statement, *Peenemünde: Schatten eines Mythos*.

ranks in the hierarchy of the institute, worked together so smoothly. I never heard a harsh word: everyone helped everyone else, and good humor reigned; in fact, it was a pleasure to work in this place."[120] A thoroughly pleasant and professional environment pervaded the workshops. This was a major factor in the technological advances made at Peenemünde. Most employees enjoyed the pleasant professionalism and intellectual respect of their comrades. It made the often arduous and stressful work a far more enjoyable and rewarding experience.

Nevertheless, some friction was unavoidable in a facility with so many employees. Most of the disagreements remained in the upper levels of the administrative hierarchy and did not filter down to the shop floor. These fissures opened because of the tremendous pressure on the leading Peenemünders to complete the A-4's development and begin mass production. Brauchitsch's order to accelerate production as well as Dornberger's unrealistic deadlines sometimes strained relations between department managers at Peenemünde.

The failure to solve difficult technological problems by the established deadlines and the frequent, sometimes disastrous testing errors were sources of friction. Dornberger fired off an angry memo to Peenemünde managers in February 1942, when many difficult design issues were coming to a head, several production deadlines had been missed, and the political battle over the missile's priority was near its peak. The exasperated General blamed many of these problems on a lack of cooperation that he felt endangered the entire effort. "Cooperation between the Test Group and the Design Bureau is totally absent," he raged.[121] The root of the problem was that longtime specialist Walter "Papa" Riedel, von Braun's old friend from the *Raketenflugplatz* days, was freezing the Propulsion group and Testing section out of the design process. Riedel's Design Bureau was also failing to produce workable design drawings from which other divisions could work.[122] The pressure created by shortened deadlines and development delays spurred Dornberger's heated memo, but the incident reveals deeper problems as well. Riedel was in over his head. The complexity of the work was beyond him. Just as bad, he had a difficult personality and resented the influence of neophyte diploma and doctoral engineers who were placed above him.[123] A few months later, this inauspicious situation, compounded by the pressure for quick experimental results, forced Riedel out of his position as head of the Design Bureau. His difficulties, combined with the increasing professional standards at Peenemünde, hastened his removal. Ironically enough,

---

[120] Wegener, *The Peenemünde Wind Tunnels*, 26–27.

[121] Dornberger Circular to Peenemünde, February 5, 1942, in Technical File "Peenemünde #2," NASM.

[122] Ibid. See also Neufeld, *The Rocket and the Reich*, 158–159.

[123] Konrad Dannenberg OHI and Arthur Rudolph OHI, NASM.

he was replaced by the able (and anti-Semitic) diploma-engineer with the nearly identical name Walther Riedel (no relation to his predecessor, but known by his colleagues as Riedel III – the second Riedel was test stand and deployment chief Klaus Riedel, also no relation).[124] Papa Riedel moved into the Production Planning group, where he was assigned to preparing production drawings – for all intents and purposes, a demotion.[125]

The Papa Riedel case is instructive for a number of reasons. There was obviously some unavoidable friction between influential people at Peenemünde. The tremendous pressure for results was one reason why. The unrealistic deadlines expected of the developers made for many long nights and several angry memorandums from Dornberger, their chief representative. The sometimes difficult Riedel was faced with a task that was beyond his talents and unsuited to his personality, a situation that made constructive cooperation – and success – a difficult objective to achieve. In fairness to Riedel, bringing the chaotic design drawing situation to heel was gargantuan project, but he did not have the ability to complete it, despite his years at the Heylandt Works and under von Braun at Kummersdorf. Riedel was a talented practical engineer, but he was not suited to the complicated task before him. His prickly nature did not help matters. Peenemünde's increasing academic and professional standards, combined with Riedel's less distinguished education, also may have played a role in his demotion. The energetic Riedel III, who had more formal academic training and who proved more willing to work collaboratively, moved into his position, and within a short time, the dysfunctional relationship between the Design group and other branches dramatically improved. The emphasis on high professional standards is clear in this case. Those who had the requisite training and who could work within the formal and informal stipulations laid out by Peenemünde authorities would flourish, whereas those who could not would flounder. Though von Braun no doubt felt some degree of personal loyalty to his longtime friend and colleague, Riedel's failure to work closely with those in other divisions, as well as his obstreperousness with other Peenemünders, forced von Braun's hand. In the emerging world of the professional rocket specialist, Riedel was unsuited to a high-ranking administrative position. Though the failures to develop the weapon on time can in no way be laid solely at his feet, its technological complexity, the intense pressure for rapid results, and heavy emphasis placed on teamwork made "Papa" a hindrance that had to be removed.

The Riedel case also offers the opportunity to examine another aspect of von Braun's leadership style as well. Von Braun's proactive, interventionist management approach, combined with his deep theoretical and technical knowledge, kept his employees in line as much as it drove his own desire to

---

[124] Fritz Mueller, OHI, NASM.
[125] Konrad Dannenberg OHI, NASM.

see the program through to success. His management philosophy was based on teamwork, cooperation, and constructive feedback, but he also incorporated rewards for particularly hard and successful workers and punishments for those who did not live up to his standards. Most employees spoke very highly of von Braun, who was quick to praise and reward, but the young aristocrat was not afraid to push disaffected employees back into line, chide managers who broke the rules or missed deadlines, and fire workers who did not live up to his very high standards. His carrot-and-stick approach went a long way toward urging the Peenemünders to their best efforts.

With a very personal managerial touch, Von Braun built an excellent rapport with his employees. According to guidance and control specialist Helmut Hoelzer, who designed one of the world's first analog computers at Peenemünde, von Braun encouraged technological debate, which could lead to some bruising staff meetings. Once von Braun had made his decision, however, no matter how difficult, he always made sure that there was no personal damage done. He might buy the loser in an argument a drink or drop in to the workshops the next day to make sure that there were no hard feelings. "Hardly anyone held a grudge against him," Hoelzer recalled.[126] Von Braun was also generous with rewards for excellent work. Some were major professional awards; others were more personal expressions of appreciation. For example, in October 1944, he awarded the title of *Oberingenieur* (Senior Engineer) to several engineers who consistently made major advances in the development and production processes.[127] The title was a mark of major distinction in the engineering profession. Bruno Helm, an assembly foreman, won a prize for improvements he made in sealing rocket combustion chambers.[128] Others might get a bottle of liquor or champagne in recognition of their work. Loyalty to von Braun was an important part of the group reality at Peenemünde. His cult of personality was widespread among his subordinates, and for good reason. Von Braun was a brilliant manager with a knack for solving difficult technical problems and dealing with people on a very personal level. His penchant for doling out rewards to the deserving encouraged individual dedication and enhanced personal identification with the undeniably complex development and production work.

Von Braun himself, who had done so much for the success of the program, received his own share of the accolades. In the summer of 1943, just before the A-4 was scheduled to go into mass production, Albert Speer asked Hitler to award von Braun with the prestigious title of Professor, a high honor usually given to senior scholars with a lifetime of work behind them. Hitler

---

[126] Helmut Hoelzer OHI, NASM.
[127] Von Braun and Storch to Dannenberg, Hackh, Heimburg, Tessmann, and Martin, October 15, 1944, RH8/v.1941, BA/MA.
[128] Bruno Helm Basic Personnel Record, Box 703, RG 165, Entry 179, File "Boston," NARA.

immediately agreed and insisted on signing the diploma himself.[129] A little over a year later, in late 1944, when hundreds of A-4s were rolling off the assembly line every month and being fired on London and Antwerp, von Braun and Dornberger both received the Knight's Cross with Swords of the War Service Cross for their efforts on behalf of the missile program.[130]

Von Braun was also quick to crack the whip on subordinates who failed to live up to his lofty standards or whose performance was a drag on development and production. In January 1943, when engineers at Peenemünde were frantically attempting to begin mass production and pressure from regime authorities for results was nearly unbearable, von Braun found himself working almost nonstop in an effort to coordinate the work of the Development and Production groups. Production drawings were in disarray and von Braun was beginning to lose his patience. In a memo to a group of engineers in charge of organizing the production drawings for electrical parts, von Braun made it clear that deadlines were not to be missed. In no uncertain terms, he informed these engineers, "If I ascertain that the deadlines have been exceeded and there has been no report of intervening difficulties, I will call the responsible people into account."[131] In another case in early 1944, von Braun arranged to have a group of development engineers shifted to the mass production facility in central Germany because they were, according to von Braun, incapable of independent hard work. Von Braun was furious about giving them instructions that were "in no way carried out to my satisfaction and that they in fact passively resisted." He sent them to the production facility "where these men will have increased supervision."[132] One way or another, von Braun was absolutely determined to get the best out of his staff.

Engineers and scientists at Peenemünde also had other perks to their work and outlets for their professional aspirations. Secrecy regulations forbade them from publishing their work publicly, but they did amass an impressive secret archive of technical documents, reports, and studies at Peenemünde itself. Subjects ranged from development of experimental equipment to processes for creating new fuels, design changes, wind tunnel tests, and assembly techniques. The technical archive was a veritable gold mine of experimental and production-related material.[133] After the outbreak of the war,

[129] Speer Meeting Minutes, July 8, 1943, T-74, Reel 192, RG-242, p. 3,405,674, NARA. See also Neufeld, *The Rocket and the Reich*, 192.

[130] Riedel to Kunze, December 11, 1944, FE 732, NASM. Two missile production directors won the award as well.

[131] Braun to Arbeitsausschuss, "Elektrische Geräte," June 26, 1943, FE 732, NASM.

[132] Braun to Rickhey, April 23, 1944, FE 694/a, NASM. The production facility was the infamous underground factory "Mittelwerk," which used slave laborers to construct the V-2. This will be examined in detail in the next chapter.

[133] For example, see the report on experiments in connection with the emptying of missile fuel tanks, FE 110, test results on new liquid propellants, FE 128, a report investigating the air flow patterns in the wind tunnel, FE 579, and a mathematical exposition on iteration

Ordnance also marginally loosened security considerations so that outside experts might be able to take some part in improving the pace of development at Peenemünde. Missile specialists held several conferences with their colleagues in the universities. The most well known of these was "Wisdom Day" in September 1939, an event in which nearly forty professors came to Peenemünde to contribute their knowledge to the increasingly intensive work.[134] These were, in short, professional conferences of the highest order, designed for the same purposes as those in other academic and professional meetings, and they went a long way toward satisfying the professional ideals of engineering and scientific specialists at Peenemünde. The work contracted to researchers at the technical universities in Darmstadt, Dresden, Stuttgart, Hannover, and Berlin was especially important. Much of it centered on parts development, but the researchers also contributed important theoretical ideas and mathematical calculations, designed to help reduce the dispersal of the missile from its target and increase its range. They also suggested theoretical techniques to assume launch and trajectory angles, as well as write important papers on the abilities of different instruments within the rocket.[135]

All of their cutting-edge work combined with the exclusivity of the technical community at Peenemünde to awaken an increased sense of professional self-worth.[136] The specialists there defined themselves in terms of their unique work and took a great deal of pride in their accomplishments. Gerhard Hufer stated that "I was immensely proud to be at Peenemünde and associated with that wonder weapon which we called the V-2. We knew all about the so-called V-1 flying bomb, but it was our rocket that was the big hope. We realized that the enemy could shoot down the V-1, but they could have no defense against our rockets."[137] Another specialist declared to Huzel shortly after his arrival on the base, "We here are super engineers!"[138] Huzel's own opinion was somewhat more modest, but still an evocative statement of his belief in his colleagues' professional worth.

---

methods applied to differential equations, FE 621, all held at the NASM archive. These are but a fraction of the hundreds of technical reports that are available to researchers at NASM and the Deutsches Museum.

[134] Neufeld, *The Rocket and the Reich*, 83.

[135] Gerhard Reisig, *Raketenforschung in Deutschland: wie die Menschen das All eroberten* (Berlin: Wissenschaft und Technik Verlag, 1999), 100–103. Reisig's Measurement section especially benefited from the connections to the technical universities. See, for example, Protokoll über die VP-Hochschultagung in Darmstadt, September 29 to October 1, 1942, RH8/v.1265, BA/MA. This particular meeting focused on how to solve several of the seemingly intractable guidance problems.

[136] William J. Goode has argued that the elite of any profession is almost always conscious of a communal identity. This was certainly the case at Peenemünde. See William J. Goode, "Community Within a Community: The Professions," *American Sociological Review* 22 (April 1957), 194.

[137] Quoted in Middlebrook, *The Peenemünde Raid*, 28.

[138] Huzel, *From Peenemünde to Canaveral*, 77.

He and his coworkers were "believing, stubborn, undaunted, hard workers" whose genius, "given unshaken belief, untiring effort, ingenuity, hard work, dedication, is capable of solving almost anything."[139] Peenemünders viewed themselves as elite members of their profession. In tackling some of the most complex technical challenges in the world, they became dauntless "super-engineers." The elite community that grew up on Usedom defined and fulfilled its members' aspirations. Their personal sense of significance, professional achievement, career development, and peer prestige were all enhanced by the work they did on the island.

In the late afternoon of October 3, 1942, the Peenemünders' hard work finally paid off. After seemingly endless development delays that forced Dornberger's bleak September assessment of the program's political support, the rocket labeled A4/V4, the fourth fully constructed experimental rocket, thundered off its launch platform and flew away east over the Baltic Sea. It climbed to an altitude of fifty-six miles before returning to Earth at nearly 3,100 miles per hour, crashing into the sea some 120 miles away. For the first time, a man-made object had reached into space. It was a titanic achievement. Dornberger, who was there to observe the launch, "... wept with joy. I couldn't speak for a moment; my emotion was too great. I could see that Colonel Zanssen was in the same state. ... We yelled and embraced each other like excited little boys."[140] News spread around the base like wildfire. Werner Rosinski remembered that after the launch, "Everyone was really excited. Everybody thought that we've got it made now."[141]

That evening, during a celebration in the officers' club, Dornberger, in language reminiscent of the Weimar rocketry days, told his colleagues "The following points may be deemed of decisive significance in the history of technology: we have invaded space with our rocket and for the first time – mark this well – have used space as a bridge between two points on the earth. ... To land, sea, and air may now be added infinite empty space and an area of future intercontinental traffic, thereby acquiring political importance." Space travel was one thing, but Germany was also fighting a war. "So long as the war lasts," he pointed out, "our most urgent task can only be the rapid perfecting of the rocket as a weapon."[142] The exigencies of Germany's war would not wait. They must not, as they knew, stop working.

Despite their impressive technological achievement, the October 3 launch was surpassingly lucky.[143] Major developmental problems still existed, and it would be many months before the Peenemünders would be able to construct a relatively reliable mass-produced weapon. Even so, the successful

---

[139] Ibid., 77–78.
[140] Dornberger, *V-2*, 12–13.
[141] Werner Rosinski Interview with sociologist Donald E. Tarter, UAH.
[142] Dornberger, *V-2*, 17.
[143] Neufeld, *The Rocket and the Reich*, 165.

launch caught the attention of many important members of the regime, and the pressure to succeed grew even more while powerful organizations jockeyed for control of the program. The Peenemünders continued on, laboring tirelessly to capitalize on their remarkable achievement. Their efforts to do so drove them into the arms of some of Hitler's most ruthless men.

4

"Production by Convicts: No Objections"

The Peenemünders' October success aroused the increased interest of powerful institutions within the Third Reich. Albert Speer's Armaments Ministry had, under Fritz Todt, been content to supply labor and technical guidance to the construction at Usedom. Speer saw that the time was right to take greater control of the program and push for the highest priority for mass-producing the nation's most advanced weapon. In addition, Heinrich Himmler, whose fascination with technical novelty was surpassed only by his ideological fervor and quiet barbarity, came to see his SS as the organization most fit to guide the burgeoning success of the missile program. Though the Army still had an influential voice in decisions about the project, it began to experience a slow but steady decline in influence over its prized program. Nevertheless, longtime civilian and military specialists at Peenemünde proved themselves to be perfectly willing to cooperate with these new but contentious allies. Despite some misgivings, the Peenemünders accepted their ideologically motivated, often intrusive masters, ultimately discovering that collaboration with these organizations could assist them in very important ways on the path to completion of their work.

Nevertheless, at the end of 1942, many obstacles still loomed in their way. Perhaps the most difficult of these was the labor supply needed to continue and expand their work. German industry in general suffered from a labor shortage, but by early 1943, those at Peenemünde felt it most acutely. The October 1942 success meant that the regime expected mass production to begin shortly thereafter, but a number of difficult development and production problems remained. The developmental issues would have to be solved by sheer determination and brainpower, but the labor problem was hugely problematic because military call-ups created a general dearth of qualified workers. In early 1943, administrators at Peenemünde solved this problem by agreeing to the use of concentration camp slave labor to mass-produce the A-4 missile. As early as the middle of 1943, Peenemünde, with its deeply ingrained institutional culture of self-interest, unavoidable National Socialist ideological messages, and steadily increasing cooperation with some of the regime's most barbaric elements, had embarked on a path that would eventually involve it in one of the regime's most infamous instances of criminal abuse of foreign labor.[1]

ENTER THE ARMAMENTS MINISTRY AND THE SS

As the Peenemünders slowly began to overcome the technical difficulties of missile development in the summer and fall of 1942, the issue of mass production loomed ever larger. The missile program had always been carried out on a crash basis, but Dornberger's and the Army leadership's incessant demands that the missile be operational as soon as possible only increased the pressure to complete development and usher in mass production. Once the missile had been successfully launched in October 1942 and began to show, at least to regime authorities, its promise as a "wonder weapon," Peenemünde's nominal independence as an Army program began to wane. Under Fritz Todt, the Armaments Ministry had earlier begun working to subsume the program under its umbrella. Todt's successor, Albert Speer, worked even more energetically to assume greater control. At the same time, the SS turned its attention to the promising developments taking place at Peenemünde.[2]

---

[1] Recent studies have done a great deal to elucidate the framework in which the decision to employ slave labor was made as well as the actual conditions of those prisoners who worked to manufacture the missile in the underground factory. However, this scholarship tends to draw too strict a division between the development engineers at Peenemünde and the production engineers from the Armaments Ministry and SS. Their arguments imply that once mass production began, with labor supplied by the SS, development specialists at Peenemünde had a minimal and uneven impact on the A-4 program generally, only suggesting technical changes to improve performance and to solve some of the more intransigent operational issues. Production engineers supplied by the Armaments Ministry and SS, they indicate, began to dominate the most important decisions made with regard to the A-4. This picture of the Peenemünders draws too great of a distinction between Peenemünde's development employees and the production engineers who plied their trade in Mittelwerk. See, for example, Michael Thad Allen, *The Business of Genocide: The SS, Slave Labor, and the Concentration Camps* (Chapel Hill, NC: University of North Carolina Press, 2002), 208–239; Neufeld, *The Rocket and the Reich: Peenemünde and the Coming of the Ballistic Missile Era* (Cambridge, MA: Harvard University Press, 1995), 167–238; André Sellier, *A History of the Dora Camp: The Story of the Nazi Slave Labor Camp That Secretly Manufactured V-2 Rockets* (Chicago: Dee, 2003); and Jens-Christian Wagner, *Produktion des Todes: Das KZ Dora-Mittelbau* (Göttingen: Wallstein Verlag, 2001).

[2] Allen has argued that these battles over large and important projects by such influential organizations were not merely efforts to gain increasing power in the polycratic National Socialist system. Rather, in the second half of the war, these organizations, especially the SS, were motivated to remake the state on the model provided by their own specific ideological vision. See Allen, *The Business of Genocide*. Karin Orth's wide-ranging study of the Nazi camp system argues less convincingly that such battles were merely a part of Himmler's "political power calculations." See Karin Orth, *Das System der Nationalsozialistischen Konzentrationslager* (Munich: Pendo Verlag, 2002; original work published 1999), esp. 162–221.

This came about in part because of Hitler's increased interest. Seven weeks after the successful launch test on October 3, 1942 (and in the midst of the cataclysmic battle for Stalingrad), he ordered Albert Speer to begin mass-producing the missile as quickly as possible. The dictator wanted to exact "vengeance" upon England for the destructive Allied bombing raids on German cities.[3] To bring the A-4 on line as quickly as possible, Speer organized what he called the "A-4 Special Committee" in early December. This group was made up of some twenty subcommittees whose members included specialists from Peenemünde, industry representatives, and Armaments Ministry officials. Its task was to coordinate the production and delivery of parts and completed missiles, organize proper transportation of raw materials, and test the mass-produced missiles for quality control. Speer, recognizing that the Peenemünde developers lacked the proper production experience, gave the leadership of the committee to Gerhard Degenkolb, a well-known and successful locomotive production engineer.[4] Degenkolb had a blustering, overbearing, and even rude personality, and he was a fanatical Nazi. He came with a reputation for ruthlessly completing his large projects with little regard for cost or human considerations. Dornberger's physical description of him is indicative of his personal distaste for the gifted but imperious engineer.

He had a well-nourished appearance. In his round, sallow face, the obliquely set, keen blue eyes darted restlessly hither and thither. Prominent swellings above his eyebrows and the clearly marked veins in his temples were evidence of a hasty temper. This was Degenkolb, one of the closest associates of our greatest adversary in the Ministry of Munitions, [Karl Otto] Saur, the all-powerful Hauptamtsleiter (Chief of the Regional Party Office). . . . [Degenkolb] had a completely bald and spherical head, his soft, loose cheeks, bull neck, and fleshy lips revealed a tendency toward good living and sensual pleasures, while the restlessness of his powerful hands and the vigor of his movements were evidence of vitality and mental alertness. He was never still. His reputation as the creator of the war locomotive stood high.[5]

In January, the energetic Degenkolb began assembling his subcommittees. Of note, he made von Braun the chairman of the subcommittee for "Final Acceptance" (*Endabnahme*). Von Braun's deputies included the chief propulsion expert Thiel and several other engineers from Peenemünde.[6] By this point, production planning was lagging badly behind. Bottlenecks in raw materials and transportation held up installation of the factory's assembly lines, and the vital production blueprints were not yet prepared either at

[3] Neufeld, *The Rocket and the Reich*, 169–170.
[4] Detmar Stahlknecht, "Protokoll über Besprechung," December 5, 1942, RH8/v.1959, *Bundesarchiv/Militärarchiv* (BA/MA).
[5] Walter Dornberger, *V-2* (New York: Viking Press, 1954), 75.
[6] Von Braun to Degenkolb, February 11, 1943, FE 732, National Air and Space Museum (NASM).

Peenemünde or at contracted subsidiary firms. With his subcommittees filled out by the middle of February 1943, Degenkolb attacked these problems with characteristic energy. He seized control of production, directing the final installation of the assembly plant and dictating production schedules for various missile components. Degenkolb's innovations to the production program streamlined subassembly, systematized communications between the Peenemünde developers, their suppliers, and other subsidiary firms, and rationalized innovations in mass production by ordering batch runs that set strict deadlines on the inclusion of technical improvements. These were vast improvements over the Peenemünders' efforts, whose constant tinkering made a hash of the process to translate development models to mass production models.

Specialists who worked with Degenkolb on these projects despised him. Degenkolb wasted no time on pleasantries and did nothing to ingratiate himself with his colleagues. Dornberger wrote of him that "He intervened brutally wherever he considered it necessary to do so, pulled all the strings he thought needed jerking for him to get his way, scrounged, dismissed, or interchanged executives without any special mandate on the strength of his position in the Ministry of Munitions. He dispensed insults, curses, and threats, and refused to go into detail. . . . He acted like a burly, endlessly threatening slave driver."[7] Many Peenemünders beyond Dornberger considered him a crude barbarian who lacked any appreciation for the complexity and importance of their technological achievements. Von Braun sometimes clashed bitterly with Degenkolb, and other engineers thought his plans were completely unrealistic.[8]

To make matters worse, Degenkolb distributed a production schedule in the beginning of April 1943 that was well nigh impossible to fulfill. He ordered a monthly output of thirty missiles by July 1943, a number that would be ramped up to 450 by November and a preposterous 900 missiles per month by December.[9] These missiles were to be produced at three sites: the production plant at Peenemünde; the Zeppelin Airship Construction Company in Friederichshafen, on the border with Switzerland; and Rax Werke in Wiener Neustadt, just south of Vienna. A previous production schedule devised by Detmar Stahlknecht, an Armaments Ministry expert who worked closely with the Peenemünders beginning in mid-1942, called for the maximum of 300 missiles per month at Friederichshafen and Peenemünde to be reached much later, by September 1944.[10] Degenkolb brusquely shoved this schedule aside in favor of his more ambitious and

---

[7] Dornberger, *V-2*, 89.

[8] Engineer Kowall, "Fertigungsplanung Gerät A4 (Sonderausschuss A4). Stand 1.2.43. Aufstellung zu Position 5b der Planungsubersicht, 1.2.32," GD 638.0.17, Deutsches Museum.

[9] Gerhard Degenkolb, Fertigungsprogramm A4, April 2, 1943, FE 732, NASM.

[10] Stahlknecht to von Braun, February 24, 1943, FE 358, NASM. Arthur Rudolph, the production chief at Peenemünde, was even less optimistic. He felt that the Peenemünde production

unrealistic plan. Critical problems still needed to be ironed out, however. The most obvious was that assembly plants were nowhere close to being ready, but the Peenemünders were also still struggling to get consistent results from their test launches. Parts lists were at best half complete, and the situation with the production drawings was still unresolved.[11] Even as Degenkolb disciplined production planning, his schedule for final assembly was a fantasy because it simply ignored these realities. Worse, the ambitious schedule created a great deal of unhappiness and dissatisfaction among the missile specialists, who felt acutely the additional strain that it placed on them.

Nevertheless, Degenkolb quickly moved to impress upon the Peenemünders that he was unwilling to brook any opposition to his schedule. On April 15, less than two weeks after he distributed his plan, the combative engineer convened a meeting at the headquarters of the A-4 Special Committee in Berlin. Present, along with Degenkolb and his deputy, Heinz Kunze, were Dornberger, Zanssen, von Braun, Rudolph, and several other important representatives from Peenemünde. Degenkolb reiterated to his audience in no uncertain terms that his production schedule stood as ordered. In an effort to make his subcommittee chairmen more accountable for their progress, he also ordered them to send biweekly status reports directly to him. All work not directly related to the missile was to cease immediately so that all available energy could be focused on rapidly completing the A-4.[12] Two weeks later, Degenkolb delivered a circular to Peenemünde that reemphasized these points, ordering development on all other projects to stop "until the development of this instrument is tirelessly brought to a conclusion." Degenkolb also warned that once production began, "any impairment of production will not under any circumstances be tolerated."[13] His impossible expectations and unrealistic demands sent many Peenemünders into a fury. According to Rudolph, Thiel was so angry that he threatened to quit Peenemünde to take a teaching position at a university.[14] Thiel had already complained to von Braun that this was no simple piece of equipment that could be moved into mass production on a whim. He argued strenuously and in vain that Degenkolb had no appreciation for the engine's technical complexity.[15] Georg Tiesenhausen stated that Degenkolb's demands created

plant would be able to produce approximately 250 missiles per month, or about 3,000 per year; see Rudolph, "Vortrag Dir. Rudolph vor den Mitgliedern des A4-Ausschusses ahnlässlich ihres Besuches am 10.3.43 in Peenemünde," FE 833, NASM.

[11] Neufeld, *The Rocket and the Reich*, 175.

[12] Gerhard Degenkolb, Aktennotiz [Memorandum] Nr. T-9/43, FE 833, NASM.

[13] Gerhard Degenkolb, Anordnung [Instruction] Nr. 3/43g, FE 732, NASM.

[14] Arthur Rudolph Oral History Interview (OHI), NASM. Dornberger noted that Thiel threatened this on a number of occasions, but in this case, he was struck by his sincerity. See Dornberger, *V-2*, 148–152.

[15] Thiel to von Braun, March 16, 1943, RH8/v.1960, BA/MA.

a "colossal strain."[16] The pressure to successfully complete development at Peenemünde, already nearly overwhelming, became unbearable under Degenkolb's demands. His impossible production schedule and ceaseless haranguing only increased the tension under which the Peenemünders had to complete their work.

Even so, despite their personal distaste for Degenkolb and the increased pressure that he placed on the Peenemünders, the leadership of the Baltic facility did their best by the A-4 Special Committee's Chairman. They despised his personality and gnashed their teeth at his orders, but they nevertheless committed themselves to his schedule. In the first place, Dornberger, probably sensing that the chickens of his earlier overly optimistic salesmanship had come home to roost, gave the Peenemünders a direct order to follow Degenkolb's demands.[17] Von Braun was also a central figure in the effort to enforce Degenkolb's schedule. He set an example to his colleagues by laboring mightily to fulfill Degenkolb's orders, working on his own initiative to help open up production bottlenecks, shorten delivery delays, and improve quality control.[18] He also used his authority to coax, chide, and push his flagging subordinates back into line. In a circular that he sent to all of his deputies at Peenemünde and to his Final Acceptance Subcommittee at the end of April, von Braun made it clear that they were to put forth their best efforts to meet Degenkolb's schedule. He stated flatly that "The published production program of A-4 Special Committee Director Degenkolb is to be seen as the only valid one for future production planning." Sensing the dissatisfaction among many Peenemünders, he required "All employees of the [Development Office], the Special Committee, and the Work Committees to support the standing precepts with all means and to take up a healthy collaboration with this position."[19] Despite his closed-door battles with Degenkolb, von Braun's public demonstration of loyalty to the project in turn helped ensure his subordinates' best efforts.

Von Braun clearly sensed the displeasure with the Degenkolb numbers and sought to make sure that the Peenemünders did their best to support the Degenkolb program. When the specialists' identification with their institutional goals began to break down in the face of an overbearing ideologue who seemingly had no idea of the difficulties involved in the missile's development, von Braun, the most influential and inspiring leader in the entire effort, was able to bring their support for the program back into line by deploying his powerful managerial and symbolic authority. To be sure, everyone at Peenemünde had no choice but to accept Degenkolb's program or perhaps face the end of their work, but this only explains

---

[16] Georg von Tiesenhausen OHI, NASM.
[17] Walter Dornberger, Aktennotiz [Memorandum], June 6, 1943, RH8/v.1210, BA/MA.
[18] Von Braun to Degenkolb, March 6, 1943, FE 732, NASM.
[19] Von Braun Rundschreiben [Circular], April 30, 1943, FE 732, NASM.

part of the dynamic at the facility in early 1943. A fatalistic acceptance of the inflated and premature production numbers by the Peenemünders would hardly help to rapidly overcome the myriad of technical problems still facing the developers. Instead, when an opportunity arose for the engineers to delay, equivocate, or simply slow the pace of their work for lack of enthusiasm and poor morale, von Braun intervened, declaring that the Degenkolb schedule would stand, but also successfully appealing to them to redouble their efforts, despite the seemingly impossible demands placed upon them. In the end, this was one of the secrets of the success of the A-4 development and production programs. Von Braun's dynamic leadership helped to restore flagging enthusiasm at Peenemünde and encouraged the specialists to make even greater efforts on behalf of the program. Many of them wavered in the face of Degenkolb's often outrageous orders, but their shared dedication to his ultimate goal, propped up by von Braun's energetic interventionism, drove them to accelerate the already breakneck pace of their work in line with the demands placed upon them by powerful regime officials.

Even though they never met Degenkolb's schedule, by early August, the Peenemünders had managed to iron out many of the prickly development and supply issues and came to an uneasy, but permanent, truce with the tyrannical A-4 Special Committee Chairman. At a meeting on August 4, Degenkolb and his deputy Kunze met with Dornberger and Zanssen in an effort to specifically lay out the terms of cooperation between the Army missile program and the Armaments Ministry representatives under Degenkolb. They agreed on a number of important points. Albin Sawatzki, a Degenkolb appointee who formerly worked for the Henschel Corporation to produce Tiger tanks (and an engineer who would later figure prominently in production) was given the responsibility for overall production planning. Degenkolb compromised with Dornberger by making Sawatzki and production managers whom he assigned to the Rax Werke in Wiener Neustadt formally subordinate to Zanssen, the Army base commander at Peenemünde. Importantly, production planning at Peenemünde fell to von Braun's deputy in development, Eberhard Rees. Dornberger and Degenkolb agreed to give Rees "full dictatorial powers," and he was fully responsible for the completion of the Degenkolb program at Peenemünde. Further, they ordered Rees to confer with Rudolph, Sawatzki, and Thiel in order to come up with a final plan for labor demands in production. The minutes for this meeting formalized the plan to carry out production by using concentration camp slaves, and Rees had the responsibility of providing accommodations for these "convicts," as Dornberger would call them.[20]

---

[20] Niederschrift über die Besprechung am 4.8.43 beim Heimat-Artillerie-Park 11, RH8/v. 1254, BA/MA. The original decision to use concentration camp prisoners is discussed in the next section.

Thus, despite initial, widespread disagreement and personal dislike on the part of many Peenemünders for Degenkolb, they were able to look past their differences in the interests of moving the program forward as quickly as possible. Peenemünde specialists forged a cooperative relationship with clearly defined circles of influence and control between themselves and representatives of the Armaments Ministry. In addition, by clarifying these areas of competence and control, their plans further involved the Peenemünde specialists in an increasingly brutal National Socialist labor policy. In one stroke, they granted a *civilian* engineer full authority over all aspects of production at Peenemünde while legitimizing the use of slave labor to carry it out by granting it an additional official seal of approval. Peenemünde specialists fully accepted the utilization of slave labor to complete their work; indeed, they had very few qualms about doing so in a regime that made high virtues out of service to the state and the exploitation of foreign enemies.[21]

BRINGING IN THE BLACKSHIRTS

At the same time that the Peenemünders were grappling with the Armaments Ministry's demands to ramp up production, the SS began to take a keen interest in the events on Usedom. For both personal and ideological reasons, Himmler had long possessed an abiding fascination for complex technology, though his knowledge of it and his organization's ability to produce it on a large scale was demonstrably subpar.[22] A program that had the size and spectacular potential of the A-4 was bound to draw his interest eventually. Nevertheless, the first meaningful contacts between his organization and the missile producers were due largely to the prodding of Peenemünde officials themselves. In December 1942, Himmler visited the facility, where he toured the grounds and witnessed an unsuccessful launch test. Less than a week later, Dornberger decided that he might be able to exploit Himmler's recent interest to enhance the program's status within the war economy. He ordered the Peenemünde Development Works Army Commander, Lieutenant Colonel Gerhard Stegmaier, who was "happy as a little child about his special greeting from the Reichsführer-SS," to pass along a message to Himmler through Stegmaier's friend, Gestapo Chief Gottlob Berger.

---

[21] Another factor in this truce was that in July, Karl Otto Saur, Speer's ruthless deputy, giddy with the possibilities of the V-2, ordered that production be ramped up to 900 in October and 1,500 by January, effectively doubling Degenkolb's program in half the allotted time. See Dornberger, Aktennotiz [Memorandum] Nr. T-21/43, FE 833, NASM. Speer recognized the absurdity of these demands and, to the relief of many Peenemünders, ordered a reversion to Degenkolb's overblown schedule, which was reasonable by comparison. See also Neufeld, *The Rocket and the Reich*, 194–195.

[22] Himmler, according to an SS business manager, "supported all inventors on principle." See Allen, *The Business of Genocide*, esp. 57–164, for the myriad of failed technological projects attempted by the SS.

Dornberger wanted to use the connection to meet with Hitler in order to pitch the aims of the program directly to him one more time.[23] The meeting never actually materialized, but Dornberger continued to develop his connection to Himmler through Stegmaier and Berger in an effort to curry favor, gain influence, and make sure the production program maintained the highest possible wartime priority.[24]

There was also another important connection between Peenemünde and the SS. Wernher von Braun himself joined the organization in 1940, although he claimed in a statement to the U.S. War Department in 1947 that he did so only after a local SS Colonel acting on Himmler's order urged the young aristocrat to enroll. After some deliberation about whether or not membership would benefit him and the missile program, von Braun agreed to join. He was not a particularly active member and joined the SS out of self-interest, not for ideological reasons. Von Braun only attended meetings periodically and was known to have worn his uniform only a handful of times, surprising some who had no idea that he was an SS member.[25] Von Braun used his SS membership only when he had to; his ideological convictions, such as they were, did not guide his actions. Rather, his career in rocketry assumed precedence over all other things. Nevertheless, the young aristocrat's membership was yet another indication that the tenets of this organization were at least not anathema to him. At minimum, he was willing to cooperate with them as long as such an activity would further his personal and professional goals. That this collaboration might also benefit the SS was a consequence von Braun did not seem to consider. Even so, his membership in Himmler's organization would have important repercussions later in 1943.

Each organization attempted to exploit its interest in the other. In the spring of 1943, a few months after his first visit to Peenemünde, Himmler made a naked attempt to reduce the Army's control of the program and gain greater influence over it for the SS. Based on a fallacious report that Berger received from Stegmaier, Himmler accused Peenemünde's commander, Leo Zanssen, of being a member of the local branch of the "Catholic Action," an anti-Nazi group largely made up of Catholic priests, and ordered him removed from his post. Zanssen was indeed a Catholic, but he was nevertheless a loyal member of the Army who, if anything, may have been growing disillusioned with the Nazi regime after a tour of duty on the Eastern Front. Even so, Zanssen's disillusionment was not so deep that he worked for the defeat of the Third Reich. Fritz Fromm, the Chief of Army Armaments

[23] Berger to Himmler, December 16, 1942, Reel 117, T-175, RG 242, National Archives and Records Administration (NARA). Neufeld, *The Rocket and the Reich*, 176.
[24] Neufeld, *The Rocket and the Reich*, 176–178.
[25] Michael Neufeld, "Wernher von Braun, the SS, and Concentration Camp Labor: Questions of Moral, Political, and Criminal Responsibility," *German Studies Review* 25 (2002), 61–62; also *The Rocket and the Reich*, 178–180.

and a chief supporter of the missile program, empowered Dornberger to investigate the charges. Dornberger temporarily assumed direct control at Peenemünde and set about uncovering the affair. He eventually managed to clear his friend's name and have a shaken Zanssen returned to Peenemünde. Stegmaier, Zanssen's duplicitous subordinate, was shipped out to command the training school for the future A-4 firing troops. Dornberger probably hoped his connections might still help the project and could at least rely on Stegmaier's loyalty to the regime. Himmler, meanwhile, quietly backed off.[26]

Though this affair resulted in a number of charged confrontations and bruised egos, Himmler made little progress in his attempt to gain influence over the program. Nevertheless, the incident could not but have shaken the Peenemünders and encouraged them to act even more strictly within what they thought were proper boundaries of behavior. It only made more apparent to them that the Gestapo may very well have been actively operating behind the scenes at Peenemünde to root out all anti-Nazi elements. If an officer as important as Zanssen could be accused of seditious activities and removed from his post, there was no telling who among the civilians might be next. Dornberger wrote that after he managed to restore Zanssen to his post in the fall, "The threat of a formidable power working behind the scenes remained."[27] This threat would rear its head again nearly a year later, but even then, the General would continue to show a remarkable proclivity to seek out the SS in order to fulfill the program's needs. Though the Zanssen affair was a rattling experience, it did not permanently poison the relationship between missile program administrators and the SS. In any case, Himmler's men were not about to go away.

FOREIGN LABOR AT PEENEMÜNDE

Despite Himmler's meddling and the near imprisonment of one of Peenemünde's key military figures, administrators at the missile base continued to remain open to the options offered by cooperating with the SS. Their automatically activated adherence to Peenemünde's institutional goals of delivering on the Army's promises to the regime dictated that they should think flexibly about their remaining problems and remain open to any possible solutions. Labor supply remained one of the most intransigent problems. Though many popular histories and memoirs of the period loudly proclaim that the SS forced slave labor upon the helpless Peenemünders, the truth is far more sinister. High-ranking officials at Peenemünde either recommended

---

[26] On the particulars of this affair, see Roll 124, T-175, RG 242, NARA. Michael Neufeld has also ably sorted out this story. See *The Rocket and the Reich*, 180–183; also Dornberger, *V-2*, 182–185.

[27] Dornberger, *V-2*, 185.

the employment of SS-controlled concentration camp labor or enthusiastically agreed to use it once the option became available. In no way did the SS compel the Peenemünders to use slave labor.[28] Once at the facility, the uses to which these prisoners were put in many ways foreshadowed their future experiences in Mittelbau-Dora.

The use of foreign labor, either forced labor, prisoners of war, or concentration camp prisoners, in the missile program reflects the general pattern that existed throughout the Nazi regime. As was the case across the country, there was no predetermined plan for its mass operation at Peenemünde. However, even though the Nazis were wary of using it at all because of ideological and security considerations, construction administrators at Peenemünde recognized very early that the deadlines established by their own optimistic projections and by regime authorities could not be reached without resorting to foreign labor. Acute manpower shortages because of the large military drain on the domestic labor pool forced their hand. In the end, foreign labor proved to be an essential element, not only because it eased pressure on the labor sector in Germany in general but also because it was central to the establishment, expansion, and completion of the technological work that was carried out at Peenemünde.[29]

Foreign labor at Peenemünde has roots that stretch back until just before the outbreak of war in 1939. The first foreign workers in Peenemünde were voluntary Czech contractors. They arrived at the base to work either on the construction of the mammoth missile production plant or at Peenemünde West, the Luftwaffe test facility. All foreign workers at Peenemünde between 1939 and 1943 labored only at construction sites around the base, not on development or production projects. It was not until 1943 that foreign workers were allowed into production. Nevertheless, because of security concerns after the outbreak of war, Army administrators removed these Czechs from the island.[30] The first forced foreign laborers to arrive at Peenemünde were those who were dragooned by the vicious occupying forces in western Poland in the middle of 1940.[31] These workers eventually numbered between 600

---

[28] Mark Spoerer has surveyed thirty-three industrial firms that used concentration camp prisoners as their labor force. He has found only one instance in which state institutions coerced factories or firms into employing slave labor against their managers' will. See Mark Spoerer, "Profitierten Unternehmen von KZ-Arbeit? Eine kritische Analyse der Literatur," *Historische Zeitschrift* 268 (1999), 61–95.

[29] Ulrich Herbert, *Fremdarbeiter: Die Politik und Praxis des "Ausländer-Einsatzes" in der Kriegswirtschaft des Dritten Reiches* (Bonn: Dietz, 1985), translated by William Templer into English as *Hitler's Foreign Workers: Enforced Foreign Labor in Germany Under the Third Reich* (Cambridge, UK: Cambridge University Press, 1997). English translation cited hereafter.

[30] Dornberger to Zanssen, August 28, 1939, FE 342, NASM; Rudolph to Speer, October 12, 1939, RH8/v.1206, BA/MA.

[31] Entstehungsgeschichte der Fertigungsstelle Peenemünde, July 2–4, 1940, RH8/v.1207, BA/MA. Hereafter cited as Entstehungsgeschichte. See also Neufeld, *The Rocket and the Reich*, 185.

and 1,000, depending upon the time of year, and augmented a German construction force of approximately 4,800 men.[32]

In short order, however, secrecy considerations also began to impinge on the use of foreign forced labor at Peenemünde. In July 1940, the Armed Forces High Command (*Oberkommando der Wehrmacht* – OKW) ordered that no foreign laborers be allowed to work in top-secret facilities.[33] The number of Polish forced laborers diminished but did not disappear altogether. The demands of quick construction dictated that they must remain in place. On July 27, Heinrich Lübke, who oversaw Peenemünde construction for Baugruppe Schlempp, informed Peenemünde administrators that he would do everything he could to retain the Polish workers.[34] In this, he was relatively successful. By December, a total of 630 Poles worked in construction projects at Peenemünde, down from a high of approximately 1,000 in August.[35] Nevertheless, throughout 1941, construction on the production plant, rail line, and other support facilities was pressed forward as quickly as possible. The only way to maintain the highest possible pace of the work was by resorting to foreign labor. For this reason, the use of foreign labor was a matter of course in construction at Peenemünde, and, outside of the demands made by OKW, there was very little discussion about its morality or disadvantages. The only questions that came up revolved around how many workers the construction administrators could procure and to what purpose they would be used.[36] Despite OKW's security concerns, nearly 1,000 Italian workers arrived in Peenemünde over the spring and summer of 1941 to help with construction projects for Peenemünde West. In April, Dornberger noted that because Italians were usefully employed by the Luftwaffe, he had no quarrel with their use at the Army facility, and Peenemünde began employing them as well.[37] In early 1942, French construction workers arrived at Peenemünde to add to the workforce.[38]

A sea change in German labor policy occurred in the winter of 1941 and 1942. The Blitzkrieg strategy, designed to capture quick victories without placing onerous demands on the home front, collapsed in the face of German defeat at the gates of Moscow. With hopes for a quick victory dashed and the reality of a long and costly war setting in, the Reich's already strained labor supply became stretched to the breaking point. To help overcome serious labor shortages, Reich officials turned in part to Soviet prisoners of war and

[32] Entstehungsgeschichte, December 2–4, 1940, RH8/v.1207, BA/MA.

[33] Entstehungsgeschichte, July 2–4, 1940, FE 830, NASM; Dornberger to Schlempp, July 24, 1940, RH8/v. 1213, BA/MA. Hitler also feared the risks of sabotage security breaches that came with the use of foreign labor. See Neufeld, *The Rocket and the Reich*, 184.

[34] Entstehungsgeschichte, July 27, 1940, FE 830, NASM.

[35] Entstehungsgeschichte, December 2–4, 1940, RH8/v. 1207, BA/MA.

[36] Aktennotiz über die Besprechung in Pee am 10.-12.2.41, February 17, 1941, FE 831, NASM.

[37] Entstehungsgeschichte, April 16, 1941, FE 831, NASM. See also Baugruppe Schlempp, Bauleitung Peenemünde, Aktenvermerk [Memorandum], July 28, 1941, NASM, FE 342.

[38] Aktennotiz [Memorandum], April 2, 1942, RH8/v. 1209, BA/MA.

other "Eastern Workers."[39] In the second half of 1942, an increasing number of these prisoners began arriving in the area around Peenemünde. Planners there began to consider using them not only as construction workers but also as assembly and production personnel. Some 400 Soviet Army officers were housed in Wolgast, on the mainland about six miles away from the base. These men had technical backgrounds and the authorities utilized them as skilled laborers on Usedom.[40] Hundreds of other unskilled Soviet prisoners were kept at a barracks camp outside of Trassenheide, approximately one mile south of the Settlement. Arthur Rudolph originally hoped to use many of these prisoners in the assembly hall, but he was forbidden to do so by regime authorities for security reasons.[41] All told, by April 1943, Army and Armaments Ministry authorities exploited more than 3,000 foreign laborers and prisoners of war on the island.[42]

During this period, foreign labor also became the common solution for labor problems at subsidiary firms that manufactured parts and assemblies for the missile. Even at the subsidiary plants, Peenemünde administrators had a strong influence on decisions about the use of these workers. Perhaps the most important subsidiary firm was the Zeppelin Company in Friederichshafen in southwest Germany. In the middle of 1941, Peenemünde developers considered using the assembly plant there to produce small parts and fuel tanks for the missile.[43] By the end of the year, they decided to expand it into a second mass production site.[44] In April 1942, Ordnance unveiled ambitious proposals to expand this plant and equip it with engine test stands, a liquid-oxygen plant, and the requisite service facilities.[45] The next month, von Braun himself traveled to Friederichshafen to assess what needed to be done to begin mass production there.

Von Braun's inspection was a thorough one. He toured the assembly halls, delivery areas, work facilities, train installations, and power supply, and he was careful to note the labor requirements for the factory. A number of skilled workers at the plant were already available, but they did not have backgrounds in missile production, and von Braun believed that they needed to be complemented by trained workers from Peenemünde. He considered transferring some VkN soldiers to Friederichshafen for this task. Von Braun

[39] Ulrich Herbert, "Labor as Spoils of Conquest, 1933–1945," in David F. Crew, ed., *Nazism and German Society, 1933–1945* (New York: Routledge, 1994), 222.

[40] Martin Middlebrook, *The Peenemünde Raid: The Night of 17–18 August, 1943* (London: Cassell, 1982), 32.

[41] Arthur Rudolph Aktenvermerk [Memorandum], February 9, 1943, RH8/v. 1210, BA/MA.

[42] Neufeld, *The Rocket and the Reich*, 185.

[43] Niederschrift über die Besprechung am 3.-4.9.41 in Friederichshafen, Entwicklungsarbeiten bei Luftschiffbau Zeppelin GmbH Friederichshafen, September 8, 1941, FE 728/B, NASM.

[44] Neufeld, *The Rocket and the Reich*, 143.

[45] Georg Thom, "Bauvolumen Gerät A4," FE 728/B, NASM. The test stand would eventually be constructed at Oberaderach.

ordered the Zeppelin Works to send four work and production planners to Usedom in order to learn the best way to run a missile assembly plant. Shop floor labor was difficult to find, but he recommended that "Construction of fuel tanks can be done by foreign workers and prisoners of war." Von Braun felt that approximately twenty German supervisors would be required to work with them and that the company could work out the details of their supervision itself.[46]

In time, von Braun's recommendations were taken to heart. In October, Detmar Stahlknecht drew up a projected number of Soviet prisoners of war needed for labor in various subsidiary firms and sent it to Fritz Sauckel, the Gauleiter (District Chairman) of Thuringia and General Plenipotentiary for Labor Supply in Germany.[47] Stahlknecht requested that firms such as Klein, Schanzlin, and Becker in Frankenthal, Ardelt Werke in Eberswalde, and Friedrichs and Company in Hamburg all receive between 25 and 130 Soviet workers. Most importantly, Stahlknecht ordered 200 prisoners for work at the Zeppelin factory. Stahlknecht had also worked out precisely which skills were necessary and the number of each set of skilled laborers that would be needed. Among other things, he requested from Sauckel 100 mechanics, ten lathe operators, and ten toolmakers for the Zeppelin Works, but he provided a precise list of skills for each of the seven firms he was requesting prisoner labor for.[48] In November, the Armaments Ministry informed Stahlknecht that the prisoners he requested were unavailable, as they were urgently required for mining operations elsewhere.[49] Ultimately, the SS provided the required labor for the Friedrichshafen plant in the form of concentration camp prisoners. In February 1943, Dachau administrators made the first shipment of prisoner labor to the Zeppelin Works in order to begin parts assembly for the missile.[50]

Thus, the transition to total war not only intensified the pressure on the Peenemünders to complete their own tasks, it also drastically cut into the available supply of German labor while increasing the compulsion with which the missile specialists would put labor to work. Well before they considered using concentration camp slave labor, managers at Peenemünde took the initiative and reached out to find whatever sources of labor they

[46] Von Braun, "Niederschrift über die Dienstreise vom 2.-5.5.42 nach Friederichshafen," RH8/v.1959, BA/MA. In his excellent history of the German missile program, Neufeld notes that the impetus for foreign labor at Friederichshafen came from Dornberger in October. See *The Rocket and the Reich*, 184.

[47] The Nuremberg Tribunal found Sauckel guilty for his supervision of forced and slave labor programs in Nazi Germany. He was hanged on October 16, 1946.

[48] Stahlknecht to Sauckel, "Ausländischer Bedarf an russischen Fachhandwerkern für Sonderprogramm Peenemünde," October 28, 1942, R41/282, *Bundesarchiv-Lichterfelde* (BAL).

[49] Letsch to Stahlknecht, November 12, 1942, R3901/20.173, BAL.

[50] Martin Weinman, ed., *Das Nationalsozialistische Lagersystem* (Frankfurt am Main: Zweitausendeins, 1990), 629.

thought might be useful for the successful completion of their work. As von Braun's evaluation of the Friederichshafen plant and Stahlknecht's labor request shows, civilian experts were central to the planning for forced labor in the missile program. How could it be any other way? No one else could evaluate the projects' needs (nor was even aware of the project), and their expertise made them indispensable for planning future work. This committed search for labor, coupled with the shortages even after the German economy turned to Soviet prisoners (well over half of the 3.3 million Soviets captured in 1941 perished by the end of the year), meant that it was only a matter of time before Peenemünde managers sought out what many across Germany believed to be the one last limitless supply of human reserves in the Reich – concentration camp slaves.[51]

The turn to slave labor at Peenemünde itself happened in early 1943. As the scheduled onset of serial production loomed closer and closer, the question of laborers to assemble the missiles became increasingly pressing. In April, one of Degenkolb's representatives, a Mr. Jaeger, the head of the Labor Operations Subcommittee of the A-4 Special Committee, recommended to Arthur Rudolph the use of concentration camp labor to assemble missiles on Usedom. On April 12, Rudolph went on a tour of the Heinkel aircraft factory in Oranienburg, which used prisoners to manufacture airplanes. Heinkel began using slave labor in this factory by requesting prisoners from Sachsenhausen in 1941. In the summer of 1942, it built a subsidiary camp of the large concentration camp Sachsenhausen on the grounds of the factory. By April 1943, nearly 4,000 detainees worked in the aircraft plant.[52]

After his inspection, Rudolph returned to Peenemünde with an enthusiastic assessment of concentration camp labor for missile production. Rudolph noted that Heinkel ordered prisoners from the SS according to professional group – an arrangement that dictated the necessity of using civilian experts to determine labor needs. One German civilian served as a supervisor for every ten prisoners. He also reported that much of the Heinkel prisoner labor force was crowded into a large locker room directly adjacent to the assembly hall. The SS guarded the prisoners and provided food, clothing, and cleaning facilities for them. Ever-present secrecy and security considerations also figured largely in Rudolph's report. He noted that prisoners of different national groups were not segregated on the shop floor, but rather that Heinkel managers integrated all of the various nationalities in the factory. "This by itself is decisive for the conduct of work," Rudolph wrote. "The

[51] Edward Homze, *Foreign Labor in Nazi Germany* (Princeton, NJ: Princeton University Press, 1967). 83. See also Herbert, *Hitler's Foreign Workers*, 133–149. In late 1941 to early 1942, the SS also undertook a dramatic reorganization of its camp administration, setting up the *SS-Wirtschaftsverwaltungshauptamt* (SS-WVHA, or SS Economic and Administrative Main Office). One of the primary tasks of this office was to provide cheap slave labor to German industry. See Allen, *The Business of Genocide*, 128–164.

[52] Orth, *Das System*, 175–176.

mixing together of nationalities has the advantage of limiting the formation of secret resistance groups." Moreover, he summarized later in the report, "The operation of detainees offers considerable advantages over the earlier use of foreigners, since all tasks not related to work will be taken over by the SS and offer greater security in terms of the demand for secrecy." Rudolph – clearly impressed with the SS's ability to everlastingly ensure that these prisoners were totally cut off from the outside world – closed his report by noting that prisoner labor was the most feasible way to equip the production plant at Peenemünde with workers. He would request that Jaeger contact the SS about providing prisoners. Rudolph would begin fencing the plant as well as the streets around it in order to make it secure for prisoner labor.[53]

Concerns about maintaining secrecy, therefore, drove home the idea that foreign forced labor was inadequate. On one hand, forced laborers were generally able to live and work together in groups, according to nationality. Some were also eligible for vacation time and could return to their native countries during periods of leave. Indeed, foreign laborers helped tip off British intelligence to the existence of Peenemünde.[54] On the other hand, SS minders ensured that the slaves beneath them were offered no such solicitude. Their policy toward slave laborers perfectly complemented the Peenemünders' secrecy considerations. After their inspection of the Heinkel factory, Rudolph and Jaeger also made it clear in June 1943 that, "for reasons of secrecy and security," they wished to exchange the French forced laborers building the production plant with prisoners who were not eligible for any vacation time. Foreign laborers who had the option of taking short trips home, they emphasized, should not be allowed on the grounds of the facility at all. Heinz Kunze, Degenkolb's deputy on the A-4 Special Committee, agreed immediately and directed that all forced laborers, not simply the French, be exchanged for prisoners who were not eligible for vacation.[55] This left concentration camp slaves as the only option as a production labor force. Thus, Peenemünde's budding relationship with the SS grew from the ranks of its senior management and was conditioned on one hand by the dearth of labor in wartime Germany and on the other hand by their overweening desire to maintain the absolute secrecy of their work. Rudolph recognized that he would solve two problems at once, and it was only after his positive assessment of slave labor and direct request for camp prisoners from the SS that the Armaments Ministry actually agreed to its utilization.

---

[53] Arthur Rudolph, "Besichtigung des Häftlings-Einsatzes bei den Heinkel-Werken, Oranienburg, am 12.4.43," April 16, 1943, RH8/v.1210, BA/MA.

[54] Middlebrook, *The Peenemünde Raid*, 38–40.

[55] Aktenvermerk über die Besprechung beim A4 – Auschuss (Arbeitseinsatz) am 2.6.43 in Berlin (Lokomotivhaus), FE 833, NASM.

In addition, quite separately from Rudolph's endorsement of slave labor, Dornberger also embraced the idea of using SS prisoners in the assembly plant. At nearly the same time that Rudolph was touring the Heinkel works and giving his assessment of its arrangement with the SS, Dornberger was inspecting the two other planned assembly facilities, the Zeppelin factory in Friederichshafen and Rax Werke in Wiener Neustadt, Austria. (In July, a fourth production facility, DEMAG Fahrzeugwerke in Berlin, was added). Dornberger noted the possibility of "a closed operation of 2200 skilled laborers from concentration camps around Rax Werke." These prisoners, he proposed, should be housed in the immediate vicinity of the factory hall. With this arrangement, both the camp and factory could be fenced in and security maintained relatively easily.[56] Like Rudolph, Dornberger believed that secrecy and security were two of the major advantages offered by slave labor. Its use meant that industrial security could be expanded, labor problems solved, and costs reigned in. All the while, projected output remained the same. The similarity to Rudolph's ideas in the pivotal month of April 1943 was based on a strong collective understanding of the goals of their common endeavor.

Peenemünde administrators and the A-4 Special Committee embraced the proposals to use slave labor and moved quickly to make arrangements with the SS. The first SS guards arrived in May to set up a stronger security cordon and watch over a group of 500 camp inmates ordered by the Luftwaffe for work at their research and testing facility at Peenemünde West. On June 17, the first 200 concentration camp prisoners to work in the missile assembly hall at Peenemünde East arrived with their complement of SS guards from Buchenwald. They were housed in the cellar of the production plant, and their first task was to build a fence around the massive assembly hall in which they lived and worked.[57] By the beginning of August, 600 skilled concentration camp prisoners were in place at Peenemünde. Base administrators had plans to build a camp just outside the assembly plant that could accommodate up to 2,500 slave laborers. Once this camp was complete, its commandant, "In direct cooperation with Herr Director Rudolph, will be able to call train after train of prisoners to Karlshagen."[58] The arrival of these prisoners on Usedom marks the consummation of the relationship between Peenemünde and the SS.

Slave labor at Peenemünde, then, emerged out of a variety of different considerations. From the earliest days of the program, pressure to show results

[56] Heereswaffenamt, Arbeitsstab A4, Aktennotiz über Reise mit dem Sonderausschuss A4 nach Friederichshafen und Wien vom 13.-20.4.1943, April 24, 1943, RH8/v. 1959, BA/MA. For the inclusion of DEMAG's facility in production plans, See Neufeld, *The Rocket and the Reich*, 193.

[57] Entstehungsgeschichte, June 17, 1943, RH8/ v.1210, BA/MA.

[58] "Niederschrift über die Besprechung am 4.8.43 beim Heimat-Artillerie-Park 11," RH8/ v. 1254, BA/MA. Quotation in Aktenvermerk über Besprechung beim A4 – Auschuss (Arbeitseinsatz) am 2.6.43 in Berlin (Lokomotivhaus), FE 833, NASM.

was omnipresent. However, the era of total war had a dramatic impact on the missile program at the base, dramatically restricting the available labor pool while making missile operations an increasing priority. The failure of German conventional weaponry gave cause for many in the regime to see Germany's salvation in the new "wonder weapons." Hitler finally fast-tracked the V-2 production program at the end of 1942, just after the major defeats in the Soviet Union and Africa.[59] Though there can be little doubt that many Peenemünders welcomed this decision, it also placed great pressure on them to finally meet their institution's goals by completing development and beginning mass production. In turn, this pressure, combined with the constant need to maintain the utter secrecy of the program and Himmler's desire to establish a presence at Peenemünde, pushed the boundaries of the possible, making the use of slave labor not only a conceivable option but also the best one. Like the rest of the German state, Peenemünde officials sought to mobilize every last drop of labor capacity available within the Reich. The victims of their technological tunnel vision would be the unfortunate mass of starving prisoners within the slave empire of the SS.

Nevertheless, these larger considerations fail to fully explain the turn into moral abomination. Cultural dynamics within the community at Peenemünde also made their unique contribution to this shift. The defining feature of life at Peenemünde was in large part the ubiquitous secrecy that permeated life there. This was obviously of paramount importance when it came to considering a labor force as well. The negative effects of secrecy also tended to shut out criticism and feedback within the community. If some Peenemünders dissented, they probably feared registering their feelings for fear of reprisal by the regime. The isolation that secrecy granted the Peenemünders also sharpened their internal focus on their own unquestioned institutional goals, fostering a climate in which an admittedly shrinking number of alternatives were a priori not even considered. Anthropologists and sociologists have demonstrated that this dynamic leads members of secret societies "to become mired down in stereotyped, unexamined, often erroneous beliefs and ways of thinking. Neither their perception of a problem nor their reasoning about it then receives the benefit of challenge and exposure."[60] This stunting of moral considerations was compounded by the general climate of racism and xenophobia that marked everyday life in the Third Reich. The Peenemünders' relative contentment with their lives and disinclination to risk parting with the comfortable advantages to living

[59] Neufeld, *The Rocket and the Reich*, 191.

[60] Sissela Bok, *Secrecy: On the Ethics of Concealment and Revelation* (New York: Vintage Books, 1989), 25. This dynamic has important contemporary examples as well. It figured largely in the 2004 controversy over the CIA's intelligence estimates regarding weapons of mass destruction in Iraq. See "Senators Assail C.I.A. Judgments on Iraq's Arms as Deeply Flawed," *New York Times*, July 10, 2004. The Senate Select Committee on Intelligence labeled this dynamic "Group Think."

on Usedom only helped to seal the matter. The pressures of the war, the ideological tenor of National Socialist Germany, and the internal cultural dynamics at Peenemünde utterly eradicated the conceptual possibility of alternatives to forced and slave labor while ensuring that opposition to its use was totally absent.

The practice of absolute secrecy also had an impact on a secondary level. Because a fundamental tenet of secrecy is the creation of boundaries and the segregation of outsiders, discrimination in one form or another is essential to its proper functioning. Peenemünders identified themselves as a cohesive community with like minds and similar interests. They most assuredly did not consider foreign workers to be a part of their elite society. The sense of separateness imparted by secrecy gave them the opportunity to segregate themselves from the prisoners. Dornberger also reinforced this distinction between Peenemünders and prisoners on Usedom. Before concentration camp prisoners arrived on Usedom, he informed the Peenemünders that "In the near future, convicts [*Strafgefangene* – Dornberger used this term instead of the commonly employed SS term *Häftlinge*, or detainees] who are to work with everyone will appear here. I say to you now directly that they are all murderers, thieves, and criminals, and every criminal will always protest that he is innocent."[61] By drawing a distinct difference between the orderly, law-abiding, "elite" Peenemünders and the dirty, poorly treated, underfed, and supposedly criminal mass of foreign workers, Dornberger's pronouncement helped to activate and reinforce the Peenemünder's group identity as well as their perceptions of the foreign labor force on the island. At the same time, it helped establish an environment in which the prisoners' priorities would mean virtually nothing to the civilians on Usedom.[62]

Indeed, one of the most remarkable features of the Peenemünde community is the absence of dissent over the issue of slave labor there. Personnel might have expressed their disagreement through minor administrative obstruction or quiet individual work slowdowns. They were quick to raise strident objections – for which they suffered no reprisal – when

---

[61] K. Friederich Baudrexl, "Als Techniker in der deutschen Rüstung," in Torsten Hess and Thomas A. Seidel, eds., *Vernichtung durch Fortschritt: am Beispiel der Raketenproduktion im Konzentrationslager Mittelbau* (Berlin: Westkreuz Verlag, 1995), 17.

[62] Social psychologists have argued that in order to know what (or who) a group actually is, it is helpful to know what (or who) it is not. Therefore, having an outgroup with which to compare one's ingroup helps to clarify the categorization process. These group-level categorizations become more prevalent in intergroup situations, like that on Usedom. According to social psychologists, once these types of group-level categories are activated, members try to differentiate their group from the comparison group. Inevitably, they argue, most intergroup comparisons favor the ingroup, and the priorities of the outgroup are virtually ignored. See R. Scott Tindale, Catherine Munier, Michelle Wasserman, and Christine M. Smith, "Group Processes and the Holocaust," in Leonard S. Newman and Ralph Erber, eds., *Understanding Genocide: The Social Psychology of the Holocaust* (New York: Oxford University Press, 2002), 146.

they disagreed with policies they felt negatively impacted their work or made impossible demands upon them. But there was no great hue and cry, or even a considered debate in Peenemünde, over the use of slave labor. Nor was there a slackening of the frenzied activity there when concentration camp prisoners and their SS masters began arriving. Employees at the base simply went along with it because of their absolute dedication to their closely knit community and its goals. Management automatically endorsed it, and, through their passivity, employees gave their tacit consent. They never weighed alternatives in the balance, and a slowdown in the work, an act whose details would only be perceptible to the employees themselves, was inconceivable, both for the Peenemünders and for their military masters. Of course, most did not have access to the levers of power at the base, and it is perhaps unfair to expect an outpouring of disagreement or anger over foreign labor on Usedom. This, however, does not mean that any dissent could not have been registered in more subtle ways. Instead, they adhered closely to the institutional goals of their community, expressing this commitment through collective, dedicated action that rapidly moved the program through the design stage to early phases of mass production. A combination of group self-interest, ideology, and the ever-present culture of secrecy, all exacerbated by the intensifying pressure of Germany's military situation, eliminated nearly any possibility of serious examination of the course of their work and the regime that sponsored it.

Interestingly, the most serious objections raised by Peenemünders and, therefore, one of the most serious threats to development and production, had nothing to do with slave labor. When Albert Speer's Armaments Ministry assumed an increased position of influence within the program by assigning the overbearing Degenkolb, thoroughly despised by everyone, to coordinate development and production, many Peenemünders complained noisily and, in some cases, even threatened to quit. In the end, what kept them in line was Peenemünde's institutional culture, which promoted group interest above all else (including concerns for the prisoners), as well as a liberal amount of managerial arm-twisting on the part of key figures such as von Braun. These important features of the missile program enabled Peenemünders to eventually see past the problematic demands (and Degenkolb's abrasive personality) that were imposed upon them by the Armaments Ministry. Despite the difficulties and complaints, they carried on as they always had, working feverishly to complete their work.

The Peenemünders' relationship to Degenkolb also illustrates an important point. When Degenkolb began making his influence on the program felt, Peenemünde specialists were able to register their displeasure with him without fear of reprisal or punishment. Engineers, scientists, and technicians replied to the Special Committee Chairman's demands with vocal, strident, and often angry responses. In this case, they clearly felt comfortable registering their dissent and displeasure. Some, like Thiel, even went to the extreme of threatening to quit the project altogether. Dissent, therefore, was not out of the question at Peenemünde, but internal hostility over practices within the program only surfaced when the designers' prerogatives themselves were threatened. Only when higher regime authorities imposed seemingly unfair policies on the Peenemünders did they act in a way that threatened to weaken the program from within. In truth, even these problems were mitigated by the deeply ingrained, automatically activated sense of loyalty to Peenemünde's mission. The acquisition of concentration camp workers promised to alleviate the program's most pressing labor problems, which a priori cut off any concern for the moral dilemmas wrapped up in slave labor. The longtime presence of foreign forced labor probably only served to inoculate German civilians against such moral concerns. Self-interest reigned at Peenemünde, promoting and justifying cooperation with some of the regime's most fearsome elements. The victims of the Peenemünders' self-interest were inevitably the dragooned foreign workers themselves.

THE LIFE OF A FOREIGN LABORER AT PEENEMÜNDE

An investigation into the life of foreign workers of any kind – forced labor, prisoners of war, or concentration camp labor – at Peenemünde presents a number of challenges. In the first place, the changing numbers of workers at the base makes it difficult to determine with certainty the amount of foreign labor used there over time. Baltic winters sometimes shut down construction, and the shifting priority level of the production program between 1939 and 1942 resulted in major fluctuations in the number of workers, including German civilians, assigned to Peenemünde.[63] It is also difficult to conceptually separate the different forms of foreign labor at Peenemünde. Forced laborers (foreigners drafted against their will into work gangs, but who could claim a limited number of extremely circumscribed rights, such as a tiny salary) worked side by side with prisoners of war and even, for a time, concentration camp laborers (who had no rights whatsoever), especially at construction sites on Usedom. Fewer foreign workers were employed in development areas because of secrecy considerations. The final difficulty involved in examining foreign labor at Peenemünde is that there is, at present, a limited amount of documentary evidence available on foreign workers involved with missile development. The majority of foreigners worked for Baugruppe Schlempp on construction projects around the base. The Army only employed them for a relatively short period.[64] Many

---

[63] In the winter of 1939–1940, poor weather forced construction activities to shut down. Work was restarted at the end of March 1940. See Godomar Schubert, "Wichtige Daten bei der Durchführung des Vorhabens Peenemünde," FE 342, NASM.

[64] In September 1940, Baugruppe Schlempp took over construction management duties from the Army Construction Office. Its local manager was Erwin Mahs, who reported to Heinrich

Baugruppe Schlempp records were destroyed or are currently unavailable. It is therefore difficult to formulate an exact picture of the conditions at the facility for a large number of foreign laborers. Nevertheless, some broad conclusions can be drawn.

The basic fact of life for foreign labor at Peenemünde was that the conditions of life on the island varied dramatically according to individual prisoners' skills and tasks.[65] Foreign labor carried out much of the backbreaking heavy construction work, such as the building of dykes, laying of roads, and clearing of forests, which required the large concentrations of mostly unskilled manpower to complete. The work was dirty, exhausting, and dangerous, and the risk of injury or death was great. Unskilled labor, which was relatively plentiful and cheap, required no training before being put into operation and could be replaced without any decline or slowdown in productivity. Therefore, Baugruppe Schlempp construction managers at Peenemünde had very little compunction to ensure that the conditions under which they labored were anything more than bare subsistence.

These unskilled, forced foreign laborers at Peenemünde found extremely difficult lives during their time at the facility. Administrators dedicated an absolute minimum of resources for their well-being. Their primary concern was the timely completion of their work. Erwin Mahs, the leader of Baugruppe Schlempp on site at Peenemünde, often met with Dornberger, Zanssen, and their deputies. According to those who knew him, Mahs "only cared about building" and ignored the needs of the construction gangs on the island.[66] Construction administrators segregated unskilled forced laborers from the rest of the German construction workers. They had to live in cramped, shoddily constructed barracks, and Nazi officials often expropriated their miniscule wages. Backbreaking, twelve-hour days at the

---

Lübke in Berlin. According to at least one report, in addition to its missile related construction work, the Baugruppe Schlempp office in Peenemünde also helped construct the labor camps and concentration camps that served as a repository for slave labor. See Max Düring Statement, AV 7/85, Bd. 25, *Bundesbeauftragte für die Unterlagen des Staatsicherheitsdienstes der ehemaligen Deutschen Demokratischen Republik* (BStU).

[65] In contrast, the German historian Ulrich Herbert has argued in several different pieces that the system of forced labor in Nazi Germany adhered to a strict set of racial guidelines that, despite some adjustments, was never significantly altered. See Herbert, *Hitler's Foreign Workers*, 1–12. According to the Nazi conceptions of race, prisoners from northern and western Europe, such as the Scandinavians or French, occupied the highest position in the racial hierarchy and were treated accordingly. However, according to Herbert, a procession down the racial hierarchy reveals progressively worsening living and working conditions. Below the northern and western Europeans on this scale were southern Europeans, followed by Slavs, especially Russians, and finally, the concentration camp prisoners and Jews, who received the worst treatment of the lot. For Herbert, "One's belonging to a specific *Volkstum*, a specific national ethnic background, determined to a pronounced degree the actual fate of the individual laborer." See Herbert, "Labor as Spoils of Conquest," 241.

[66] Franz Brauns Statement, AV 7/85, Bd. 25, BStU.

construction site were normal, and construction managers scarcely considered safety precautions for these workers.[67] The Italians, who were citizens of Germany's closest ally and, technically speaking, voluntary contract workers, might have expected decent conditions. Instead, they found the situation at Peenemünde unbearable. In October 1941, they staged an uprising that Army security forces rapidly quelled. Several of the Italians were arrested and the rest went back to work, but shortly thereafter, construction administrators removed them from Usedom.[68] A 1942 report noted that French workers, who had arrived only earlier that year, were sick and exhausted from overwork.[69] In the beginning of 1943, over three hundred Dutch laborers who had arrived four months earlier found the conditions at Peenemünde so difficult that they refused to return from their Christmas vacation.[70] In October 1943, a typhus outbreak ravaged the population of foreign workers; 1,300 Poles were unable to work for almost a month, pushing back deadlines for the ongoing building programs planned by Baugruppe Schlempp.[71] Without question, the furious pace of construction along with the extremely difficult conditions under which the unskilled forced laborers lived and worked took a heavy toll.

Unskilled concentration camp laborers, who began arriving in the middle of June, experienced even worse suffering at Peenemünde. Many showed up at Peenemünde in terrible condition, the victims of malnutrition and long train rides in overstuffed railroad cars.[72] While at the facility, these workers endured grueling conditions. According to Paul Baader, a VkN soldier who worked on materials testing in the development workshops, unskilled concentration camp prisoners always received the worst and dirtiest work.[73] Workers on the construction brigades ran the ever-present risk of serious injury or death. Werner Rottleb was a camp prisoner sent from Neuengamme (near Hamburg) to Peenemünde in 1943 and set to work on various arduous construction projects in both Peenemünde East and West. The food in his camp was terrible and of insufficient quantity. SS captors beat and even shot several workers at the worksite.[74] Many prisoners who did not work at the construction sites unloaded trains or ships in exhausting and brutal transport kommandos. Karl Krüger worked with a civilian group

---

[67] Middlebrook, *The Peenemünde Raid*, 31–32.
[68] Entstehungsgeschichte, October 16, 1941, FE 831, NASM.
[69] Aktennotiz [Memorandum], April 2, 1942, RH8/v. 1209, BA/MA.
[70] L. H. Jahnke Statement, "KZ Peenemünde – Bericht der VdN-Forschungskommission Rostock," AV 7/85, Bd. 32, p. 2, BStU.
[71] Entstehungsgeschichte, October 19, 1943, FE 873, NASM.
[72] Walter Grewe Statement, "KZ Peenemünde – Bericht der VdN-Forschungskommission Rostock," AV 7/85, Bd. 32, p. 6, BStU.
[73] Paul Baader Statement, AV 7/85, Bd. 33, BStU.
[74] Werner Rottleb Statement, "KZ Peenemünde – Bericht der VdN-Forschungskommission Rostock," AV 7/85, Bd. 32, p. 2, BStU.

in Peenemünde harbor that marked and recovered test rockets in the Baltic. He saw how camp prisoners slaved to move cement sacks from ships to waiting trucks and trains that would carry them off to construction sites around Peenemünde. The prisoners had to carry their loads over impossible distances, and SS guards beat or shot whoever could not bear the work.[75]

Indeed, SS guards, in the various camps as well as in the work gangs, did their best to treat their charges in the murderous tradition of their organization. Many unskilled laborers were killed during their time at Peenemünde, but a direct figure is virtually impossible to estimate, given the paucity of sources. Most prisoners were beaten or worked to death in the construction and transport kommandos; many others were shot. According to one German witness, all of the concentration camp prisoners working at either the Army's facility or at Peenemünde West were "underfed, always hungry, totally weakened."[76] A former prisoner at Peenemünde stated that

The SS guards treat the inmates most cruelly and inhumanely. Food is very insufficient, and most of the inmates appear to be hardly more than skeletons covered with skin. The SS guards refer to their prisoners as bone men, or the blue-white football club – blue-white because of the color of their uniforms, and football because they kick the heads of those who faint from weakness.[77]

The explanations of deaths given by the SS, such as being shot "for resistance," hung "on the order of the Reichsführer-SS [Himmler]," or the ubiquitous phrase "shot while attempting to escape" (used by camp administrators to explain away the multitude of random acts of arbitrary violence that SS tormentors engaged in), only obscured the truth of the matter.[78] Many prisoners who were no longer able to work were murdered by lethal injection by an SS medical officer, who "proudly claimed that he had killed in this manner more enemies of the Reich than many soldiers in the front lines with their machine guns had."[79] The remains of many prisoners killed on Usedom were incinerated in a crematorium in Greifswald on the mainland.[80]

---

[75] Karl Krüger Statement, AV 7/85, Bd. 25, BStU.

[76] Karl Dachner Statement, AV 7/85, Bd. 25, BStU.

[77] Horst Lukat Interrogation, Box 647, RG 165, Records of the War Department General Staff, Entry 179, Enemy POW Interrogation File (MIS-Y), 1943–1945, NARA.

[78] See the collection of death reports in AV 7/85, Bd. 26, BStU.

[79] Horst Lukat Interrogation, NARA.

[80] "Auszug aus den Totenlisten des Krematoriums in Greifswald," AV 7/85, Bd. 26, BStU. Such generally vicious treatment may have inspired a resistance movement on Usedom. In 1965, Theo Franz, a former civilian construction worker at Peenemünde then living in East Germany, wrote to Albert Norden, the infamous SED apparatchik, that a member of a prisoner resistance group told him that a group of Soviet officers was shot in the woods around Greifswald and buried in a mass grave. The resistance group, purportedly known as the White Guard, was made up of Belgian, Polish, and Russian prisoners. See Theo Franz to Albert Norden, February 5, 1966, AV 7/85, Bd. 32, BStU.

In contrast, technically proficient foreign workers often fared much better than their unskilled compatriots. Skilled workers assigned to Peenemünde East found more comfortable accommodations as well as much easier working conditions during their time on the island. Frenchman Michel Fliecx arrived in Peenemünde from Buchenwald in the early summer of 1943. On his second day at the facility, he and his fellow inmates had to line up inside the factory grounds, where a civilian specialist who wore a party pin asked for their technical qualifications. According to Fliecx, the Peenemünder singled out welders, lathe operators, mechanics, and other skilled laborers. Fliecx, along with many other prisoners, was a university student, but luckily he managed to convince the civilian overseers that he was in fact a technically skilled laborer.[81] Work allocations in the assembly facility were assigned on the basis of technical qualifications. Once given a position in the factory, a German prisoner, Willy Steimel, noted that working there alongside civilians was generally not hard. He testified after the war that prisoners' technical skills "Forced the civilian management to value the prisoner as a specialist and also to treat him accordingly, namely as a human being. This brought about a partly bearable situation."[82] Before the concentration camp prisoners arrived at Peenemünde, Dornberger had also enjoined the civilians who would work with them to "Lead them in their work. Show them what a German can do. But do not engage in any chicanery with them."[83] For Fliecx's group, workdays were from 7:00 in the morning to 5:30 in the afternoon, with a thirty-minute break for lunch. At the end of the shift, prisoners even had the opportunity to clean themselves and lounge on warm evenings under the spruce trees inside the factory fence.[84]

Conditions outside of work also augured well for skilled prisoners. Fliecx considered the food "adequate, but we were nevertheless hungry."[85] Steimel, who also made the two-day journey from Buchenwald, noted that the food was much better than at his former camp.[86] The prisoners were given individual bunk beds with two blankets each, as well as their own washbasins. The sixty SS overseers, except for a sadistic Romanian-German guard, a *Volksdeutscher*, nicknamed "Moustache," generally treated them well. When Moustache did go hunting for victims on which to take out his frustrations,

---

[81] Michel Fliecx, "Wegen des Vergehens der Hoffnung – Zwei Jahre Buchenwald – Peenemünde – Dora – Belsen," AV 7/85, Bd. 32, BStU.

[82] Willy Steimel Testimony, Roll 4, M-1079, RG 242, U.S. vs. Kurt Andrae et al., NARA. According to Neufeld, Steimel's testimony is not entirely trustworthy because he was probably an SS informant. See Neufeld, *The Rocket and the Reich*, 189. Nevertheless, most of his observations are confirmed by Michel Fliecx's important memoir, which was unavailable to Neufeld when his work was published.

[83] Ibid., 8–10, 13.

[84] Michel Fliecx, "Wegen des Vergehens der Hoffnung," BStU.

[85] Ibid.

[86] Willy Steimel Testimony, NARA.

civilian managers were able to complain about him to his commanding officer and limit his arbitrary brutality.[87] Steimel also noted that SS guards "Could not use the usual methods of the concentration camp because of the fact that a lot of civilian and military workers were present and did not permit it."[88] Under these conditions, the health of the prisoners generally improved, but this only cast into more stark relief the differences between skilled and unskilled foreign workers at Peenemünde.

Administrators of the missile program explicitly sought skilled laborers to fulfill their needs. In July 1943, 400 unskilled French prisoners arrived at the missile factory from Buchenwald, and production administrators immediately attempted to exchange them for trained workers. The exchange never took place because more prisoners arrived on the heels of this transport, and they were put to work in the necessary areas.[89] Rudolph and Jaeger catalyzed the process. They made requests, divided by skill, through Gerhard Maurer, a high official in the SS organization with the bland-sounding title of SS Economic and Administrative Main Office (*SS-Wirtschaftsverwaltungshauptamt* – SS-WVHA), who assigned prisoners from all corners of the Nazi empire accordingly.[90] This system worked efficiently. Whatever its other failings, the SS promptly satisfied the Peenemünders' demands, and by early August, final plans to inaugurate the missile's mass production could be laid.

After years of painstaking development, the final stages of the A-4 program seemed to be taking shape. Production schedules (however burdensome and irrational) were set; factories were either completed or being converted to accommodate the missile's requirements; and foreign slaves helped ease the difficult labor shortage. The A-4 was far from ready for mass production – design problems vexed the developers into 1944 – but by the middle of 1943, Peenemünders could more clearly imagine what the final outlines of their space-age work would be. Despite the technical difficulties, they were closer to success than ever, but what they did not know was that just over the horizon loomed a nightmare of suffering, murder, and misery.

---

[87] Fliecx, "Wegen des Vergehens der Hoffnung," BStU.

[88] Willy Steimel Testimony, NARA. Even so, Fliecx notes that some guards did beat prisoners whom they caught sleeping and sometimes made them perform physically torturous exercises during roll calls.

[89] Entstehungsgeschichte, July 11, 1943, RH8/v. 1210, BA/MA.

[90] Aktenvermerk über die Besprechung beim A4 – Auschuss (Arbeitseinsatz) am 2.6.43 in Berlin (Lokomotivhaus), FE 833, NASM. For transport lists of prisoners to Peenemünde, see National Archives Captured German Records Collection, ACC1996.A.0342, Reel 161, located at the United States Holocaust Memorial Museum.

5

"At the Limits of Existence"

The secret of Peenemünde could not last forever. A pivotal moment in its history occurred the night of August 17–18 when the Royal Air Force descended on Peenemünde. The massive air raid conducted that night had a profound effect on life on Usedom. It scattered much of the work to sites across the Third Reich and brought the missile specialists into increased contact with some of the most barbaric conditions within the Nazi empire. Those Peenemünders who were displaced carried on much as before, bending all of their effort toward successfully producing a usable weapon that could help reverse German fortunes in the war. Slave labor continued to be an explicit part of the bargain, a situation that led to the bloodiest chapter in the checkered history of the German ballistic missile program.[1]

By the summer of 1943, employees at Peenemünde could consider themselves extraordinarily lucky. Their work, though strenuous, was as rewarding as anything they could have imagined in their professional lives. At home, they either lived the lives of happy singles, started new families, or raised their children in a small community that was knit together by deep and durable and social bonds. The utter isolation and absolute secrecy of Peenemünde kept them hidden from prying enemy eyes, and they experienced its deprivations perhaps less than any other community in Nazi Germany. Though daily life on Usedom was not without its stresses, the Peenemünders had

---

[1] Michael Allen notes the strong loyalties between the newcomers to the program, but not the old guard of Peenemünders. According to Allen, these loyalties were based almost exclusively on National Socialist ideological motives. See *The Business of Genocide: The SS, Slave Labor, and the Concentration Camps* (Chapel Hill, NC: University of North Carolina Press, 2002), 221. Likewise, Michael Neufeld, *The Rocket and the Reich: Peenemünde and the Coming of the Ballistic Missile Era* (Cambridge, MA: Harvard University Press, 1995), also necessarily emphasizes the struggles between individuals such as Dornberger and Kammler. In his *Produktion des Todes: Das KZ Mittelbau-Dora* (Göttingen: Wallstein Verlag, 2001), Jens-Christian Wagner was the historian to explicitly note the juxtaposition of a so-called clean Peenemünde and the criminal enterprise of Mittelbau-Dora. In statements made at Dora war crimes trials in 1947 and 1967, former Peenemünders went to great lengths to distance themselves from Mittelwerk and point out their differences with the SS. They often either steadfastly maintained that they had little to do with slave labor in the factory or attempted to emphasize their efforts to help prisoners imprisoned by the SS. Neither argument holds much water.

every reason to count themselves among the fortunate in a nation at war. Much of this irrevocably changed when the war finally came to Peenemünde in the middle of August 1943, a pivotal event in the history of the facility and its employees, whose lives had become so deeply intertwined with its existence.

PARADISE LOST: THE BRITISH STRIKE PEENEMÜNDE

Peenemünde authorities were woefully unprepared for the attack. The peaceful life at the base lulled nearly everyone there into a false sense of security. The war, such a "long, dim way off," had not yet intruded on their lives, and they had yet to be touched by any real deprivation. Air raid warnings occurred often enough as bomber units and reconnaissance aircraft passed by Peenemünde on their way to Berlin, but the frequent sirens did not concern the employees much. Even for combat veteran Peter Wegener, life was so pleasant and free of difficulty at Peenemünde that "It never occurred to me that I lived in a most attractive location for an enemy air raid. ... Apparently my delight in the altered lifestyle kept me from pondering the future of the laboratory."[2] According to Dieter Huzel, virtually no one on the base took the periodic air raid warnings seriously.[3] Many Peenemünders were totally immersed in the work, and their own naiveté about the war gave them the false sense that it posed no danger for the base.

Army officials at the facility only made rudimentary preparations for potential air raids, and these were well short of what was necessary. The Settlement was built with an eye toward form and function, not air raid protection, and in many locations on the island the only reasonable air raid bunkers were the cellars underneath the homes. Planners had made few considerations for protection. Elsewhere on the base, a meager number of *Splitterschutzgraben*, splinter-proof trenches, were the only air raid accommodations.[4] Nevertheless, the increasing frequency of warnings throughout the summer of 1943 forced officials to come up with some kind of plan to deal with an attack. In early August, they drew up a plan that gave mostly Army personnel, but some civilians as well, specific assignments in the event of an air raid. Civilians became responsible for fighting fires in the woods around Peenemünde and protecting sensitive missile hardware, though how this second task was to be accomplished was left up in the air. Others were responsible for relocating nonessential personnel, mostly wives and children,

[2] Peter Wegener, *The Peenemünde Wind Tunnels: A Memoir* (New Haven, CT: Yale University Press, 1996), 19, 63.

[3] Dieter Huzel, *From Peenemünde to Canaveral* (Englewood Cliffs, NJ: Prentice-Hall, 1962), 51.

[4] Middlebrook, *The Peenemünde Raid: The Night of 17–18 August, 1943* (London: Cassell, 1982), 133–153.

into more widely dispersed (and, presumably, safer) quarters. The soldiers at Peenemünde, including the VkN, had other responsibilities. These included ensuring that enough extra food was on hand for a three-day period and securing crossing the points over the Peene River in order to maintain the facility's security. Army authorities assigned soldiers from the Ninth Company of the VkN to help Army and SS men guard both the prisoner of war camps and concentration camps on the island in the event of an attack.[5] These rushed measures were only elementary plans that barely got off the ground in the days before the raid. On August 17, there was no respectable fire brigade available that was big enough to combat the effects of an attack. There was also virtually no large or well-organized medical establishment that could deal with mass casualties, and there remained a dearth of air raid shelters and other proper measures to truly protect lives against the coming onslaught.

The day before the raid was like any other at the base. In the afternoon, development chiefs held a meeting with Dornberger to protest against the demands that the accelerated production program was making on their work.[6] Many employees relaxed on the beach or in the ocean. That night, some Peenemünders gathered for drinks in the officer's club or strolled in the comfortable night air in Zinnowitz. It was an altogether pleasant and typical summer's evening at Peenemünde. Inge Holz, a secretary in the Development Works, remembered "It was a very happy evening for the girls. ... At about eleven p.m., we all went home. As we girls walked back to our home, we sang a little as we went, and we talked of the pleasant time we had had."[7] As Holz and her friends walked home, over 600 Stirling, Halifax, and Lancaster bombers of the Royal Air Force formed up over the North Sea, turned east, and made their way across Denmark on the way to Usedom.

Four years earlier, just after the outbreak of the war in September 1939, an anonymous informant left a package of documentation on the activities at Peenemünde for British intelligence personnel in Oslo.[8] The British initially doubted the veracity of their find. The idea of a rocket-powered weapons of any kind was simply too fantastic. Other bits of intelligence gathered later

---

[5] Stichwortartige Zusammenstellung der bei der Befehlsausgabe am 3.8. vormittags 9 Uhr im Offizierheim ausgegebenen Richtlinien über luftschutztechnische Massnahmen, die sofort unter Zurückstellung aller anderen Aufgaben durchführen sind, August 4, 1943, FE 833, National Air and Space Museum (NASM). In the event, the scientists, engineers, and technicians of this unit never carried out this responsibility.

[6] According to Dornberger, Thiel, Rees, and even von Braun briefly threatened to quit over the rush into production. Though this behavior certainly fits a pattern for serially despondent Thiel, I have found no evidence that Rees or von Braun actually threatened to do so. Nor is there any documented proof of this meeting. The problems Dornberger describes were common enough, though. See Walter Dornberger, *V-2* (New York: Viking Press, 1954), 149–151. See also Chapter 4.

[7] Printed in Middlebrook, *The Peenemünde Raid*, 135.

[8] Ibid., 35.

in the war seemed to confirm the Oslo Report's basic outlines: advanced German ballistic missile research was taking place at what was once an isolated fishing village, but what had been transformed into a cutting-edge development facility. British intelligence pieced this information together from prisoner of war interrogations, secret informants, and from forced laborers who worked at Peenemünde but were allowed to return home. Polish resistance members forwarded information they were able to gather about test launches to intelligence agents in Sweden, who then passed it on to London. This intelligence led the Royal Air Force to conduct reconnaissance missions over Usedom in May and June 1943. High-altitude photography confirmed the exotic work going on there. At the end of June, the British made Peenemünde a target of urgent priority and began to lay preparations for the attack. When the Royal Air Force pilots took off from their airfields in August, none of them had ever even heard of Peenemünde until the night they climbed into their planes to bomb it.[9]

The first British bombers of "Operation Hydra" arrived at Peenemünde shortly after 1:00 in the morning, while residents were asleep. Their targets were the test stands, industrial facilities, and employee settlement. The attackers intended to strike a mortal blow to the missile program by crippling development and production while killing the experts working on the project. When the bombers began arriving over Usedom, a string of targeting blunders caused many bombers to miss their assigned objectives. The foreign workers locked up in the shacks of the Trassenheide labor camp bore the brunt of these mistakes. The camp, with its closely packed wooden barracks, barbed wire fence and single exit gate, was a death trap. Between 500 and 600 foreign workers died in the attack, which lasted just under an hour. Peenemünde planners did not concern themselves with constructing air-raid shelters within the camp, and there was virtually no fire-fighting equipment on the grounds. Tragically, several air raid trenches were located just beyond the camp's fence, and though some prisoners made it out of the camp, none of them were able to reach the trenches.[10]

At the Settlement, the situation was only slightly better, but still terrifying. When the air-raid sirens and approaching engine noises jolted them from their sleep, many families at least had the chance to get to shelter. Even so, the attack was a jarring experience. Arthur Rudolph's family and their neighbors barely made the hundred-foot sprint into the shelter before the bombs began exploding around them. In the shelter, a shower of sparks from a phosphorous bomb nearly set Rudolph's young daughter's hair on fire. His wife patted out the sparks with her hands. The outer door of their

[9] See David Irving, *The Mare's Nest* (Boston: Little, Brown, 1965) and Middlebrook, *The Peenemünde Raid*, esp. 35–45.
[10] Walter Reuss, "Erfahrungsbericht über die Bombennacht vom 17. zum 18.8.43," August 30, 1943, FE 833, NASM.

shelter was blown away by a near miss.[11] Another civilian in a different
shelter recalled that "I had experienced raids in Berlin, but I had never
experienced such intense bombing and, this time, I felt that we really were
going to die."[12] By the end of the raid, almost three quarters of the dwellings
in the Settlement were destroyed and 178 of its inhabitants lost their lives.[13]
In an area that housed nearly 4,000 people, this is a surprisingly small
number, but it perhaps could have been even smaller. A report filed after the
bombing raid indicated that there was not enough fire-fighting equipment
in the Settlement and that those in charge of air-raid countermeasures had
failed to fill many of the water tanks that were to be used to fight fires in
case of an attack. The Peenemünders' ability to limit the extensive damage
and loss of life was severely limited by this oversight.[14] Despite relatively
light loss of life, the development chiefs found out later that they suffered a
major blow: Walter Thiel, the head of the propulsion group who had done
so much to improve A-4 engine performance, perished with his entire family
when their shelter suffered a direct hit.[15]

Development and production facilities were luckier. The most important
buildings survived the raid. The targeting errors, anti-aircraft fire, and the
individual pilots' tendencies to "pile on" the destruction already created by
earlier attack waves meant that more bombs fell on the Trassenheide camp
and the Settlement. Some planes did find their mark, however. Some thirty-
five buildings were either destroyed or damaged in the raid. The assembly
plant received only minor damage, and the Luftwaffe facility, liquid-oxygen
plant, and Aerodynamics Institute were untouched.[16]

In the wake of the British raid on Usedom, Peenemünders became more
reflective about their jobs. The attack forced many to confront the reality
of war and their own participation in it. Scientist Siegfried Winter stated
that "I began to realize that here I was, possibly sitting on the end of
an English bomb, yet during the day, I was working at preparing exactly
the same thing, in rocket form, to send to the English. . . . It forced me to
take stock of what I was doing in my own work – but life took over as
normal the next morning."[17] This is as powerful as statement as any about
the inertia that the project developed and the automatic adherence with
which Peenemünders clung their institution's goals. Nevertheless, they were
generally uneasy about the prospect of another raid. "We had been stung
once," Huzel wrote. "An air of intense expectancy prevailed. The bombers

[11] Thomas Franklin, *An American in Exile: The Story of Arthur Rudolph* (Huntsville, AL: Christopher Kaylor, 1987), 71.
[12] Printed in Middlebrook, *The Peenemünde Raid*, 142.
[13] Ibid., 144. See also Dornberger, *V-2*, 168.
[14] Reuss, "Erfahrungsbericht," NASM.
[15] Neufeld, *The Rocket and the Reich*, 198.
[16] Middlebrook, *The Peenemünde Raid*, 150–152.
[17] Ibid., 142.

would certainly be back."[18] The British also destroyed the base's aesthetic beauty. Much of the damage was left in place in an effort to convince Allied reconnaissance that the raid had done its job.[19] Administrative offices had to be relocated, and all air-raid warnings were ever after taken very seriously.[20] From the British perspective, the raid was only partially successful, but for the Peenemünders, it obliterated the naïveté and carefree attitudes about the war that were such a part of the fabric of life at Peenemünde.

FROM PARADISE ON THE BALTIC TO PERDITION IN THE HARZ

From the perspective of development and production, a far more important result of the bombing raid was the wide dispersal of the people and facilities at Peenemünde. Families abandoned much of the Settlement and relocated to the other villages scattered on the island. The relocation of the production facilities themselves was the most fundamental change at Peenemünde, however. Peenemünde administrators agreed with Reich officials on this subject. Less than a week after the attack, Himmler convinced Hitler that the facilities on Usedom should be moved to more secure locations.[21] Hitler declared that development should be moved to the Waffen-SS camp at Bliszna in the General Government, German-administered Polish territory where the Nazis slaughtered millions of Jews. The production plant was to be relocated to an as yet unspecified, bombproof factory. Much like Rudolph did in his report on slave labor in the Heinkel Works, Himmler argued that in order to maintain the strictest secrecy around the work, the assembly lines should be fully manned by concentration camp prisoners.[22] This proved to be his most effective gambit in his ongoing efforts to gain influence over the A-4 program.

Most Peenemünde administrators resisted the wholesale relocation of their work, but there was little major opposition to Himmler's suggestion to remove production to an even more secure site. At the very least, there was a growing opinion among many administrators that the program should be both restructured and relocated. On August 23, Georg Thom, undoubtedly

---

[18] Huzel, *From Peenemünde to Canaveral*, 92.

[19] Neufeld, *The Rocket and the Reich*, 205.

[20] Huzel, *From Peenemünde to Canaveral*, 61–63.

[21] On June 21, British bombers damaged the production site at the Zeppelin factory in Friederichshafen, though they did not know that it was used for this purpose, and on August 13, the U.S. Air Force bombarded Rax Werke in an effort to halt airplane construction there. These two attacks, combined with the one at Peenemünde, convinced Hitler and his paladins that the secret program had been discovered. See Heinz-Dieter Hölsken, *V-Missiles of the Third Reich: The V-1 and V-2* (Sturbridge, MA: Monogram Aviation Publications, 1994), 90, and Neufeld, *The Rocket and the Reich*, 199.

[22] Hölsken, *V-Missiles of the Third Reich*, 97; Neufeld, *The Rocket and the Reich*, 200.

on orders from Dornberger, sent a proposal to General Friederich Fromm that Dornberger be given responsibility for managing the missile program. He argued that "A condition for success is the eventual centralization of all measures under the strictest military leadership. In the future, it must be reckoned that the enemy will try to destroy development and production, and therefore all protective and counterintelligence measures must be adjusted to a new reality."[23] Thom had another objective in mind as well. Despite Degenkolb's streamlining efforts, Thom argued that the twin structure of the German Army High Command (OKH) and Armaments Ministry was too cumbersome and did not move the program any closer to its goals. His case for centralizing the program would have effectively given Dornberger control over the querulous Degenkolb. Dornberger clearly sensed that change was in the wind and sought to manage and direct it as much as possible. He still retained, with good reason, a proprietary interest in the missile program and was convinced that it could only achieve its lofty goals with his stewardship. If change were to occur, it should do so under Army auspices.

The most important of these imminent changes was the relocation of the production facilities. Dornberger was not informed of Hitler's decision to move much of the work out of Peenemünde until August 25. It is likely, though not entirely clear, that he did not at first fully support such a radical change because it ran counter to his "Everything under one roof" concept.[24] Development was the sticking point in discussions about relocation. In a hastily assembled meeting on August 25, chaired by von Braun (with several of his key deputies in attendance), Peenemünde managers decided flatly that the development work should stay on the base. The minutes of their meeting note that "The technical work, that is, operational assembly, operational testing and operational measurement, of individual parts must remain in close proximity to Peenemünde." Eventually, only the launch tests were moved to Bliszna. In contrast, however, the managers were prepared to remove mass production and began laying plans for dispersing it and its prisoner labor force to different sites in Germany. It was an early demonstration of the Peenemünde developers' willingness to cooperate with assembly specialists in the Armaments Ministry and SS to bring their operation into line with the regime's larger goals for the program.[25]

At the same time, there was a groundswell of opinion among many other individuals in other sectors of the program that at least production should be restructured and relocated. Immediately after the raid, several employees openly wondered if the factory should not be rebuilt, but rather set up

---

[23] Thom to Fromm, August 23, 1943, RH8/v. 1211, *Bundesarchiv/Militärarchiv* (BA/MA).
[24] Neufeld, *The Rocket and the Reich*, 202.
[25] Niederschrift über die Besprechung in Karlshagen am 25.8.43, FE 732, NASM.

somewhere else.²⁶ This sentiment reached into rather high places in the administration. In his survey of the bomb damage after the raid, Arthur Rudolph, who did not participate in the August 25 meeting, also came to the conclusion that production must be moved. Though he filed his own report after Hitler's order to move production, Rudolph's language indicated that he had no doubts about the wisdom of the decision to do so, even before the Führer's decision. He wrote that "Despite [the construction design of the production plant], it appears that there is no way to guarantee production if it is located above ground. I believe that it is necessary in this case to make all facilities absolutely bomb-proof by moving them to underground locations. However, this does not apply only to the factory facilities, but rather the accommodations for all employees must also be secured against air raids so as not to disrupt the course of daily life as well as manufacturing."²⁷ For Rudolph, there was no question of the viability of moving everything associated with production, including the labor force, underground. He wanted to go even further than Hitler, arguing that even civilian employees should be housed in secure underground facilities as well. Rudolph's highest priority was to ensure that production goals could be met as quickly as possible. Part of this process was the need to keep manufacturing centers safe from attack, a major concern of many regime officials. In all likelihood, his new vision included the use of the concentration camp labor already on hand at Peenemünde to help accelerate the relocation. He would become a central figure in the installation of the underground factory that began shortly thereafter.

The effort to relocate mass production introduced to the program one of the most capable, energetic, and vicious figures in Himmler's entire murderous organization: SS Brigadier General (*SS-Brigadeführer*) Dr. Hans Kammler. The head of Office "C" (Construction) of the SS Economic and Administrative Main Office, Kammler was a dashing, brilliant officer who held a doctorate in civil engineering. Dornberger was duly impressed with Kammler's appearance: "One's first impression was of a virile, handsome, and captivating personality. He looked like some hero of the Renaissance, a *condottiere* of the civil wars of Northern Italy. The mobile features were full of expression."²⁸ Ideologically, Kammler was the perfect embodiment of the "reactionary modernist," combining equal doses of technological expertise with National Socialist fanaticism and romanticism.

²⁶ Wernher Brähne, "Die Mittelwerk GmbH. Eine Chronik über Firma und Werk," unpaginated, Gericht Rep. 299, Bd. 582, Nordrhein-Wesfälisches *Hauptstaatsarchiv Düsseldorf, Zweigarchiv Schloss Kalkum* (HStaD-ZA Kalkum).
²⁷ Arthur Rudolph, "Erfahrungsbericht über den Feind-Angriff vom 17. zum 18.8.43," FE 833, NASM.
²⁸ Dornberger, *V-2*, 198.

Born in 1901, Kammler did not fight in the First World War, but he did fight with the paramilitary Freikorps immediately after the war. He joined the Nazi Party in 1932 and held a number of administrative posts in the Air and Agriculture Ministries, volunteering his services part time to the SS. In 1941, he joined the SS full time. Oswald Pohl, the head of the WVHA, almost immediately assigned him some of the SS's most important and secret work – constructing the gas chambers at Auschwitz-Birkenau and Majdanek.[29] His murderously effective office also seized control of the slave labor industry within the SS and deployed unfortunate prisoners in mobile construction brigades with a ruthlessness that was unmatched in its efficiency and scope.[30] Given an order from Himmler to do everything possible to hasten mass production and deployment of the A-4, Kammler unleashed his limitless energy on the missile project. In doing so, he came to rely heavily upon key members of the missile program in order to meet his goals.

The process of transferring production to a bombproof facility was entirely improvised. The site selected by Kammler, Degenkolb, Dornberger, and Karl Otto Saur, Albert Speer's fanatical deputy, was a tunnel complex in the southern Harz Mountains in Thuringia, near the town of Nordhausen. Originally, the company Ammoniak, a subsidiary of I.G. Farben, dug the tunnels into the face of a mountain known as Kohnstein and mined it for calcium sulfate. In 1938, I.G. Farben struck a deal with a government-owned corporation known as "Wifo," an acronym for *Wirtschaftliche Forschungs-gesellschaft* (Economic Research Company). In exchange for paying a share of the mining expenses, Wifo received large underground storage areas for strategic gasoline and oil reserves.[31] The tunnels themselves consisted of two parallel main lines that snaked north to south, with forty-four perpendicular galleries linking them. Each main tunnel was just over a mile long, and each gallery was a little less than 500 feet long.[32] At the end of August, the Armaments Ministry and SS took them over for the purpose of missile production. On August 28, a mere ten days after the British struck Peenemünde, the first 107 prisoners arrived from Buchenwald to begin expanding the tunnels in preparation for factory installation. The new underground camp was given the code name "Dora." The future factory would be named Mittelwerk, and the factory/camp complex was known as Mittelbau-Dora.[33]

[29] Allen, *The Business of Genocide*, 140–148.
[30] Ibid., 140–239.
[31] Georg Rickhey Statement, U.S.A. vs. Kurt Andrae, et al., Roll 4, M-1079, National Archives and Records Administration (NARA).
[32] Manfred Bornemann, *Geheimprojekt Mittelbau: Vom zentralen Öllager des Deutschen Reiches zur grössten Raketenfabrik im Zweiten Weltkrieg* (Bonn: Bernhard & Graefe, 1994), 11–20.
[33] Yves Béon, *Planet Dora: A Memoir of the Holocaust and the Birth of the Space Age* (Boulder, CO: Westview Press, 1997), xii.

Kammler's main concern was the expansion of the tunnels so that they could accommodate the large assembly line necessary for missile production. As expansion proceeded, heavy equipment was shipped from Peenemünde and other assembly plants for installation in the tunnels. At Kammler's order, thousands of prisoners continuously streamed into Dora to complete this work. By the end of September, nearly 3,000 prisoners labored in the tunnels; by the end of November, there were some 8,000 slaves working in barbaric conditions underground. In this period, the majority of the prisoners arrived from Buchenwald, but in the middle of October, most of the slave laborers at Peenemünde departed with their SS guards to Dora, though some did remain behind.[34] When the new year arrived, the SS had fully proven its value as a labor supplier, however murderous. It managed to deliver nearly 10,000 prisoners to work in the tunnels.[35]

As the daily transports rolled in to provide their human cargo to this apocalyptic mining and construction project, the SS made no effort whatsoever to care for the prisoners. The level of maltreatment engendered by the Nazi idea of extermination through work [*Vernichtung durch Arbeit*] set new standards of inhumanity. The wretched slaves used dynamite, jackhammers, and hand tools to bore into the mountain, filling the tunnels with dust and ammonia fumes that burned throats and lungs. Prisoners removed rocks and boulders by hand, a dangerous job because the SS, keen to push the work forward, drove the prisoners into the rock pile without regard for loosened, only partially collapsed parts of the wall. Falling rocks crushed many prisoners.[36] The prisoners loaded the stones onto rail carts, which they pushed outside for disposal. Kapos (prisoner functionaries who supervised small groups of internees) and SS men drove the pace of the work to breakneck speed and reigned in the tunnels with wanton brutality. Yves Béon paints a terrifying picture worthy of Bosch:

The air inside, oppressively thick with choking dust, fumes of burnt oil, and humidity, engulfs the newcomers. Here are hills of gravel, there valleys filled with water, and throughout the cave, pools of light alternate with suspicious areas of shadow. Gray beings shovel, hollow out, and tear away at surfaces. Narrow hoppers loaded with stones and trash roll through a narrow passageway, pushed by men in filthy rags. In the unnatural light, lines of ghostly figures carry pieces of carpentry on their shoulders. Others push, pull, and drag insane loads. Shouting and swearing, the SS, Kapos, and Vorarbeiter [foremen] rush among them, whipping and clubbing the

---

[34] Entstehungsgeschichte der Fertigungsstelle Peenemünde, October 13–15, 1943, FE 873, NASM. Hereafter cited as Entstehungsgeschichte.

[35] Wagner, *Produktion des Todes*, 186–187; Neufeld, *The Rocket and the Reich*, 209–210. See also André Sellier, *A History of the Dora Camp: The Story of the Nazi Slave Labor Camp That Secretly Manufactured V-2 Rockets* (Chicago: Dee, 2003), 55–57.

[36] Xavier Delogne Interview, Fortunoff Video Archives (FVA), Yale University.

terrified prisoners. In the distance, the sound of mine blasting adds to the chaos, and the air resounds with a thousand clamors.[37]

When it was finished, the prisoners had managed to enlarge the former Wifo tunnels so that the entire installation contained 1,270,000 square feet of floor space.[38]

The work away from the mine face was no easier. Large transport kommandos made up exclusively of prisoner labor unloaded freight trains outside of the tunnels and often carried their cargo in by hand, though sometimes a train did bring in supplies. Installation of factory equipment usually involved simply manhandling large and heavy machinery into place. Most of the time, the prisoners had no mechanical help, and their efforts were made even more difficult by their barbarous overseers, who beat them senseless if they worked too slowly or fell out of line. Even so, the transport work continued. By the end of February 1944, according to one estimate, this exhausting, deadly "Warenannahme" kommando had unloaded nearly 1,300 freight cars worth of material.[39] By the end of December, most of the production equipment from Peenemünde and the Rax Werke in Wiener Neustadt had arrived at the Dora tunnels.

The short time away from the murderous work offered no respite. Czech survivor Wincenty Hein estimated that the prisoners had approximately eighteen hours of activity per day and often less than six hours of rest.[40] During this rest time, prisoners rarely emerged from the tunnel. The SS gave the construction of freestanding barracks outside of the tunnels the lowest priority. Instead, the sleeping facilities that the SS allowed the prisoners were bunk beds in a cross-tunnel that was located relatively close to the mine's face. Rest and sleep were impossible, as the din from jackhammers, pickaxes, and explosions continually rang through the tunnel. Jean Michel, a French prisoner in the tunnel at Dora, wrote in his memoirs that "The noise bores into the brain and sheers the nerves. ... Over a thousand despairing men, at the limit of their existence and racked with thirst, lie there hoping for sleep which never comes."[41] Water seeped from the walls, creating a dank chill in the galleries. André Rogerie, who arrived at Dora in November, recalled

---

[37] Béon, *Planet Dora*, 4.

[38] "Production and Disposition of German A-4 (V-2) Rockets," Box 772, RG 18, Records of the Army Air Force, Entry 1, Air Adjutant General, Bulky Decimal File, 1946–1947, NARA.

[39] Wincenty Hein Testimony, ZM 1625, Bd. 23, Akte 35, *Bundesbeauftragte für die Unterlagen des Staatssicherheitsdienstes der ehemaligen Deutschen Demokratischen Republik* (BStU). Also see Wagner, *Produktion des Todes*, 87.

[40] Wincenty Hein, "Lebens- und Arbeitsbedingungen der Häftlinge im Konzentrationslager 'Dora'- 'Mittelbau' und ihre Folgen," ZM 1625, Bd. 22, Akte 34, BStU.

[41] Jean Michel, with Louis Nucera, *Dora*, trans. by Jennifer Kidd (New York: Holt, Rinehart, & Winston, 1979), 68.

that the dust could be so thick that prisoners could not see from one end of the sleeping tunnel to the other.[42] Construction supervisors divided up the prisoners into two shifts of twelve hours each, meaning that the sleeping quarters were always occupied and crawled with filth, vermin, and disease. There were no cleaning facilities and only makeshift latrines, which were made out of oil drums that were cut in half and periodically sprinkled with chlorine. Brutal kapos or SS guards sometimes pushed the already dysenteric and miserable prisoners into the barrels for sport. Outbreaks of tuberculosis and pneumonia swept mercilessly through the prisoner population, and the corpses of those who died in the night were piled up at the entrance to the sleeping tunnels.[43] One prisoner remembered laconically, "I dreamed about Buchenwald like it was Heaven when I was in Dora."[44]

The underground camp had an atrocious death rate: 172 prisoners died in November. In January 1944, that number increased to 669. In March, 721 prisoners, an average of 24 per day, were worked to death in the tunnels.[45] To this number must also be added several outbound transports of prisoners whom the SS deemed "unfit for work" and were therefore likely candidates to be murdered elsewhere. In the month between the beginning of January 1944 and early February, three transports of over 2,000 prisoners went to the massive slave and extermination camp Majdanek in the General Government. On April 8, the SS sent another 1,000 prisoners to Bergen-Belsen in Lower Saxony.[46] Most of those who survived the trip to Majdanek were then transferred to Auschwitz and murdered there. Those whom the SS sent to Bergen-Belsen were crammed into shoddy, disease-ridden barracks and left to die. The death rate became so bad at Dora in the winter of 1943 to 1944 that the inbound prisoner transports from Buchenwald could barely keep pace with the catastrophe unfolding under Kohnstein. Of the 17,000 prisoners shipped to Dora between August 1943 and March 1944, 6,000 died in the course of expanding the tunnels and installing the missile factory – a death rate of well over one third of all prisoners.[47] Only when the expansion and installation work was completed, coupled with the construction of

---

[42] André Rogerie Interview, FVA, Yale University.

[43] Sellier, *A History of the Dora Camp*, 59–60. See also the André Rogerie Interview, Yale University.

[44] Ben Giladi Interview, FVA, Yale University.

[45] Wagner, *Produktion des Todes*, 647. Yves Béon writes of a prisoner named Jacky whose task it was to roam the tunnels with a cart, searching for dead bodies to take to the morgue. See Béon, *Planet Dora*, 11–12.

[46] See the collection of transport lists in RG 04.006M, Nazi Concentration Camp Records, 1939–1945, Reel 18, United States Holocaust Memorial Museum (USHMM). See also Wagner, *Produktion des Todes*, 492.

[47] Manfred Bornemann and Martin Broszat, "Das KL Dora-Mittelbau," in *Studien zur Geschichte der Konzentrationslager, Schriftenreihe der Vierteljahreshefte fur Zeitgeschichte* (Stuttgart: Deutsche Verlags-Anstalt, 1970), 166–171. See also Wagner, *Produktion des Todes*, 188–190.

prisoner barracks outside the tunnels in the spring of 1944, did the death rate finally begin to decline.[48]

Mittelbau-Dora was not the only location where concentration camp prisoners slaved and died for the sake of the missile program. SS and Army officials approved other underground locations. In September, "Papa" Riedel and Godomar Schubert surveyed a site east of Salzburg for the installation of the development works underground. The site was to be supplied by prisoners from concentration camp Ebensee, a subsidiary camp of Mauthausen. Construction began in November, but this project, code named "Zement" (Cement), suffered from conflicts between the Army and SS as well as extremely high cost.[49] The development works never relocated there, but a number of section chiefs at Peenemünde, including Papa Riedel, temporarily transferred to Zement to assist in the work.[50] Other test areas were constructed at Lehesten, in central Germany, and at Redl-Zipf in the Austrian Alps, just north of Ebensee.[51] All of them used slave labor in their construction and, in the case of Zement, mirrored on a smaller scale the horrors of Mittelwerk.[52] However, the new factory under Kohnstein became the focal point of the missile program until the end of the war. The number of staff at Peenemünde was dramatically smaller, and the base itself was reduced to a pure research and development facility.

There is no question that the person who bore overall responsibility for this "empire of horror," as Michael Neufeld has appropriately called it, was Hans Kammler.[53] The SS General successfully mobilized concentration camp labor for the project and brought his considerable resources to bear in

---

[48] Wagner, *Produktion des Todes*, 647.

[49] See Florian Freund, *Arbeitslager Zement: das Konzentrationslager Ebensee und die Raketenrüstung* (Vienna: Verlag für Gesellschaftskritik, 1989). In the summer of 1944, Armaments Ministry officials scrapped the plan to relocate the development works and instead drew up plans to use Zement to assemble tanks as well as house an underground oil refinery.

[50] Entstehungsgeschichte, November 2–16 and 25, 1943, FE 833, NASM.

[51] See Dorit Gropp, *Aussenkommando Laura und Vorwerk Mitte Lehesten: Testbetrieb für V2-Triebwerke* (Bad Münstereiffel: Westkreuz Verlag, 1999) and Florian Freund and Bertrand Perz, *Das KZ in der Serbenhalle: Zur Kriegsindustrie in Wiener Neustadt* (Vienna: Verlag für Gesellschaftskritik, 1987). Lehesten, code-named "Mitte," was used to calibrate rocket engines and as a liquid-oxygen facility. It was supplied with laborers from Buchenwald and its subcamps. The Redl-Zipf facility, code-named "Schlier," was used in the same capacity and received prisoners from Mauthausen. Originally, authorities planned to subsume them under DEMAG's authority, but in December 1943, they were made a part of Mittelwerk GmbH. See Niederschrift über die 1. Sitzung des Beirates der Mittelwerk GmbH am Freitag d. 10. Dezember 1943, December 14, 1943, R121/405, *Bundesarchiv-Lichterfelde* (BAL).

[52] Army officials' attempts to maintain control over the missile program meant that they themselves would be willing to manage slave labor. At the end of December 1943, Thom laid a proposal before Kammler that spelled out plans to put Zement under the control of the Army Ordnance Office. See Entstehungsgeschichte, December 28, 1943, FE 833, NASM.

[53] Neufeld, *The Rocket and the Reich*, 209.

order to expand the size and scope of Germany's missile program. Kammler's organization had skills that the war economy demanded, and it specialized in managing slave labor at far-flung locations across the Reich.[54] Dora was Kammler's awful triumph. Kammler, however, was forced to rely on like-minded individuals who perhaps did not share his ideological vision, but who could at least come to quick agreement with him on technical matters. The engineers in the A-4 program were well suited to this task. Their particular expertise, combined with their unique zeal for the success of the missile, perfectly complemented Kammler's own ruthless drive to set up the production facility, and they became vital cogs in the machinery of destruction under Kohnstein.

THE "FACTORY COMMUNITY": CIVILIANS AT MITTELBAU-DORA

While the prisoners labored furiously to expand the tunnels in the winter of 1943 to 1944, factory installation proceeded apace. Factory managers were able to install machinery in the tunnels so quickly that, on New Year's Eve, the first missiles rolled off of the assembly line. These weapons were so badly flawed that they were returned almost immediately to the factory, and serious developmental issues remained to be ironed out. Even so, it was an important symbolic achievement, despite the barbarity with which it was completed. In any case, technical problems and the transfer of production from Peenemünde to Mittelwerk delayed the original production schedule by several months. In May 1944, the factory managed to turn out 253 missiles, but a raft of technical problems caused output to drop precipitously throughout the summer. Only in September did Mittelwerk begin to produce anything like the high numbers that were originally planned, usually between 600 and 700 missiles per month.[55] Overcoming major technical obstacles required the close cooperation of specialists in both Peenemünde and Mittelwerk.

A company known as Mittelwerk GmbH (Central Work, Ltd.) managed the carnage at Mittelbau-Dora in cooperation with the SS. At the end of September 1943, Gerhard Degenkolb had moved to streamline the production operation, which sometimes struggled under the Army's ungainly bureaucracy. Under his supervision, the A-4 Special Committee created this company to manage missile production, and the company officially came

---

[54] Allen, *The Business of Genocide*, 202–206.

[55] Neufeld, *The Rocket and the Reich*, 213. The most pressing and difficult of the challenges faced by the developers was the issue of "air bursts," in which inbound missiles broke up during reentry. The problem took months to solve, and it was only in late 1944 that it was finally reduced. The issue was that the outer skin of the missile, weakened by heat friction during reentry, tore off of the body, resulting in the missile's breakup. See Neufeld, *The Rocket and the Reich*, 220–230.

into being on October 7.[56] A state-run corporation that was organized along the lines of a private business, Mittelwerk was financed by the Armaments Ministry and placed under its umbrella firm, Rüstungskontor GmbH.[57] Originally, Degenkolb himself chaired the company's advisory board, whose members included Dornberger and Heinz Kunze, Degenkolb's deputy.[58] Its board of directors was made up of industry men Kurt Kettler from Borsig, Otto Bersch, who previously worked for an automotive firm in Breslau, and Dora camp commandant Otto Förschner. Kammler himself directly placed Förschner on the board in an effort to maintain a prominent role for the SS in policy-level factory decisions, even though the SS man spent his entire adult life as a career soldier.[59] Officially, Förschner was in charge of security and countersabotage. In theory, he could also participate in business decisions, but in reality, his utter lack of managerial experience meant that he had no input at all in daily determinations regarding factory operation.[60] Förschner also relied on his subordinates and prisoner functionaries to run daily camp operations, keeping his distance from the prisoners and usually, but not totally, refraining from abusing them. At the same time, he did nothing to

---

[56] Grundungseintrag Mittelwerk GmbH, 10.7.43, Reel 12, M-1079, NARA.

[57] See Rainer Fröbe, "KZ Häftlinge als Reserve qualifizierte Arbeitskraft: Eine späte Entdeck-ung der deutschen Industrie und Ihre Folgen," in Ulrich Herbert, Karin Orth, and Christoph Dieckmann, eds., *Die Konzentrationslager – Entwicklung und Struktur, Bd 2* (Göttingen: Wallstein Verlag, 1998), 636–681. Michael Allen has pointed out that National Socialist polycracy did not just lead to internecine struggles, as is so commonly assumed. It also made possible useful business arrangements among like-minded individuals in different organiza-tions, a necessary precondition for the use of the slave labor services offered by the SS. See Allen, *The Business of Genocide*, 168–171.

[58] Vermerk über Besprechung im Reichsministerium für Rüstung und Kriegsproduktion, Gen-eralreferat Wirtschaft und Finanzen, betr. Mittelwerk GmbH., am 21.9.43, R121/405, BAL. See also Protokolle der Mittelwerk-Gesellschaftsversammlungen am 24.9.1943, R121/544, BAL.

[59] Born in 1902, Förschner joined the Reichswehr when he was twenty years old. Immediately after leaving the Reichswehr in 1931, he entered the SS, and three years later he enrolled in its officer candidate school at Bad Tolz. In late 1938, his career was marred by an incident in which he fathered a child out of wedlock with the girlfriend of a subordinate named Hugo Hochaus. Förschner, who married in 1931, paid Hochaus twenty-five Reichsmarks per month to falsely claim paternity of the child. After a few months, Hochaus attempted to blackmail Förschner into paying him even more money. Not surprisingly, the plot unraveled when Förschner, whose drinking habit got him into this mess in the first place, drunkenly divulged the secret to an SS colleague. Förschner was demoted to the rank of private (*SS-Mann*) and then nearly thrown out of the SS. The SS expelled Hochaus and he spent eight months in Sachsenhausen for his trouble. However, Förschner's long service record and Hitler's aggressive war planning in 1939 saved his career, as his unit was desperately short of officers. The SS eventually restored his status, and Förschner went on to serve in the SS Viking division on the Eastern Front, later acting as the chief officer in charge of the SS guards at Buchenwald before coming to Dora in 1943. See Otto Förschner Dossier, SS Officer Files, Reel SSO-214, RG 242, NARA.

[60] Heinrich Detmers Testimony, U.S.A vs. Kurt Andrae, et al., Roll 5, M-1079, NARA.

alleviate their suffering. His generally laissez-faire attitude and reliance on prisoners for many functions led to the rise of a substantial resistance organization at Mittelbau-Dora, which the Gestapo would combat in part by relying on information received from civilian engineers in the tunnels.[61]

The most active figure behind tunnel expansion was a young, hard-nosed engineer named Albin Sawatzki. Born in 1909 in Danzig, Sawatzki was a determined, ambitious diploma engineer with a mean streak and a propensity for violence.[62] Before coming to the missile program, Sawatzki worked as a production engineer for the Henschel Corporation, where he made a name for himself in tank production. Degenkolb recruited him from Henschel to run the A-4 Special Committee's subcommittee for serial production. After the August bombing raid at Peenemünde, he left for Thuringia to manage installation and production at Mittelwerk.[63] Sawatzki was neither on Mittelwerk's board of directors nor a member of the SS, but Kammler gave him full authority inside the factory. He had the power to request and assign prisoners and became Kammler's trustee for all problems concerning A-4 production. Sawatzki was fully independent of the factory hierarchy, but he worked closely with it to ensure that it met its production goals.[64] In December 1943, Kammler, probably sensing that Förschner was in over his head on the board of directors but also eager to maintain as much control as possible over Germany's crown jewel weapons system, attempted to place Sawatzki on the board.[65] He was rebuffed, but Sawatzki retained his position as Kammler's special envoy to Mittelwerk. In May 1944, he would become the director of the Production Planning division, officially an employee of Mittelwerk GmbH.[66]

Another key civilian engineer joined the board of directors in April 1944, just as the frenetic pace of tunnel expansion was beginning to wind down.

---

[61] For Förschner's reliance on prisoner functionaries at Dora, see Wagner, *Produktion des Todes*, 301–307. The camp commandant also relied heavily on his SS subordinates to supply labor to the factory. See Wilhelm Simon Testimony, U.S.A. vs. Kurt Andrae, et al., Roll 12, M-1079, NARA. Simon worked in the SS Labor Allocation Office in Dora, which assigned prisoners to Mittelwerk.

[62] Albin Sawatzki Dossier, Gericht Rep. 299, Nr. 430, HStaD-ZA Kalkum.

[63] Brähne, "Die Mittelwerk GmbH," HStaD-ZA Kalkum. Sawatzki met Brähne shortly after Brähne was transferred from Peenemünde to Mittelwerk. Sawatzki immediately took a liking to Brähne and gave the gifted technical illustrator what he called a "Shooting License" (*Jagdschein*) to roam wherever he wished in the factory. Sawatzki wanted him to document "in both words and pictures" the events in Mittelwerk under his leadership. Sawatzki's self-serving vision was to publicize in the future his own role at the plant after the war was brought to a successful conclusion. To his credit, Brähne, who recognized very early the callous inhumanity of Mittelbau-Dora, did exactly that, providing to posterity some of the most stark and disturbing illustrations of life and work in the factory. Sawatzki's deep belief in the ultimate victory of Nazi Germany, even at the end of 1943, is also noteworthy.

[64] Georg Rickhey Statement, NARA.

[65] See Niederschrift über die 1, December 14, 1943, R121/405, BAL.

[66] Direktionsanweisung zur MW-Gesamtorganisation, May 26, 1944, R121/405, BAL.

Georg Johannes Rickhey, a purchasing specialist who previously worked for the massive industrial firm DEMAG in Berlin, was installed as the General Director of the operation, a position that gave him a decisive voice in meetings of the board of directors. Rickhey went on trial for war crimes in Dachau in 1947 because of his connections to Mittelwerk, but he successfully avoided conviction at the hands of American prosecutors by convincing the court that he was merely an apolitical technocrat who was at the mercy of political forces beyond his control. This was hardly the case. A talented diploma-engineer, Rickhey was born in 1898 and joined the Nazi Party in 1931. In 1940, Rickhey worked as the chief technical advisor in the Main Office for Technology (*Hauptamt für Technik*) for the Gauleiter of Essen, SA Lieutenant-General (*Obergruppenführer*) Josef Terboven, a Nazi official with close ties to Hitler (who attended Terboven's wedding in 1934). In this position, he helped streamline the heavily industrialized district's war production measures and inaugurate more efficient use of its labor resources. While working under Terboven, Rickhey displayed an absolute dedication to his task and couched his work in ideological terms that insisted upon unquestioning service to the Nazi state. In a conference with his senior deputies in February 1940, he demanded that his colleagues across Gau Essen turn all of their resources to the war effort, telling them that "All tools, machines, and the laborers necessary for them must, from the smallest workshop to the largest W- [weapons] and Rü [armaments] operations, be engaged one hundred percent in production. It is the task of the representatives of the Gau's Office of Technology, and, therefore, the Party, to make exact [technical] recommendations and to uncover additional suitable areas in which machinery can be set up and brought into operation as quickly as possible. ... Total war demands the utmost exploitation of all means of production on hand and the strenuous effort of all available workers." He informed his deputies that because his office alone could not effect an increase in manufacturing productivity in the entire Gau by itself, it would be choosing "Factory managers, engineers, technicians, and work Meisters *who at the same time are members of the Party or are political leaders* ... to examine the suitability of factory facilities on hand and to make proposals, either on their own or in cooperation with the Gau and Kreis representatives, about Armaments Kommandos or [factory] conversion" [emphasis in original].[67] For Rickhey, ideology was the primary motivating factor in the work that he assigned. The Nazi party was to be the vanguard in his early

---

[67] Bericht über Besprechung auf dem Gauamt für Technik in Frankfurt/M. Friedenstrasse 2 am 20.2.40, Georg Rickhey Dossier, Gericht Rep. 299, Nr. 411, HStaD-ZA Kalkum. At his trial, prosecutors also raised the possibility that Rickhey might have been a central figure in the removal of the Jewish head of a local, private technical society called Haus der Technik in Essen in 1933, but they were unable to verify the claim. Heinz Kunze Statement, U.S.A. vs. Kurt Andrae, et al., Roll M-1079, NARA.

efforts to bend Germany's industrial might entirely to the service of the war effort.

At Mittelwerk, purchasing had proven to be one of the most difficult problems in completing the setup of the factory. In addition to his duties as General Director, Rickhey took over these functions as well as responsibility for personnel issues, quickly introducing a number of reforms that were designed to improve the purchasing and production processes. Rickhey's arrival also signified a reorganization of the corporate administration. The primary result was that Förschner's duties became limited to counterespionage. Though he remained on the board of directors, this was in effect a demotion for the camp commander.[68] The SS eventually transferred him to a command at Kaufering, a subcamp of Dachau, and his position on the board of directors remained unfilled. The SS was left without a formal representative on that body. Though there would be minor adjustments in the structure of the corporation, there were no further major reorganizations or additions to its board of directors. Civilian managers, therefore, not the SS, set forth the directives and guided the policies of the Mittelwerk GmbH. The SS, of course, ran camp Dora, but its influence on the policy decisions and much of the daily operation of the factory was strictly limited. Though Kammler set the overall conditions at Mittelwerk, he relied on civilian managers to carry out the tasks necessary to establish a mass production facility under Kohnstein. Civilians were responsible for employment and handling of prisoners inside the factory itself.

Although no one from Peenemünde served on the Mittelwerk board of directors, many development and production engineers from Usedom received positions in upper and middle management in the factory. The most important of these men was Arthur Rudolph. As the individual responsible for erecting the production facility at Peenemünde, Rudolph naturally was heavily involved in the disassembly of the factory and its relocation to the Harz Mountains. His official title with the Mittelwerk GmbH was Factory Director, and he was responsible for missile assembly and production, but his first task was managing the transfer and installation of machinery. He arrived in Mittelbau-Dora in September 1943, and in this capacity he worked hand in glove with Sawatzki.[69]

A number of Rudolph's deputies and lower-level managers from Peenemünde left the facility on the Baltic to assist Rudolph with this work, and the production engineer relied heavily on them to complete many of the major tasks. Many arrived with Rudolph in September.[70] They were among

---

[68] Direktionsanweisung zur MW-Gesamtorganisation, May 26, 1944, NS-4 Anh., Nr. 16, BAL.
[69] Entstehungsgeschichte, September 8, 1943, FE 833, NASM.
[70] Arthur Rudolph Office of Special Investigation (OSI) Interrogation, printed in Franklin, *An American in Exile*, 221.

the thousands of people who left Usedom for Mittelwerk that autumn. On November 16, for example, Albert Speer himself directed that Peenemünde had to give up at least twenty percent of its skilled personnel, or approximately 1,145 people, by Armaments Ministry estimates, for the project in Mittelwerk. Engineers, technicians, master craftsmen, secretaries, and other skilled laborers, such as joiners, electricians, and welders, streamed out of Peenemünde and into Mittelwerk.[71] By December 1, 1943, 386 people had been transferred, including 128 engineers, technicians, and craftsmen. Two weeks later an additional 347 people were transferred to Mittelwerk, 97 of them engineers and high-level technicians. Hundreds more would follow them in the weeks after.[72]

This major relocation of personnel generally proceeded smoothly and with only minor problems. Walther Riedel (Riedel III), von Braun, and Sawatzki directed much of the transfer process. Cooperation between administrators at Peenemünde and Mittelwerk was obviously essential in order to smooth the process and efficiently respond to Mittelwerk's needs. The only delays were caused by limitations placed by the Armed Forces High Command (OKW) on train travel, which caused moderate delays in personnel transfer. These delays were overcome after the Peenemünders used their influence to acquire special travel passes from military authorities.[73] There was also some inevitable disagreement over where certain important specialists should work, either at Peenemünde or at Mittelwerk, but in general the transfer of personnel from Usedom to the Harz Mountains was very smooth because von Braun labored mightily to ensure that the division of personnel between Peenemünde and Mittelwerk was equitable. By the middle of November 1943, the staff at the missile base had shrunk to just over 7,200 employees.[74] In April 1944, Sawatzki was able to determine that all of the positions needed for civilian labor had been filled, but that another 1,850 prisoners still needed to be placed in the plant.[75] Once the underground missile factory had been completed and steady operations began, approximately 3,000 German civilians found themselves working there alongside some 5,000 concentration camp slaves. The number of prisoners would multiply almost exponentially later in the war, as more and more of Germany's

---

[71] Entstehungsgeschichte, October 13, 19, and 20, 1943; also November 11 and 16, 1943; FE 833, NASM.

[72] Undated Reisinger Report, FE 694, NASM.

[73] Von Braun to Lindenberg, December 20, 1943, FE 694/a, NASM.

[74] Von Braun to Kettler, April 2, 1944, FE 694/a and Von Braun to Sawatzki, April 12, 1944, FE 694/a, NASM. See also Neufeld, *The Rocket and the Reich*, 206.

[75] Albin Sawatzki, "Fertigungsumfang und – aufwand A4 Mittelwerksanteil Stand 1.5.1944," FE 694/b, NASM. The list of recipients of this memorandum is also indicative of the links between administrators at Peenemünde and Mittelwerk. Of the twelve recipients, five, including von Braun and Rudolph, were Peenemünders. Others addressees included Kammler, Rickhey, and Bersch.

armaments industries began shifting their operations underground in the region around Mittelwerk.[76]

Security at the new production facility was extremely tight, but it also increased dramatically as the tunnels expanded and were filled. The *Sicherheitsdienst* (Security Service, or SD) and Gestapo operated out of offices in Nordhausen (near Mittelwerk) and other towns around Kohnstein. They also established a strong presence in Mittelwerk itself. SS Lieutenant-Colonel (*Obersturmmbahnführer*) Helmut Bischoff, the sadistic head of security for the A-4 program, coordinated their activities. Bischoff received his orders directly from Kammler himself.[77] At the end of May, 1944, the Armaments Ministry designated the area within a thirty-kilometer radius of Dora as a relocation zone for heavy industry, which was being bombed to rubble by air raids on Germany's cities. This region was known as "Sperrgebiet Mittelbau," and Bischoff also assumed responsibility for security for the entire area. Those who did not work or live in the area needed special permission from the Gestapo to enter it.[78] A large motorized police unit operated out of Nordhausen and another motorized infantry unit were also assigned to help secure the area, especially Mittelwerk itself.[79] The factory's officials established a security detail [*Werkschutz*] that served as the guard troop for Mittelwerk. Its members patrolled the entrances to the tunnels and maintained security checkpoints inside the factory.[80] The SD and the Gestapo also assumed responsibility for security at Mittelwerk. They maintained organizationally distinct offices for combating sabotage and so-called terror actions, espionage, and offenses committed by civilian workers. They also ran a network of informants in the factory whose activities were coordinated by the SD office in Niedersachswerfen, a town near the north entrance to the tunnels.[81] With all of this in place, civilian employees faced a dizzying array of security measures in their daily activities at Mittelwerk.

In the fall and winter of 1943 to 1944, people who transferred to the factory from Peenemünde were first ordered to the town of Ilfeld, the seat of the Mittelwerk GmbH's local headquarters, about ten kilometers north

---

[76] Wagner, *Produktion des Todes*, 549. Major corporations such as Junkers and Askania opened operations in the tunnels in 1944. Mittelwerk GmbH also received contracts to produce the V-1 cruise missile and the Heinkel He-162 "People's Fighter" [*Volksjaeger*].

[77] Michael Allen, in his thoughtful book *The Business of Genocide*, mistakenly writes that Kammler placed Bischoff in this position because of Bischoff's supposed engineering background. Actually, Bischoff had no technical training and was a lifetime SS police official. See Allen, p. 226. For a more accurate picture of Bischoff, see Wagner, *Produktion des Todes*, 524–528.

[78] Wernher Haack Statement, ZM 1625, Bd. 40, Akte 168, BStU. How strictly this could be enforced is questionable.

[79] Heinrich Detmers Testimony, NARA.

[80] Helmut Bischoff Testimony, Gericht Rep. 299, Nr. 23, HStaD-ZA Kalkum.

[81] Adolf Häser Statement, Gericht Rep. 299, Nr. 253, HStaD-ZA Kalkum. Häser was the Chief of the Gestapo in Niedersachserfen and Nordhausen.

of the factory. There, they received instructions and several days' worth of training in the factory's secrecy regulations, counterespionage, and how to handle incidents of sabotage. When they departed Ilfeld and arrived at their new temporary residences (sometimes military-style wooden barracks – the area lacked enough proper accommodations to absorb such a large influx of personnel), Gestapo officer interviewed them, photographed them, issued them passes into the tunnels, and eventually led them into the mountain.[82] As at Peenemünde, this introduction to Mittelwerk served several functions. It recalled the Peenemünders' initiation into the world of secrecy shrouding the project, gave them entry into this exclusive world, and also brought them face to face with the oppressive mechanism of the Nazi state. The Mittelwerk employees felt nothing if not self-conscious in the knowledge that the Gestapo now had a file on them that included their name, address, and a photograph. The sense of coercion around the project only grew, but it was only one of many factors ensuring the dedication of the civilians to the missile effort.

More positive considerations also helped maintain their loyalty. The engineers, technicians, and craftsmen who moved into the area around Mittelbau-Dora in late 1943 and early 1944 found a factory system in place that emphasized their importance to the German war effort in both word and deed. The articles of incorporation for Mittelwerk GmbH, written by Degenkolb himself, attempted to manufacture a sense of community that was based both on the tasks ahead of them as well as Degenkolb's own vision as to how the operation should run. The articles recall Dornberger's speech delivered to the Peenemünders just a few months earlier, if only cast through an increasingly warped ideological prism. Consciously attempting to maintain a sense of communal interest around the work and referring to the Mittelwerk's employees as "work comrades," Degenkolb began by noting that "[Factory] Operations will be carried out in the spirit of a factory community [*Betriebsgemeinschaft*]." All of the firm's managers as well as each employee on the shop floor made up a strong, productive community of common interest. Because of the factory's overriding military importance, employees of Mittelwerk all had equal stake in the project's success. Degenkolb wrote that this common interest flowed explicitly from a sense of being a part of the national community [*Volksgemeinschaft*]. Accordingly, only "Those who possess German blood can be a member of the factory community." For him, the Nazi *Volksgemeinschaft*, based as it was in large part along racial lines and on the common welfare of all Germans regardless of class, was both a model and wellspring of the

---

[82] Wernher Haack Statement, ZM 1625, Bd. 40, Akte 168, BStU. Haack lived near Woffleben, a camp on opposite side of Kohnstein from Dora. When he arrived in the tunnels in December 1943, he remembered that there was not much light underground, there was no air circulation, and the air reeked of sulfur and ammonia.

community of missile specialists in the tunnels. Like the Nazi *Volksgemein-schaft*, Degenkolb envisioned a factory community that was to be bound together by "a spiritual commitment and reciprocal feeling of responsibility." He demanded that "The destiny of the whole operation is the destiny of the entire factory community. This destiny is therefore the communal task of all work comrades, who must uphold operations with their last reserves of strength and productivity. ... The supreme principle of the National Socialist Party, 'Communal interest before personal interest,' is the highest goal of the factory community."[83] As at Peenemünde, the success of one of Germany's most modern factory production lines was to be based not simply on the skill of its laborers, its speed, or its efficiency, but also on the active identification of its workers with the goals for which it was put to use. The sense of involvement in a project that was somehow larger than the sum of its parts was an important factor at both locations. The only thing different for Degenkolb was the motivation. His ideas embodied reactionary modernism in its ultimate form, embracing antimodern notions of race and the German spirit as the central factors in the success of one of the world's most advanced weapons. Though his rhetoric differed from that espoused at Peenemünde, his message was similar. Personal interests should be set aside and communal interests embraced so that the work of defending the nation could be completed. Every worker's last effort should be bent toward achieving this goal.

Degenkolb then went on to discuss the means by which it could be realized. After his resounding call to adhere to the tenets of a racially defined factory community centered on common national interest, most of these directives were mundane by comparison. Nevertheless, they were important for the emphasis they laid upon factors that defined the Peenemünde community of specialists. For example, Degenkolb ordered that all employees receive a copy of the articles of incorporation and give their signature to confirm that they would abide by the rules it contained. In addition, he demanded that employees behave according to the strictest rules of secrecy regarding the plant's operation. Degenkolb encouraged all employees, no matter what rank, to proactively seek out improvements that could be made in their individual sectors and for management to be flexible in responding to these suggestions. Factory managers received full authority to hire or requisition the necessary workers for their individual section. Degenkolb also noted that employees could be summarily fired for such transgressions as

---

[83] Betriebsordnung Mittelwerk GmbH, December 23, 1943, R121/405, BAL. Michael Allen labels this effort to manufacture not only technology but also the National Socialist spirit – an act that supposedly would "yield up the German soul" (and relegated profit to secondary status) – "productivism." There is little reason to doubt that this could be true, but Allen fails to note how easily such rhetoric could correspond with the less-ideologically inclined efforts to manufacture a sense of common interest among factory employees. See Allen, *The Business of Genocide*, esp. 165–239.

"Offenses against National Socialist principles," "Serious offenses against the laws of the Reich," and "Disrupting ongoing work." Depending on the transgression, punishment ranged from an administrative wrist-slapping to the passing of the case to the SD.[84] However, most of these specific orders were not so different from the rules employees had to live under at Peenemünde, and the transition to a different style of management in Mittelwerk was minimal.

This is evident in a letter written to Degenkolb in January 1944 by Heinz Schmid-Lossburg, an executive with Rüstungskontor GmbH, the umbrella firm that financed Mittelwerk GmbH. After reading Degenkolb's missive, he noted a number of concerns that it brought up. He opened by stating dryly that "The section concerning the factory community is somewhat unclear," and laid out a number of concerns about compensating employees, issues pertaining to overtime pay, and other financial questions. However, Schmid-Lossburg also went on to state, among other things, that much of what Degenkolb emphasized in the articles of incorporation was not necessary. Most employees had long been living and working under such rules at Peenemünde and were still bound by them. Nearly all of the employees at Mittelwerk were already obligated to follow the strictest secrecy guidelines because of their previous work, and most managers already understood that the task of looking after the employees belonged "fundamentally to the factory managers." Their roles as facilitators of improvement were clear, and they had always been open to suggestions from a range of employees. Factory managers already understood that the importance of the project required that they have confidence in their (civilian) workers. They had learned to take care of these issues during their time at Peenemünde.[85] Degenkolb's call to bring together the factory community was actually redundant, and had already been widely embraced by the employees at Mittelwerk.

Those civilians who came to Mittelwerk to join Degenkolb's "somewhat unclear" concept of a factory community received substantial compensation for relocating their homes, workplaces, and often their families. Mittelwerk GmbH agreed to give its senior managers up to 600 Reichsmarks (RM) to cover moving expenses, whereas other employees received up to RM 500 to cover expenses.[86] Because of the hasty and fully improvised transfer of production, many employees were also unable to find suitable accommodations immediately. To help overcome the difficulties associated with such a rapid move, Mittelwerk GmbH paid employees an extra per diem of RM 2.50 for the first six months they were in the area. Benefits provided by the corporation were also quite generous. Employees who worked overtime did so at a time-and-a-half rate. The board of directors also recognized that

[84] Betriebsordnung Mittelwerk GmbH, December 23, 1943, R121/405, BAL.
[85] Schmidt-Lossburg to Degenkolb, January 5, 1944, R121/544, BAL.
[86] Niederschrift über die 1, December 14, 1943, NS-4 Anh., Nr. 3, BAL.

working in the factory created a special strain on civilians in the tunnels. Tellingly, every three months, employees received a generous "allowance for difficult working conditions" [*Erschwerniszulage*], which amounted one quarter of the monthly salary. This benefit was available to employees only after they had been in the factory for two months.[87] Karl Otto Saur, Speer's deputy, also showed an important interest in the well-being of Mittelwerk's civilian employees, insisting that they were "especially burdened" and should receive extra vacation time "in view of the difficult working conditions."[88]

The employees' social welfare was also a central consideration for the board of directors. Married workers received an allowance for family-related expenses [*Unterhalts-Beihilfe*]. Single employees also received a similar allowance, though it was about half of what the married workers collected. However, if a single employee got married, he and his wife received a one-time gift of RM 150 from the corporation. On top of the family allowance, Mittelwerk GmbH awarded a one-time payment of RM 100 for the first child and a one-time payment of RM 50 for the second. No German government before the Nazis, incidentally, had ever enacted this benefit. The company also offered a subsidy of RM 10 per month for each child that married couples had beyond their second. This benefit would be in effect until the child was sixteen. Life insurance, no small matter in a country being systematically razed by the Allied bombing campaign, was also a staple benefit.[89] The Mittelwerk board of directors also proposed to offer forty percent of the company's stock to its employees, but it was an idea that never came to pass.[90] In the end, Mittelwerk GmbH was obviously keen to look after the social welfare of its employees, an official concern that could only have helped enlist and maintain employees' willingness to put forth their best efforts on the program's behalf.[91]

Perhaps more importantly, individual employees' salaries rose markedly upon transferring to Mittelwerk GmbH. Generally speaking, engineers and technicians who occupied positions in the middle and upper-middle management levels at Peenemünde earned between RM 10,000 and RM 12,000 per year, depending upon education, experience, and seniority. These numbers rose dramatically when civilian managers transferred to Mittelwerk. For example, Erich Ball, an assembly specialist at Peenemünde who earned approximately RM 10,000 per year between 1937 and 1943 (though no

[87] Bericht [Report] Dr. Kettler, December 10, 1943, R121/405, BAL.
[88] Niederschrift über die 1, December 14, 1943, NS-4 Anh., Nr. 3, BAL.
[89] Bericht [Report] Dr. Kettler, December 10, 1943, R121/405, BAL. See Götz Aly, *Hitler's Volkstaat: Raub, Rassenkrieg und nationaler Sozialismus* (Frankfurt am Main: Fischer Verlag, 2005).
[90] "Production and Disposition of German A-4 (V-2) Rockets," NARA.
[91] Götz Aly has shown that, under the Nazis, German citizens received better social welfare benefits than under any other government before it. Aly, *Hitler's Volkstaat*.

records exist for individual years, one should assume that his salary was somewhat higher in 1943 than in 1937 because of periodic raises and cost-of-living adjustments), earned a salary of RM 18,000 in his position as assembly line manager at Mittelwerk. Technician Günther Haukohl's salary numbers are nearly identical. Rudolf Schlidt, a technician who worked in materials testing at Peenemünde and helped assemble exhaust jet vanes in Mittelwerk, saw his salary increase from RM 6,000 to RM 10,000 per year. Perhaps the most dramatic salary increase was Arthur Rudolph's. The talented production supervisor, who held a two-year degree from a vocational school and who once had to subsist on just over RM 7 per week, earned a whopping RM 29,900 in his position with Mittelwerk GmbH, up from his salary of just over RM 10,000 at Peenemünde. The once impoverished technician had done very well for himself under the Nazi regime.[92] Based on the documentation available, it appears that most employees who transferred from Peenemünde likely received pay raises of between thirty and forty percent – a substantial increase by any standard. Admittedly, late in the war, the money may not have gotten them any farther in an economy in which everything was rationed, but it was at minimum a statement of the commitment that the regime was willing to make to them as well as a psychological boost to the former Peenemünders themselves.

Increased material awards, then, were used to buttress the civilians' personal dedication to the program's success and to help override any lingering personal reservations about the exploitation of slave labor in the concentration camp system. Those who came from Peenemünde and elsewhere received a generous amount of money to help overcome relocation expenses as well as several different types of inducements to work efficiently in Mittelwerk. They were well compensated for their work and they could be certain that their families would be looked after by the state-owned corporation, even if they should suffer an accident or death. In addition, the handsome benefits package offered by Mittelwerk GmbH showed that its administrators were aware of the arduous nature of employment in the factory. It was not an easy place to work, and conditions were, to borrow Saur's word, "difficult," to say the least. Management consciously attempted to alleviate the stress of working underground in a high-pressure environment in which the employees were confronted daily with the concrete reality of slavery under the Nazi regime. By offering generous pay, excellent benefits, and increased vacation time, even during the most radical period of the regime's existence, the company hoped to mitigate the strains that it knew existed in a place that must have been as difficult to work in as Peenemünde was exciting. The rupture from their comfortable lives on the Baltic was no doubt unpleasant, but missile program administrators did their best to overcome any remaining

---

[92] Erich Ball, Günther Haukohl, Rudolf Schlidt, and Arthur Rudolph Basic Personnel Records, Box 703, Entry 179, RG 165, File "Boston," NARA.

sense of dislocation. Even the bombastic Degenkolb, with his bizarre articles of incorporation, got into the act. Though his missive may have been opaque to many employees, the tangible material benefits of working and living around Mittelbau-Dora were perfectly clear. Even if life there was not as rewarding as in Peenemünde, employment under Kohnstein offered other advantages, such as higher salaries and excellent social welfare benefits.

Another major benefit was the potential for professional advancement. Personnel who had previously been employed at Peenemünde often assumed positions in the Mittelwerk factory that were essential to its successful operation. For many, transfer to the underground facility meant a great deal of upward professional mobility. Many employees who were deputies and assistants in Peenemünde became section chiefs and division managers in Mittelwerk. According to Dornberger himself, nearly every working group in the factory was headed by an engineer who had originally been employed at Peenemünde.[93] Sensibly, engineers at Peenemünde who had experience in a particular area at their former base went on to become the experts in this same field in Mittelwerk. Several individual cases as well as entire factory divisions within the Mittelwerk factory serve as useful examples of the importance of the Peenemünders to the efficient functioning of the operation.

Engineer Erich Ball, who helped plan assembly at the production plant in Peenemünde, arrived from Usedom in September 1943 and assumed the position of assembly line chief at Mittelwerk. Günther Haukohl, a skilled technician at Peenemünde who worked on the installation of the assembly line, helped plan extended manufacturing facilities and a repair shop in Mittelwerk.[94] Engineer Firnrohr (first name unknown), a deputy group leader in the assembly plant at Peenemünde, became the division head of the department responsible for assembly of the center section of the missile at Mittelwerk. Engineer Busselt, a deputy leader in the division responsible for testing the wiring in experimental missiles at Peenemünde, became the head of the group responsible for testing the missile's wiring at Mittelwerk.[95] Many other engineers and technicians from Peenemünde followed this career arc, which established direct connections between the research and development station on the Baltic and the assembly plant in the Harz Mountains.

Perhaps the most important section in this regard was the Labor Operations division (*Betriebsarbeitseinsatz*), which allocated both civilian and

[93] Walter Dornberger Statement, ZM 1625, Bd. 44, Akte 189, BStU.
[94] U.S.A. vs. Kurt Andrae, et al., Roll 4, M-1079, NARA. The notoriously brutal kapo "Big Georg" Finkenzeller worked under Haukohl.
[95] My sincere thanks to Torsten Hess of the KZ Gedenkstätte Dora-Mittelbau for his help in working through a number of the connections that individual engineers had between Peenemünde and Mittelbau-Dora.

prisoner labor inside the assembly halls. As Factory Director, Rudolph ultimately was in charge of this division, but former Peenemünde specialists supervised its daily activities. The two men in charge of this section, Engineers Broszat and Weckbrodt, were former Peenemünders who worked closely with Rudolph in designing the production plant there. In the Mittelwerk factory, they worked with the SS-run Labor Allocation Office in concentration camp Dora in order to assign semiskilled prisoners to the proper assembly and transport details. Starting in the autumn of 1943 and continuing through March 1945, civilian managers in factory labor operations received daily prisoner strength reports that detailed how many slave laborers came into Dora, how many had died, and how many total prisoners worked in the tunnels. These reports made the staggering death rate in the tunnels abundantly clear, and from them, Broszat and Weckbrodt were able to request and allocate more prisoners as needed.[96] In addition, Engineer Raschdorf, Broszat's deputy in this division, also previously worked at Peenemünde. In late 1944, Hellmut Simon arrived from Peenemünde to take Broszat's place. The Factory Labor division was also the civilian office in charge of supervision and control. Engineer Stuhlfauth, who ran this section, was not a Peenemünder, but his deputy, an engineer named Kuhlmann, had transferred from Usedom in November.[97]

A second important example of the centrality of civilian experts from Peenemünde in daily factory operations was in the Production Quality Control division (*Fertigungsaufsicht*). This division was created before the August air raid and subsequent dispersal of the base's facilities and personnel. It remained under the administrative control of von Braun's development group at Peenemünde even after its main office relocated to Mittelwerk in May 1944. The Quality Control division was the embodiment of cooperation between missile specialists both in development and production. Von Braun and Sawatzki worked very closely to ensure that this division functioned smoothly at Mittelwerk. In April 1944, von Braun traveled to the factory in order to discuss with Sawatzki how to improve and simplify the group's work. Both men agreed that one of the major problems confronting them was the difficulty of creating easily mass-produced, readily installed assemblies out of the often custom-made pieces of equipment made for test firings at Peenemünde. The experience of mass production had shown both of them that "the questions that still remain to be solved must be worked out by cooperation between development, subsidiary firms, and assembly." These questions would be tackled by the Quality Control group, which would function as a technical "storm troop" [*Stosstrupp*] in attacking

---

[96] For examples of these strength reports, see Veränderungsmeldungen, ZM 1625, Bd. 69, Akte 348, BStU.
[97] Mittelwerk GmbH Betriebsabteilungen, December 12, 1943, NS-4 Anh., Nr. 16, BAL.

problems as they came up while also coordinating the functions of all Army developers, production specialists, and private industry.[98]

The group was made up of approximately fifty people from Peenemünde, engineers who both supervised the incorporation of design changes in subsidiary firms and tested assemblies and subassemblies in the factory. These civilian engineers also employed skilled prisoners as assistants. Hans Lindenberg, the chief of this very important group, was one of von Braun's deputies when Lindenberg worked at Peenemünde. For his tasks at Mittelwerk, he cooperated directly with Sawatzki on questions of development and production, but he was technically still directly subordinate to von Braun.[99] Even though he lived and worked in Ilfeld, not far from Mittelwerk, Lindenberg often sought out von Braun's support on numerous production and design questions. Moreover, in addition to coordinating development and production, the Quality Control group was also charged with "removal of specialists (foreigners) [from the assembly line] who are not particularly qualified" [parentheses in original].[100] The fate of prisoners who were not on the factory assembly line was common knowledge. Most were worked to death in the myriad of SS-managed construction projects in the Nordhausen area in 1944 and 1945.[101]

Importantly, Wernher von Braun helped write the service instructions for the Quality Control group and was central in the setup of this office.[102] Throughout the early part of 1944, while the tunnels were still being expanded and more parts of the factory were being brought on line, he remained a central figure in defining the mission of the Quality Control group. In February, he wrote a stern circular to all of the members of the Quality Control group reminding them of the importance of their work. In von Braun's typical carrot-and-stick approach, he wrote that he would unerringly support any specialists who rejected flawed items even if their decisions set back production quotas. However, he also pointed out that he would call to account those members of the group who frivolously ignored the demands of production or who could not clearly justify why production numbers were not being met. All members of the group were to give their forthright cooperation to all of the firms involved in the manufacture of the A-4 so that high volume, steady output could be achieved. "I don't need to mention to you," he wrote, "that attaining a high output of instruments as soon as possible is everyone's dearest wish." Von Braun reminded the old

---

[98] Albin Sawatzki, Aktenvermerk über die Besprechung am 12.4.1944 im Mittelwerk, April 16, 1944, FE 694/a, NASM.

[99] Ibid.

[100] Dienstanweisung für die Fertigungsaufsicht, May 15, 1944, NS-4 Anh., Nr. 23, BAL.

[101] See Wagner, *Produktion des Todes*, 359–367 and Allen, *The Business of Genocide*, 222–232.

[102] Wernher von Braun Statement, February 7, 1969, ZM 1625, Bd. 60, Akte 268, BStU.

Peenemünders that in order to produce this high volume of serviceable missiles, the developers and producers had to closely coordinate their activities. For him, this was an overriding concern: "We have no time to lose! The fastest possible introduction and adaptation of [test] results in the [office responsible for coordinating technical changes] is the decisive demand upon which the success or failure of our entire project depends."[103] Cooperation, then, was to be the watchword for this group, as indeed it was for all Peenemünders who came to work at Mittelwerk. As at Peenemünde, the fate of the entire project depended on the positive interaction of a motivated, cooperative German workforce. The Quality Control group, made up almost exclusively of Peenemünders, was a lynchpin to this success.

The links between engineers at the distant locations were founded on more than professional grounds. Away from the shop floor and in higher management echelons, the social bonds between leading development and production engineers were close. For example, Rickhey hosted a number of parties, complete with cognac and cigars, for Mittelwerk and Peenemünde managers.[104] One occasion between the Peenemünders and Mittelwerk management even revealed a certain level of indifference to the crimes committed in the name of missile assembly. In December 1944, von Braun, Dornberger, and Heinz Kunze (Degenkolb's deputy) won the Knight's Cross for their efforts to develop and produce the A-4. A number of leading Peenemünders and production engineers, including Rickhey and Sawatzki, were invited to the celebration, which was organized by Riedel III.[105] The atmosphere at the event was pleasant and relaxed, with drinks flowing and the conversation lively. As part of the festivities, guests received cartoon illustrations that made light of each other's particular eccentricities and foibles, but that also revealed a callous attitude toward the fate of the prisoners at Dora. Dornberger's illustration depicts the General on the hunt for a buck, his favorite pastime, reckoning a five-meter dispersal on his shot and declaring wistfully, "Those were the days!" [*Das waren noch Zeiten!*]. Von Braun's drawing shows the young engineer reclining in an armchair while he dreams of using a solar reflector to cook his breakfast. More troubling was Sawatzki's illustration. In it, a mass of prisoners struggles to pull his car out of ditch while a kapo looks on. In Rickhey's drawing, inmates labor to pull an A-4 out of a tunnel, observed at a distance by a man in a crisp suit, presumably Rickhey himself.[106] Any levity these illustrations may have added also reveals a distinct unconcern on the part of the Peenemünders with the struggles of the prisoners who labored in the tunnels of Mittelwerk. Illustrations

---

[103] Wernher von Braun, Rundschreiben an sämtliche Angehörigen der Fertigungsaufsicht, February 17, 1944, RH8/v. 1980, BA/MA.
[104] Julius Bouda Testimony, U.S.A. vs. Kurt Andrae, et al., Roll 7, M-1079, NARA.
[105] Riedel III to Kunze, December 11, 1944, FE 372, NASM.
[106] Dinner card draft illustrations, FE 855, NASM.

FIGURE 8. A Peenemünder who had knowledge of events at Mittelbau-Dora drew these illustrations in late 1944. Dornberger's light-hearted illustration has him aiming at a buck and declaring, "Those were the days!" In contrast, Sawatzki's depicts him observing concentration camp prisoners as they pull his car out of a ditch; Rickhey's has him overseeing slave labor operations to produce the missile. [Fort Eustis Collection, FE 855, National Air and Space Museum, Smithsonian Institution]

aside, these social occasions also helped to reaffirm the engineers' identification with each other and their willingness to cooperate in the project. Such events were important for maintaining an efficient, collegial relationship between developers and producers. In short, they reinforced the highly developed sense of solidarity among the stressed and beleaguered missile specialists.

The division, therefore, between specialists at Peenemünde and production engineers at Dora was not sharply defined. Some historians and many
rocketry enthusiasts have, for very different reasons, implied that one group
had little to do with the other except for the occasional exchanges of correspondence or odd meeting. A weapon as radical and advanced as the A-4,
however, dictated that the developers remain in close contact with production engineers in order to iron out the inevitable wrinkles that would appear,
not only in a weapon that was being rushed into operational use but also
in the transition from experimental to mass production. The transplanting
of final assembly from Peenemünde only served to complicate this transition, but it also maintained the link between development and production
personnel. Moreover, many Peenemünders, who had internalized the missile
center's central message of service to the nation in time of war as well and its
deeply ingrained culture of group self-interest, moved into the Mittelwerk
factory and became key figures in the management of missile production.
Indeed, this line between the two groups was not a solid barrier at all, but
rather a semipermeable membrane in which people and information could
be easily moved back and forth. Former Peenemünders inhabited the middle-
and upper-level management strata of Mittelwerk GmbH, an arrangement
that assisted the exchange of information between the two sites. These managers also occupied positions in which their decisions would have the most
impact upon the daily lives of concentration camp prisoners forced to work
in the service of the missile program. Every day at work, they were confronted with the reality of slave labor. The decisions that they made from
moment to moment on the shop floor reflected the deeply internalized culture of their former institution as well as the increasingly radicalized political
climate in the last eighteen months of the National Socialist regime.

SLAVE LABOR AND THE CIVILIAN WORKFORCE
AT MITTELBAU-DORA

German civilian specialists played a part in the systematic persecution of
slave laborers in Mittelwerk. However, direct individual involvement in the
crimes associated with slavery in the tunnels varied widely. The civilians'
behavior was clustered into two basic groupings. Passive facilitators who
efficiently did their jobs in support of the war effort made up one general
group. By their hard work, they created the essential precondition for more
extreme acts to be perpetrated by their more vicious and ideologically motivated colleagues. They tolerated Nazi brutality either by turning a blind eye
to it or, worse, by turning it to their own advantage. A much smaller but
more brutal group of engineers made up the balance of the civilian staff.
This group included people like Sawatzki, who embraced slave labor and
showed an astonishing combination of cruelty and technological rationality that served the twin goals of A-4 production while subjugating those

perceived to be enemies of the Nazi state. These men were more heavily involved in Nazi crimes, though their numbers in the factory were much smaller. Their direct and more criminal participation in the excesses of the Nazi regime was one of the results of the radicalized political atmosphere and the specialists' narrowed ethical outlook that only made room for the goals of mass production. In any case, almost everyone involved was at least indifferent, or at least oblivious, to the tremendous human suffering that they caused.

The period of factory installation described above was certainly the most brutal and terrifying for the prisoners at Dora. Civilians who worked among these prisoners treated them in a variety of ways. Most of the civilians involved in setting up the factory came from Peenemünde or were sub-contracted employees of subsidiary firms. Some of these civilian engineers physically mistreated the prisoners in the tunnels despite an SS order forbidding the practice. In December, Förschner circulated a notice that prohibited civilians from having any contact with prisoners except for the purposes of explaining their work. He made it clear that the SS was formally the only group responsible for disciplining the prisoners.[107] Most civilians complied, but some outright ignored this order. Sawatzki himself was perhaps the worst offender in this regard. Eddie Verheyn, a French prisoner, testified in the U.S. Army Dora war crimes trial that Sawatzki would roam the tunnels in the morning, "calling us French swine and kicking several of us here and there."[108] Gerhard Hobert recalled clearly that civilians beat prisoners, accusing in particular Sawatzki, as well as an engineer named Siegel and Karl Seidenstucker, a section chief on the assembly line. "It is not only true that the directors were bandits," he testified, "but the section leaders were just as bad."[109] Finally, Engineer Jakob, a former Peenemünder who was in charge of tail assembly on the production line, was also accused of abusing prisoners. According to one prisoner, he purportedly "... took pleasure in beating people and having people beaten."[110] Both Seidenstucker, who was described by one prisoner as a "sadist," and Jakob were two of Rudolph's deputies.[111]

For the most part, however, civilian specialists in the tunnels left the disciplining of prisoners to the kapos and SS.[112] They still had little sympathy

[107] Förschner Aktennotiz [Memorandum], December 30, 1943, NS-4 Anh., Nr. 3, BAL.
[108] Eddie Verheyn Testimony, U.S.A. vs. Kurt Andrae, et al., Roll 4, M-1079, NARA.
[109] Gerhard Hobert Testimony, U.S.A. vs. Kurt Andrae, et al., Roll 4, M-1079, NARA. Willi Burgdorf, a prisoner who worked directly for Seidenstucker, offered that the engineer "was a bad fellow."
[110] Eddie Verheyn Testimony, NARA; Willy Steimel Testimony, U.S.A. vs. Kurt Andrae, et al., Roll 5, M-1079, NARA.
[111] Tadeusz Kahl Testimony, 28/83 USA, BStU.
[112] Sellier, *A History of the Dora Camp*, 87. Sellier also notes that of the thousands of deaths in this period, very few were because of outright executions.

FIGURE 9. In January 1944, concentration camp inmates watched by a civilian specialist work to expand the tunnels under the Kohnstein Mountain before factory machinery is installed. This is the only known extant photograph depicting civilian supervision of factory installation. [Ullstein Bild – W. Frentz]

for the prisoners suffering through their troglodyte existence. In the terrible winter months of 1943 to 1944, the civilians were wholly indifferent to the misery around them. Yves Béon remembers civilians who acted as if the prisoners were not even there, "continually measuring the galleries according to the plans they carry. They move about, climbing the piles of rubble, going around machines and reels of cable, past turning concrete mixers, but never looking at the tattered men around them, nor even hearing the shouts, the vicious clubbings, or screams of pain. Quietly, they indicate location points desired for machines, for junctions, for joints and fixing points for the electrical and pneumatic air ducts."[113] Factory installation, not prisoner care, was the first priority for most civilian specialists in the winter of 1943 to 1944. Dr. Karl Kahr, the SS physician at Dora, testified in 1947 that the tempo of the work was one of the primary causes of fatigue and accidents. He placed equal blame for this on civilian employees, kapos, and the SS.[114] Those prisoners who did not or could not contribute their full energy to achieving the goal of quickly installing the factory were easily expendable and readily replaced. A drumbeat of instructions from above reinforced these

[113] Béon, *Planet Dora*, 24.
[114] Karl Kahr Testimony, U.S.A. vs. Kurt Andrae, et al., Roll 8, M-1079, NARA.

ideas, warning about sabotage and delays while pointing out, for example, that "The fast and programmatic execution of our work of production and the guarantee of the working reliability of our plants and manufacturing equipment are at present our first requirements."[115] No mention was ever made of the care of the prisoners.

Though most civilians abstained from direct abuse of the sort Sawatzki and others doled out, they could still be directly implicated in the *system* of maltreatment in Mittelwerk. Clement van Hamme, a Dutch prisoner, explained that "The civilians who were with us did nothing but watch the work and command. They beat us little or not, but denounced the men who did not work to the SS."[116] In the early stages of tunnel expansion and factory construction, some German specialists blatantly (and, it must be said, illegally, even in the Nazi context) mistreated prisoners whom they deemed too lazy or too slow while most cared little about the prisoners' suffering. In the course of doing their jobs properly, however, they became involved in the structure of abuse in the factory by reporting misbehaving workers, replacing those who could no longer function efficiently, and utterly failing to look out for their well-being.

In the late spring and early summer of 1944, slave labor in the tunnels under Kohnstein began to take on a new aspect. As the expansion of the galleries was completed and the installation of the factory assembly line began receiving its finishing touches, the ferocious pace of the work as well as the catastrophic death rate under the mountain began to abate. By April 1, the primary tunneling work was complete, the tunnels cleaned, air conditioning installed, and machines readied for operation.[117] The SS also finally began constructing barracks outside of the tunnels in January 1944. They began moving the prisoners into them shortly afterward. The last of the prisoners emerged from the tunnels to take their places in the barracks in June.[118] Camp Dora would eventually be made up of over fifty barracks. It contained its own crematorium as well as a gallows and separate prison that came to be known as "The Bunker." Almost all of the barracks were equipped with wash rooms and latrines, which helped limit the spread of disease.[119] The SS also built a medical barracks so that there was at least the possibility of medical care for the prisoners. The food provided to the prisoners improved as well. Those fortunate enough to have the requisite skills to work on the assembly line experienced a drastic

[115] Kettler Sonderdirektionsanweisung D, January 8, 1944, NS-4 Anh, Nr. 3, BAL.
[116] Clement van Hamme Testimony, U.S.A vs. Kurt Andrae, et al., Roll 2, M-1079, NARA.
[117] Wincenty Hein Testimony, BStU.
[118] Monatsbericht [Monthly Report], May 23, 1944, Gericht Rep. 299, Bd. 562, HStaD-ZA Kalkum.
[119] Wagner, *Produktion des Todes*, 192–194. Wagner also points out that the lowered death rate at Dora was also the result of shipping several thousand prisoners who were no longer able to work to Dora's growing network of subsidiary camps.

improvement in conditions, and the death rate dropped dramatically that spring.[120] The accommodations above ground, although not ideal, were nonetheless a monumental improvement over the cold, wet, and disease-ridden sleeping tunnels in the mountain.

Attitudes toward the prisoner labor force also fundamentally changed. Until the end of March, 1944, most of the prisoners in the Mittelbau-Dora complex worked at the tunnel face, blasting the rock and transporting it out. Many others worked in so-called *Fachkommandos* as joiners, electricians, handworkers, and other skilled positions, and a small number of prisoners, perhaps five percent, were employed as secretaries, clerks, and stenographers in the business offices.[121] As the work in the tunnels began to require fewer high concentrations of unskilled manpower for digging, transport, and construction work, the job sites underground began to be populated increasingly by skilled laborers whom civilian managers and SS officials alike viewed as a much more valuable commodity. Assembling the missiles required a competent workforce that was at least moderately familiar with modern technology. Electricians, welders, metalworkers, and mechanics were especially valuable professions for prisoners. Most often, the SS housed these skilled workers in the main camp of Dora itself, where the skilled labor pool could be most easily accessed and prisoners could move relatively easily to and from work.[122] Skilled inmates in the main camp generally received better treatment as well.

For example, in May 1944, an outbreak of typhus struck Dora and threatened to devastate the camp. The SS medical staff ordered that "skilled prisoners who are important for the factory" receive inoculations against the disease so that they could continue their work.[123] Starting in July, prisoner doctors conducted twice-weekly health inspections, especially with the aim of limiting the spread of fleas. The availability of food and water improved as well. These improvements did not come about out of humanitarian concern but rather out of the camp administrators' narrow technical self-interest. Unskilled prisoners were often banished to Dora's increasingly dense network of subsidiary camps, where conditions were much worse. They had to live with other, often fatal, disadvantages that skilled workers sometimes did not have to face. Many of them lacked adequate shoes, especially in the transport kommandos, where, according SS doctor Joachim Mrugowsky,

---

[120] Monatsbericht [Monthly Report], Häftlingskrankenbau Dora, May 23, 1944, Gericht Rep. 299, Bd. 562, HStaD-ZA Kalkum.

[121] Anklageschrift gegen Bischoff, Busta, Sander, Konzentrationslager Mittelbau/Dora, ZM 1625, Bd. 5, Akte 15, BStU. See also Wincenty Hein Testimony, BStU.

[122] Wagner, *Produktion des Todes*, 367–368; Neufeld, *The Rocket and the Reich*, 225; Sellier, *A History of the Dora Camp*, 149.

[123] Monatsbericht [Monthly Report], May 23, 1944, Gericht Rep. 299, Bd. 562, HStaD-ZA Kalkum.

"abrasions [on the feet] are a consequence of the work performed."[124] Such a mundane oversight as proper footwear might seem harmless, but in writing of his experience in Auschwitz, Primo Levi reminds his readers that "Death begins with the shoes; for most of us, they show themselves to be instruments of torture, which after a few hours of marching cause painful sores which become fatally infected. ... To enter the hospital with a diagnosis of 'swollen feet' is extremely dangerous, because it is well-known to all, but especially to the SS, that there is no cure for that complaint."[125] So-called incurable prisoners were no longer of any value to the SS and were murdered. Skilled prisoners who lived in Dora ran the same risks, but the chances of infected feet were mitigated by Förschner's move in the early summer of 1944 to provide them with leather shoes. Other camp officials and medical personnel improved the prisoners' access to better clothing.[126] Those who labored in positions requiring less skill but more physical exertion, and who, consequently, were most in need of proper shoes and new clothes, received nothing.

Civilian specialists had virtually no contact with the prisoners in the camp, but they worked side by side with them in the factory every day. Technical considerations were the dominant factors in decisions made about the treatment of prisoners. Arthur Rudolph admitted that all of the civilians who worked in the factory were keenly aware of the A-4's importance to the war effort and that the primary objective at Mittelwerk was to mass-produce missiles quickly and efficiently.[127] Rapid mass production of the A-4 required permanent and trained groups of workers. Continuous training of new workers was expensive and hampered the pace of production, as did a never-ending stream of prisoner abuse. It was far easier and made much more technical sense to keep skilled workers alive and in the workshops than it was to train a new one every time a worker died or suffered injuries requiring his removal from the assembly line. Former prisoner Wincenty Hein pointed out that "Since qualified work groups were more valuable [than unskilled labor], the treatment of the prisoners during their work time in the factory improved."[128] At Rickhey's direction, administrators introduced a premium system in which prisoners could be rewarded with extra rations and cigarettes from the camp in exchange for excellent work.[129]

Improved conditions for the prisoners had little to do with the engineers' humanitarian concerns and much more to do with producing functional

[124] Mrugorsky to Kammler, July 4, 1944, Roll 1, M-1079, NARA.
[125] Primo Levi, *Survival in Auschwitz* (New York: Simon & Schuster, 1996), 34–35.
[126] Mrugorsky to Kammler, NARA.
[127] Arthur Rudolph OSI Interrogation, printed in Franklin, *An American in Exile*, 218.
[128] Wincenty Hein Testimony, ZM 1625, Bd. 23, Akte 35, BStU.
[129] Wagner, *Produktion des Todes*, 393–394.

FIGURE 10. Civilians working with a skilled prisoner laborer from concentration camp Dora test an A-4's guidance equipment in Mittelwerk. The prisoner is French physicist Charles Sadron. [Ullstein Bild – W. Frentz]

missiles and their own narrow self-interest. Because it was in the missile specialists' interests to keep skilled labor alive or uninjured, technically trained prisoners stood a much better chance of survival. Engineer Willibald Feier, who worked at Peenemünde from 1941 until his transfer to Mittelwerk in 1943, remarked that "Since a huge death rate ruled among the prisoners at this time, we civilians appealed to the SS guards with the goal to reduce the death rate. This was necessary for us, since we were concerned about having unskilled workers and it took a long time for us to train them."[130] Feier's candid statement indicates that he was conscious of the horrors in the early months of working in the tunnels, but it also shows the sharp narrowing of civilians' ethical sensibilities when it came to considerations of the prisoners' conditions. Skilled workers had value; unskilled workers did not. Civilians viewed the prisoners through the lens of technical function. They assigned no value to those unlucky enough to be without the requisite skills. Their personal identification with the goal of mass production, combined with pressure from above, manifested itself in a willingness to think not in terms of the human cost of their work but rather the potential output of human labor.

[130] Willibald Feier Statement, Gericht Rep. 299, Bd. 600, HStaD-ZA Kalkum.

The circumstances created by wartime events also cast the engineers' considerations in a revealing light. In the spring and summer of 1944, Allied air attacks increasingly began to create major problems for firms that manufactured A-4 parts and assemblies. Many of the production sites were relocated to Mittelwerk; by the end of the summer, the entire tail assembly, the servomotors, and the central section of the missile were being put together in the tunnels.[131] The decision to move servomotors into Mittelwerk is an important example of the missile program's managers' narrow but strong identification with the goals of their work. Servomotors had been manufactured by the firm Brown, Boverie, and Cie at locations in Saarbrucken and outside of Paris. Because of the intensity of Allied air attacks, the program administrators, including Dornberger, von Braun, Rudolph, Sawatzki, and Rickhey, agreed to move the production sites to Mittelwerk. However, camp Dora did not have enough skilled workers necessary to man the assembly positions in the tunnels. The workforce had to be relocated as well but of course would not come along voluntarily. The only way to make the French workers come to Mittelwerk was to arrest them, transport them to Dora, and put them to work as concentration camp prisoners under Kohnstein.[132] No one participating in these decisions objected to forcing the French workers into slavery, though there is also no evidence that the arrests ever took place.

The general improvement in working conditions did not necessarily translate into freedom from abuse at the hands of the SS, kapos, or even civilian specialists. An SS guard named Erwin Busta roamed the halls of the factory and dispensed fearsome beatings. Busta, whom the prisoners nicknamed "Horsehead" [*Pferdekopf*] because of his elongated features, was a sadist who meted out ferocious abuse to skilled and unskilled prisoners alike. He was often seen hunting for prisoners that he felt were not working hard enough or fast enough. Busta committed a litany of brutalities in the tunnels, whipping a prisoner to the point of unconsciousness and then shooting him in the head (August 1944); shooting two Russian lathe operators for no apparent reason while they were at their work stations (December 1944); and beating another prisoner to death with an electric cable (winter 1945).[133]

---

[131] Prüfbericht Mittelwerk GmbH zum 30.9.44, R121/303, BAL.

[132] Protokoll der Besprechung im Büro Rickhey am 6.5.44, FE 694/b, NASM. Whether or not these workers were actually arrested and brought to Dora is unclear. In von Braun's defense, the young engineer had himself been arrested by the Gestapo and released only two months earlier. He lived with the threat of rearrest for the rest of the war, and it is unreasonable to expect him to explicitly protest against slave labor after this episode. Neufeld, *The Rocket and the Reich*, 213–220. I examine this incident in further detail in the next chapter.

[133] Anklageschrift gegen Bischoff, Busta, Sander, Konzentrationslager Mittelbau/Dora, BStU. See also Neumann Testimony, "Verläufiges Ergebnis der Ermittlungen," ZM 1625, Bd. 4, Akte 14, BStU.

At least one civilian who worked in the tunnels recalled after the war that even Germans in Mittelwerk feared Busta.[134] Even so, they were not above using the SS man for their own ends. Vadim Bykadorov, a Russian engineer who arrived at Dora in the summer of 1944, testified in 1967 that "German civilians who tested the quality of the work intimidated the prisoners with the threat that they would report cases in which the quality of work was not high to 'Pferdekopf.' . . . The results of such reports were that the prisoners were beaten or taken out of the kommando and never seen again."[135] Some prisoner kapos also vented their anger on the prisoners they supervised on the shop floor.[136] Skilled or not, few prisoners were completely exempt from every form of arbitrary abuse, though the pace and number of incidents slackened considerably in 1944.

One of the principal reasons for this was that technical considerations forced the SS to defer to the authority of civilian engineers and administrators. Though the SS did indeed set up the framework in which the factory functioned, civilian engineers were fundamentally entrusted with its daily operation.[137] The conversion from construction to mass production enhanced the authority of the civilian missile specialists in the factory while weakening the authority of the SS on the shop floor. Ernst Dutzmann, the former head of the Army Acceptance Office (*Heeresabnahmestelle* – responsible for testing and delivering finished missiles and other materials to the missile battalions), stated that "I did not see prisoners who were employed in assembly get mistreated by SS guards while they worked. The German expert employees were their direct supervisors during work."[138] Another civilian engineer remarked that "I myself only saw a few SS men in the underground factory. They were occasionally in the long tunnels. I never saw SS men in the side tunnels [where subassemblies were put together]."[139] On the shop floor, civilians could, within their areas of expertise, exert authority even over SS men. Some were even able to keep the vicious "Pferdekopf" in check.[140]

However, the civilians, who were under tremendous pressure to produce results, still witnessed scenes of arbitrary brutality. Worse, they sometimes participated in the mistreatment of prisoners. According to one former prisoner, approximately eighty percent of the civilians treated the prisoners normally and did not abuse them, but the other twenty percent had no qualms about slapping, kicking, or otherwise harming prisoner labor.[141] The civilian board of directors was acutely aware of this behavior and under no

---

[134] Luise Speiss Statement, ZM 1625, Bd. 34, Akte 19, BStU.
[135] Vadim Bykadorov Testimony, Gericht Rep. 299, Nr. 171, HStAD-ZA Kalkum.
[136] Heinz Hilgenböcker Statement, Gericht Rep. 299, Nr. 274, HStAD-ZA Kalkum.
[137] Hubert Tacke Dossier, ZM 1625, Bd. 35, Akte 131, BStU.
[138] Ernst Dutzmann Testimony, Gericht Rep. 299, Nr. 210, HStaD-ZA Kalkum.
[139] Heinz Krause Statement, Gericht Rep. 299, Nr. 188, HStaD-ZA Kalkum.
[140] Sellier, *A History of the Dora Camp*, 139.
[141] Josef Klaes Testimony, U.S.A. vs. Kurt Andrae, et al., Roll 5, M-1079, NARA.

circumstances condoned it. In a memorandum circulated through the Mittelwerk factory in the early summer of 1944, factory management pointed out that

The camp doctor has repeatedly determined that detainees who work in the offices or on the factory floor have been beaten by company employees because of this or that offense, or have even been stabbed with sharp instruments to the point that they must be given medical treatment. Such interference with the camp commanders' authority on the part of Mittelwerk employees must cease under all circumstances. If a prisoner is guilty of a punishable offense... a written report must be submitted to camp commandant Förschner. ... A copy of the report is to be sent to executive factory management. Further punishment will then be undertaken by the commander against the prisoner.[142]

Despite an order previously circulated by Förschner in his position as the chief of factory security, some civilian specialists still saw fit to beat prisoners working under them. Nevertheless, most incidents were relatively isolated. In a factory of over 3,000 civilians and 5,000 prisoner laborers, incidents of direct maltreatment were the exception rather than the rule. The majority of civilian workers behaved less violently, though their reasons for this had less to do with humanitarian considerations than they did with technical reasons.

Importantly, some civilians sympathized with the prisoners and treated them as well as they could, given the circumstances. Many prisoners went to great lengths after the war to recognize the foreman or engineer who gave them a piece of bread or helped shield them from the SS.[143] For example, French prisoner Georg Soubirous recalled that "For a short time I was in an electrician's kommando named 'König.' At the head of this kommando was a German engineer named König who did everything to make our lives easier. He did his utmost to make sure that we did not become 'Strafhäftlinge' [prisoners who were punished in the Bunker in Dora]."[144] After the war, many civilians who were employed in Mittelwerk claimed to have done everything they could to improve the lot of the prisoners, giving rise to what Jens Wagner has called the "Myth of the Bread Givers."[145] Most of this testimony, offered during war crimes trials, lacks any corroboration and documentation. There is no evidence that most civilians on the shop floor actually attempted to pass food to prisoners. It is clear, however, that a bare minority of civilian workers were willing to take such risks in order to help individuals in the small groups of laborers under their control.

[142] Rickhey and Kettler, Sonderdirektionsanweisung, June 22, 1944, NS-4 Anh., Nr. 3, BAL.

[143] Sellier, *A History of the Dora Camp*, 137–138.

[144] Georges Soubirous Testimony, Gericht Rep. 299, Nr. 590, HStaD-ZA Kalkum.

[145] See, for example, Wernher Haack Statement, ZM 1625, Bd. 40, Akte 168, BStU; Indrak Statement, Gericht Rep. 299, Nr. 599, HStaD-ZA Kalkum; and Walter Kash Dossier, ZM 1625, Bd. 30, Akte 88, BStU. See also Wagner, *Produktion des Todes*, 552.

These individual cases of civilian assistance to the prisoners throw the utter inaction on the part of the vast majority of civilians into stark relief. Employees who helped the prisoners illustrate the fact that there was indeed a choice to be made in the tunnels of Mittelwerk. Civilian missile specialists were not totally constrained by fear and repression, as so many would like posterity to think. There were alternatives to simply going along with the orders of the SS. Instead, individual engineers and technicians were able to make decisions within their own areas of individual responsibility that had a profound impact on the conditions of the prisoners working under them.[146] The vast majority of them chose to neglect the difficulties faced by the prisoner population, whereas some even elected on their own accord to mistreat the prisoners, despite strict orders from the SS to avoid this at all costs. This willing indifference pervaded Mittelwerk and helped condone the wider actions that the SS saw fit to carry out against supposed enemies of the state in a political environment that was becoming ever more radical.

The enhanced authority of civilian managers on the shop floor at Mittelwerk may have lessened the incidents of abuse, but it also meant that individual specialists were more deeply implicated in the structure of abuse in the factory. The civilian factory managers, rather than the SS, had direct influence over prisoner allocation. Specialists on the shop floor requested allocations of prisoners or made changes in the staffing of individual prisoners.[147] These requests were sent to Rudolph's Factory Labor Operations Office and then on to Dora administrators, who acted upon them.[148] When allocation problems arose, these civilians sometimes even had recourse to the SS officials in the camps that assigned labor to the factory. One floor manager wrote directly to the Labor Office in Klein-Bodungen, a subsidiary camp of Dora that supplied skilled laborers, informing the office that the prisoners assigned to him "are unusable for 'Dora assembly.' I intend to exchange them with the following prisoners. ... These prisoners are skilled people and work in the transport section. According to the [prisoner] foreman 02708 in 'Dora Assembly,' they are industrious workers. I ask your permission that these three prisoners be trained in Dora Assembly."[149] It is worth recalling that unskilled laborers who were deemed "unusable for Dora assembly" faced the prospect of slaving to death in the construction kommandos dotting the area around the Mittelwerk.

Civilians in Mittelwerk also had a large part of the responsibility for recognizing and reporting incidents of sabotage. The A-4 was a rather delicate

---

[146] Wagner, *Produktion des Todes*, 553.

[147] Hubert Tacke Statement, ZM 1625, Bd. 55, Akte 264, BStU.

[148] See, for example, Herbst to Stärfl, November 17, 1944, Gericht Rep. 299, Nr. 561, HStaD-ZA Kalkum.

[149] Büttner to SS-Arbeitseinsatz, Klein-Bodungen, October 16, 1944, Gericht Rep. 299, Nr. 561, HStaD-ZA, Kalkum.

FIGURE 11. These two prisoners were sent from Dora to the subcamp of Boelke Kaserne in Nordhausen. Boelke Kaserne was one of the many subcamps to which prisoners were relegated if they could not work on the assembly line in Mittelwerk. [National Archives and Records Administration]

weapon, and even simple acts could prove fatal to its performance. Most commonly, prisoners purposely soldered weak welds, tightened nuts too much or too little, or engaged in surreptitious work slowdowns, known to factory officials as *Arbeitsbummelei*. One of the most common stories about sabotage incidents, possibly apocryphal, has Russian prisoners urinating in the missiles' engine blocks or instrumentation.[150] Civilians had strict orders to be on the lookout for these activities. In January 1944, Förschner and

---

[150] A resistance group operated briefly in Mittelbau-Dora as well, before it was brutally smashed by the SD and its leaders murdered. For a full accounting of the resistance organization and sabotage activities at Mittelwerk, see Manfred Bornemann, *Aktiver und Passiver Widerstand im KZ Dora und im Mittelwerk: Eine Studie über den Widerstand im KZ Mittelbau-Dora* (Berlin: Westkreuz Verlag, 1994) and Peter Hochmuth, *Der illegale Widerstand der Häftlinge des KZ Mittelbau-Dora: Dokumentation* (Schkeuditz: GNN Verlag, 2000). For a first-hand account of the resistance in Dora, see Béon, *Planet Dora*.

industry man Kurt Kettler ordered factory managers to take concrete measures to detect and deter sabotage.[151] The lack of technical experts among
the various security elements in the factory required the vigilance of trained
civilians. It has already been noted that upon their introduction to the factory, the SD instructed the civilians to be on the lookout for incidents of
espionage and sabotage.[152]

It is very likely that the Factory Labor division (*Betriebsarbeitseinsatz*),
under Rudolph and run by his two deputies, both Peenemünders, was one of
the civilian bodies that was directly implicated in the effort to eliminate such
acts in the factory, an endeavor that only led to the torture and murder of
slave laborers in Dora. In its capacity as the section responsible for the supervision of labor operations, sabotage reports filed by lower-level civilian managers against prisoners likely flowed through this office and into the hands
of the SS camp administration or SD, which would see to it that prisoners
were severely punished, often killed.[153] Hannelore Bannasch (another transfer from Peenemünde – Bannasch worked as von Braun's secretary there)
testified in 1947 that she often heard individuals verbally report instances of
sabotage to Sawatzki, but more importantly, she also stated that she saw a
number of written reports. She recalled that "These reports were handled by
the factory management and Mr. Sawatzki heard of them only as they were
passed on by the factory management [Rudolph]. ... If anybody had signed
[a sabotage report] at the Werke, it would have been Mr. Rudolph."[154] In
addition, Otto Förschner's secretary revealed after the war that production
managers, the factory's security detail [*Werkschutz*], and even middle-level
managers all submitted sabotage reports to the SS. The reports, sometimes
countersigned by civilian section chiefs (*Abteilungsleiter*), came to the Commandant's Office, which then forwarded them to the SD for disposal of the
case.[155] Another civilian engineer confirmed that factory management often
gave sabotage reports to factory security, which also delivered them to the
SD.[156] Despite all of this testimony, these statements must be treated very
carefully, given the paucity of documents that can directly attest to them.
Even so, within the authority structure in Mittelwerk, they make logical
sense. The retreat of the SS from the shop floor meant that civilians were the
first line of defense against sabotage, and many understood their importance
in this matter. However, the civilians were also forbidden from punishing

---

[151] Kettler and Förschner, Sonderdirektionsanweisung, January 8, 1944, NS-4 Anh. 3, BAL.
[152] See also Karczewski Testimony, Gericht Rep. 299, Nr. 589, HStaD-ZA Kalkum.
[153] Historians must be very careful with such claims because sabotage reports have not been
   located in the archives. They are left with the testimony of those who were in a position to
   actually see such reports or claimed to handle them personally.
[154] Hannelore Bannasch Testimony, U.S.A. vs. Kurt Andrae, et al., Roll 10, M-1079, NARA.
[155] Soddemann Testimony, Gericht Rep. 299, Nr. 589, HStaD-ZA Kalkum. This practice likely
   began some time in the middle of 1944.
[156] Molsen Testimony, Gericht Rep. 299, Nr. 589, HStaD-ZA Kalkum.

prisoners directly and therefore had to report these incidents up the chain of command until they reached suitable authorities. Former Peenemünder Willibald Feier stated that "Our most serious responsibility was to immediately report it [sabotage] to the SS."[157] Most civilians in Mittelwerk were aware of sabotage, understood their duties to avert it, and took action when they discovered it.

The hunt for saboteurs usually yielded results in the civilian-directed equipment tests. The primary objective of these tests was to ensure quality control, but they were also designed to discover potential cases of sabotage at several stages of the assembly process. Civilian groups tested parts and subassemblies several times before they handed the manufactured products off to the Army Acceptance Office for even more testing.[158] Former Peenemünders, many of whom were VkN engineers, and officials from Rax Werke and Luftschiffbau Zeppelin staffed the Army Acceptance Office. The office was originally made up of 120 scientists and engineers, but it grew later to nearly 200 men, and its members tested parts, subassemblies, and large assemblies once they were complete.[159] At each stage, quality control supervisors signed a certificate indicating that the part or assembly passed inspection, and then put a stamp on the equipment.[160] Every part, subassembly, and general assembly had a specific portion of the factory where it was put together. If there was a problem with a particular assembly, test engineers knew ahead of time precisely where that assembly was mounted and who was doing the work.[161] In this way, quality control could be carefully monitored at every step of the process so that if sabotage was suspected, it could be more easily traced back to its source. To complete the process, finished missiles were tested by the Army at its firing range at Bliszna, the converted Waffen-SS camp in Poland.[162] Thus the first line of defense against

---

[157] Willibald Feier Statement, HStaD-ZA Kalkum. Actual punishment for sabotage was most often meted out by the powerful and ruthless SD Office that Kammler assigned to Mittelwerk. SD officials either murdered the prisoners outright in The Bunker in Dora or tortured their victims before hanging them. Public hangings in the camp began in the autumn of 1944 after the SD uncovered the communist-led resistance movement. In March 1945, they ordered mass hangings from a crane inside the Mittelwerk tunnels as an intimidation tactic against saboteurs. See Wagner, *Produktion des Todes*, 350–357.

[158] Heinz Krause Statement, HStaD-ZA Kalkum. Hubert Tacke Statement, ZM 1625, Bd. 55, Akte 264.

[159] Three prisoners performed secretarial duties. Von Braun to Kettler, January 20, 1944, FE 694/a, NASM. Von Braun to Heereswaffenamt Amtsgruppe für Endabnahme, November 26, 1943, FE 694/a, NASM. Ernst Dutzmann Testimony, HStaD-ZA Kalkum. Heinz Hilgenöcker Statement, Gericht Rep. 299, Nr. 174, HStaD-ZA Kalkum. Hilgenöcker was himself a former VkN member who worked in the engine testing section at Mittelwerk.

[160] Heinz Krause Statement, HStaD-ZA Kalkum.

[161] Fritz Kunig Testimony, Gericht Rep. 299, Nr. 589, HStaD-ZA Kalkum; Hans Joucks Testimony, ZM 1625, Bd. 55, Akte 264, BStU.

[162] Neufeld, *The Rocket and the Reich*, 220–222.

FIGURE 12. Beginning in 1944, the A-4 was manufactured in the Mittelwerk factory. The missile's center section, containing the fuel tanks, was assembled here, in Tunnel B. Secrecy was very strict at Mittelwerk. The sign at the bottom left warns that anyone who did not work in this area was not allowed passage. [National Air and Space Museum, Smithsonian Institution (SI 79–12324)]

sabotage was not the SS or SD, but rather the civilians, most of whom had come from Peenemünde. The institutional inertia that was a result of their indoctrination there played a large part in narrowing their focus to their own priorities and ignoring those of others.

CIVILIAN MOTIVATION IN MITTELWERK

Those who came to Mittelwerk from Peenemünde carried on in much the same way as they had in previous years. The conditions in which they lived and worked every day were far more severe, but the work that they carried out at Peenemünde prepared the specialists for what they would discover at Mittelwerk and helped overcome any dislocation they may have felt. Many of the factors that had direct bearing on the work in the tunnels were very similar to those that the Peenemünders had dealt with on Usedom.

Secrecy and security practices, which formed the cultural bedrock of their identity as Peenemünders, were very similar, if more intensely followed, at Mittelwerk. In 1947, Georg Rickhey recalled that upon the move to Mittelbau-Dora, "The top secret rules, which were extremely strict anyway, were made even stricter."[163] A former employee of both facilities recalled, "If secrecy was emphasized with all means in Peenemünde, then it was handled even more sharply in the tunnels. . . . Secrecy went on ad absurdum."[164] When the Peenemünders arrived in Sperrgebiet Mittelbau, they had their photos taken and assignments noted by the SD. To enter the factory itself, they had to hold a special pass (*Werkausweis*) to get past security. Rather than a unique badge, as was the case at Peenemünde, the factory passes had a picture of the employee on the front of them as well as a special mark indicating where in the factory the employee worked.[165] Only high-ranking officials or those with special permission carried the so-called shooting license (*Jagdschein*) that gave them permission to go anywhere in the factory.[166] Factory security checked the passes once at the tunnel entrance and several times again inside tunnel.[167] Strict rules also governed the treatment of documents in the factory. All correspondence was classified as secret or top secret, and all documents were to be locked in safes when not in use. There was to be no mark on any document that indicated the location of the Mittelwerk factory. Incoming and outgoing letters all bore a generic address in Halle.

The effect of these measures was the same as at Peenemünde. Entrance became a symbol of honor, trust, and hierarchy. It was coupled with intrusion of state security into the employees' everyday lives inside and outside the tunnels. While the employees were at work, these measures contributed to a strong sense of isolation from the larger society in which the employees existed outside of the tunnels. As Werner Brähne recalled, "The V-Weapon factory was like a state within a state, and it was totally shut off from the outside world."[168] Again, employees found themselves adhering automatically to the rules governing secrecy. Many refused to walk to places in the factory that their pass did not give them entrance to, and others complained vociferously when secrecy regulations were broken.[169] Civilian engineers assiduously made sure that document and correspondence secrecy was maintained at Mittelwerk. No marks of origin, except those in code, appeared on

---

[163] Georg Rickhey Statement, NARA.

[164] Bericht eines nicht genannten "Peenemünder," veröffentlicht in der Wochenzeitung "Christ un Welt" im Juni 1950, Gericht Rep. 299, Bd. 158, HStaD-ZA Kalkum.

[165] Erich Dänicke Factory Pass, Erich Dänicke Dossier, Gericht Rep. 299, Nr. 186, HStaD-ZA Kalkum.

[166] Wernher Brähne Statement, Gericht Rep. 299, Nr. 158, HStaD-ZA Kalkum. These passes were easily recognizable by security personnel, as they had three large brown buttons affixed to their face.

[167] Heinz Krause Statement, HStaD-ZA Kalkum.

[168] Wernher Brähne Statement, HStaD-ZA Kalkum.

[169] Ewald Wenzel Testimony, Ewald Wenzel Dossier, ZM 1625, Bd. 51, Akte 259, BStU. Wenzel worked as a plumber and truck driver in Peenemünde from 1941 until 1944, when he relocated to Mittelwerk.

letters and delivery crates.[170] They assiduously reported to security officials the concerns that they had about transgressions against secrecy considerations. Most often, the transgressions came in the form of correspondence from outside companies that used the Mittelwerk address directly, rather than its generic address in Halle, or other such lapses that would expose the location of the factory. One engineer who received a letter addressed to "Elektromechanische Werk Ilfeld/Harz" (Elektromechanische Werk was the corporate name given to Peenemünde in 1944) wrote to the factory security that "Since such a designation totally contradicts the rules for secret correspondence, we request that the necessary measures be employed to deal with this problem."[171] Again, as in Peenemünde, civilian employees remained strict custodians of official secrecy.

Other security activities also increased in scope and intensity, sharpening an already present sense of coercion in the atmosphere around Mittelwerk. SS and Gestapo officials regularly intercepted mail and screened it for content. In one incident, factory security intercepted two postcards written by a civilian mechanic named Johannes Mrosek on which Mrosek gave his family instructions for reaching him directly in Ilfeld, site of the factory headquarters. Security officials gave him a stern warning never to do this again.[172] The SD also searched the accommodations of factory employees. In some cases, it discovered major offenses, such as in the case of engineer Fritz Schweinberger, formerly of Rax Werke, who was caught with several secret and top-secret documents in his apartment.[173] In another case, an employee left secret documents out on a workstation instead of locking them up in the safe when he left there. He was arrested by the Gestapo and spent two weeks in the Nordhausen prison.[174] The SD also operated a network of informants in the factory from its office in the local town of Niedersachswerfen.[175] Both civilian and prisoner informants were ubiquitous, leading to increased tension, especially between civilians and prisoners. Willy Steimel, probably a prisoner informant himself, stated in 1947 that "A higher grade of mistrust

[170] Unknown Prisoner of War Testimony, File "V-2 (A-4) Missile (Germany, WWII), Intelligence Interrogations," NASM.

[171] Heinemann to Abwehrbeauftragten, March 9, 1945; NS-4 Anh., Nr. 4, BAL. Factory officials routinely sent letters to both Bischoff and Förschner (who left Dora in January 1945 and was replaced by former Auschwitz Commandant Richard Baer) that they had collected from Mittelwerk employees complaining about insecure correspondence practices. Bischoff had been attempting to straighten out this issue since early 1944. See Bischoff Rundschreiben, March 7, 1944, RH8/v.1265, BA/MA.

[172] Schwohn to Bischoff, date unclear, NS-4 Anh., Nr. 4, BAL. Schwohn sometimes asked Bischoff to discipline the offender, which was often a far worse fate. Also see Schwohn to Bischoff, March 22, 1945, NS-4 Anh., Nr. 4, BAL.

[173] Bischoff to Kammler, Series 2, Folder 2, RG 10.228.0002, USHMM. Schweinberger's fate is unknown.

[174] Georg Rickhey Dossier, Gericht Rep. 299, Nr. 411, HStaD-ZA Kalkum.

[175] Häser Statement, Gericht Rep. 299, Nr. 253, HStaD-ZA Kalkum.

of the prisoners [than there was at Peenemünde] existed and went like a red thread through all of the happenings of the plant from the beginning to the end of the camp. The reasons for this must be found in the appearance of the SD and the Gestapo."[176]

More problematically, Nazi ideology began to play an ever-increasing role in the world populated by the missile specialists.[177] Years of intense propaganda against Germany's enemies had resulted in a population that was radically nationalist, if not xenophobic. The German populace, which was perfectly aware of many of the atrocities committed by the Nazis in their name, feared the concentration camp prisoners in their midst and searched for ways to impose some kind of order on their slowly disintegrating society. According to Jens Wagner, this quest for public security, at least in central Germany, drew the population closer and closer to the only organ that could successfully impose it – the Nazi police apparatus.[178] In the missile program itself, indoctrination, both into the regime's ideology and into the program's central ideals, only increased the tendency to look past violence against prisoners or increased individual willingness to carry it out. Once the moral proscription against violence had been broken, the abuse of the prisoners in one form or another became a more conceivable act and easier to carry out again and again. When the act was done, there could be no going back.

The veil of secrecy in the underground tunnels of Mittelbau-Dora only worsened this situation. The utter isolation produced by the tunnels increased the chances that mistreatment of the prisoner labor force would occur. Security measures had a powerful, socially binding effect, but also a darker side.[179] When civilian power over the prisoners was joined to secrecy in the tunnels of Mittelwerk, the threat of prisoner abuse increased exponentially. Pressure, power, professionalism, fear, and secrecy all coalesced in the tunnels under Kohnstein, with disastrous results for the prisoner labor force.

---

[176] Willy Steimel Testimony, NARA.

[177] See Norbert Frei, "People's Community and War: Hitler's Popular Support," 59–78, and Hans Mommsen, "The Indian Summer and the Collapse of the Third Reich: The Last Act," 109–127, both in Mommsen, ed., *The Third Reich Between Vision and Reality: New Perspectives on German History, 1918–1945* (New York: Berg, 2001).

[178] Wagner, *Produktion des Todes*, 554–559. Robert Gellately also shows that support for the regime remained strong, even late in the war, because of the balance that the police apparatus struck between order and oppression. See *Backing Hitler: Consent and Coercion in Nazi Germany* (New York: Oxford University Press, 2001).

[179] Sissela Bok has argued that one of the most insidious effects of deep secrecy is that it debilitates character and judgment. "[Secrecy] can also lower resistance to the irrational and the pathological," she writes. "It then poses great difficulties for individuals whose controls go awry." Secrecy carries some risk of corruption for everyone, but when it is combined with extraordinary powers over others, with no accountability to those whom it affects, the temptation for abuse is great. Sissela Bok, *Secrecy: On the Ethics of Concealment and Revelation* (New York: Vintage Books, 1989), 25, 110.

Ironically, as Wagner points out, after the war, secrecy enabled individuals in the missile program, especially those who had worked at Peenemünde, to argue that they had no idea of the criminal activities going on at Mittelbau-Dora.[180] Secrecy served as the ultimate enabler, erecting a framework in which the crimes could be committed, and subsequently creating a plausible basis for deniability after the acts had been uncovered. The actual activities of those left behind at Peenemünde prove otherwise.

[180] Wagner, *Produktion des Todes*, 562.

6

"We Still Had a Fatherland to Fight For"

For those who stayed behind at Peenemünde, the last eighteen months of the war continued to present huge challenges. In addition to ironing out the A-4's remaining technical bugs (even as mass production commenced at Mittel-werk), they worked feverishly to improve the missile's performance and capabilities, expand the A-4's operational versatility, and develop other anti-aircraft missile systems. As the war situation became progressively worse for Nazi Germany, the Peenemünders responded with prodigious activity that, while unsuccessful, clarified their level of commitment to their work while making important contributions to the conceptualization of future weapon systems, including the surface-to-air missile and the submarine-launched ballistic missile. Much of this furious work required their continued cooperation with Kammler and the SS, and they proved capable of striking a mutually accommodating relationship with Himmler's men. This was true even at the upper levels of the program's administration, where there were a number of personal clashes but also a conscious effort to overcome any acrimony between individuals for the good of the program. These months were characterized not by internal dissension and collapse, but rather by technical creativity and managerial cooperation.

After the war, the story told by Peenemünders and their supporters about this period was dominated by themes of resistance, dissent, and distrust of Hitler and his regime. They painted the SS as an unstoppable marauder, plundering and subjugating all projects that were not yet in its ken. According to their own mythmaking, the Peenemünders feared Himmler's and Kammler's designs on their work, and they had no love for the regime. All of the work that they did in this period was supposedly done purely in the name of spaceflight.[1] Such sentiments reflected similar postwar claims made by Albert Speer, who asserted that the SS conspired to infiltrate Armaments Ministry projects and that he actually never embraced a cooperative

---

[1] See Wernher von Braun, "Reminiscences of German Rocketry," *Journal of the British Interplanetary Society* 15 (May–June 1956), 125–145; Walter Dornberger, *V-2* (New York: Viking Press, 1954); Dieter K. Huzel, *From Peenemünde to Canaveral* (Englewood Cliffs, NJ: Prentice-Hall, 1962).

relationship with Himmler's organization.[2] These claims by Peenemünders ignore the reality of their daily tasks.[3]

At Peenemünde, missile specialists showed an extraordinary willingness to invest all of their time and creative energies, to the point of transcending liberal, "Enlightened" standards of behavior, in order to fulfill their wartime work. This last spasm of activity drove them to design weapons that were well beyond their own technological capabilities while also resorting to technologies that were of limited value and had questionable chances for success. The Peenemünders did not, as they and their enthusiastic supporters so often assert, merely satisfy the demands of the State and SS in the last year and a half of the war. Nor did they carry out their projects because they were forced to do so. Rather, their feverish activities during this period are evidence of their willingness to defend the Third Reich from its increasingly powerful enemies. In their own way, they contributed to the atmosphere of increasing desperation and radicalization that characterized the last year of the Nazi regime.[4]

CRISIS AND REORGANIZATION: FROM CONFLICT TO COOPERATION

The years 1944 and 1945 were tumultuous ones for the group of Peenemünders who remained on Usedom, and they struggled on many fronts to see their operation through to completion. Administrators had to cope with the departure of thousands of experts to Mittelwerk and elsewhere. Those who stayed behind lived in fear of more air raids like the one of August 17–18, 1943. Their fear was justified. U.S. Army Air Forces struck Peenemünde three times in the summer of 1944.[5] In addition, technical

---

[2] See esp. Albert Speer, *Infiltration*, trans. by Joachim Neugroschel (New York: Macmillan, 1981).

[3] Such claims made by the Peenemünders also fail to stand up to the historical record. Michael Neufeld's examination of this period in the program makes clear that there were indeed serious conflicts at the highest levels of the Reich over administering the program. At the same time, however, he notes that the Peenemünde developers themselves continued their work on a number of radically advanced projects that were not affected in any way by the SS. His approach, which focuses explicitly on the actual technological work that went forward, indicates that the Peenemünders continued to do their duty. However, it does not help to explain the level of the Peenemünders' commitment to completing their duty.

[4] Karl-Heinz Ludwig, in his seminal work *Technik und Ingenieure im Dritten Reich* (Düsseldorf: Droste Verlag, 1974), coined the term "self-mobilization" (*Selbstmobilisierung*) to describe the activities of technical specialists in the years of Hitler's regime. Self-mobilization, as Ludwig conceives it, was the willing involvement of individuals who went far beyond the call of duty to advance the objectives of the regime. This term is particularly well suited to understanding the commitment of the Peenemünders during the entire war, but it is particularly appropriate to their situation between 1943 and 1945.

[5] Michael Neufeld, *The Rocket and the Reich: Peenemünde and the Coming of the Ballistic Missile Era* (Cambridge, MA: Harvard University Press, 1995), 247–248.

problems with the missile were not ironed out in any acceptable way until late 1944. Finally, the program itself was buffeted by external conflicts between the Army, Armaments Ministry, and the SS over who would manage it, resulting in fundamental changes in the facility's administration and leadership.

A great deal of strain had been growing between the various organizations involved in the A-4 program in the first half of 1944. As the three powerful institutions all jostled to assume control of the program, the areas of authority for each organization became increasingly hazy. In practice, this meant that the first nine months of 1944 were marked by a substantial amount of personal and administrative friction at the highest levels of the program. However, the general state of cooperation at the middle and lower levels of the program attenuated this friction, and it had little impact on the activities occurring on the shop floor. The compromise solution struck at the highest levels – more through necessity than by desire – reflected the cooperative realities of the work going on at the level of middle and shop floor management. Despite the steadily worsening wartime pressure, by the beginning of autumn 1944, the three organizations had come to an uneasy truce just in time for the onset of mass operations against Allied targets in the West.

Heinrich Himmler's longtime desire to see the A-4 program put under his leadership was the catalyst to the conflicts of early 1944 and led to several dramatic battles over personnel. From the Peenemünders' perspective, the most worrisome of these was the arrest of Wernher von Braun, his brother Magnus (a chemist who arrived in Peenemünde in 1943), Klaus Riedel (the chief of ground equipment development for the A-4 and one of von Braun's colleagues from the *Raketenflugplatz*), and Helmut Gröttrup (Ernst Steinhoff's deputy and liaison to Dornberger). The documentary record of this incident is incomplete and unclear, but historians have pieced together what probably happened to the young development chief and his colleagues. In February 1944, von Braun, a major in the SS, received an order to report to Himmler's headquarters of Hochwald, near Grossgarten (Pozezdrze) in East Prussia. During their meeting, Himmler offered von Braun all of the resources at his disposal to speed up progress on the A-4. Sensing that Himmler was making a power play for the program, von Braun politely rebuffed the SS Chief. The development head was loyal to Dornberger, with whom he had worked for nearly all of his adult life, and felt little compunction to abandon Army control for the SS. His rejection of Himmler's offer forced the Reichsführer to try a more aggressive approach. A month later, in late March, Wernher von Braun and his brother, Riedel, and Gröttrup were awakened by late-night visits from Gestapo officials, who arrested them all, probably with Himmler's blessing. The Gestapo was supposedly tipped off by an informant who claimed to have overheard the developers discuss at a party earlier in the month that they were only interested in traveling to space, not in building weapons for the Nazis, and felt that Germany was bound to lose the war. These statements were, in no uncertain terms, construed as acts of high treason, even if they could not be corroborated by anyone else. All four men languished in a jail in Stettin (now Szczecin, Poland) for almost two weeks (Magnus was released a few days earlier) while Dornberger attempted to work through Field Marshall Wilhelm Keitel and Gestapo Chief Heinrich Müller to secure their release. Finally, using his connections to the *Abwehr*, he was able have the engineers released into his custody.[6]

After the war, von Braun's arrest proved to be a blessing for those Peenemünders who were in the United States. Many of them held it up as direct evidence that they fundamentally disagreed with the regime's motives and attempted to resist SS intrusions at all cost. This argument conveniently ignores the documentary record, which makes it clear that the program's administrators had earlier *sought out* cooperation with the SS. Their regular and meticulous deception about this was done to conceal any evidence of their own willingness to tolerate Himmler's organization and serve their own interests. It is true that von Braun had a long history of fascination with civilian space travel and that he probably engaged in surreptitious idle chatter with his colleagues at Peenemünde about the idea. However, there is no evidence that he failed to put forth his best effort to build a missile that could be used in wartime. Throughout his time at Peenemünde, he made it clear to his subordinates that he expected their utmost effort to make their instrument work reliably (including adding measures to improve its range and accuracy). At least in his written correspondence, he betrayed no sign of being troubled by the introduction of slave labor to the program in the early summer of 1943, and he was long an advocate of the use of forced labor to help development and production achieve its goals.

Doubtlessly, his arrest caused a great deal of worry for von Braun and his colleagues at Peenemünde. According to Dornberger, they were all aware that they were arrested because of reports filed by Gestapo informants in Zinnowitz.[7] The engineers' brief incarceration made it even more readily apparent that the totalitarian state had an eye fixed on their activities and was not afraid to exercise its arbitrary power, even if it meant arresting the program's most important figure on questionable charges. As with the Zanssen affair, it became plain to the Peenemünders that the Gestapo could exercise its power any time and anywhere it chose, and that none of them were safe from its reach. The sense of being watched at all times could only have increased. Nearly all of von Braun's activities for the rest of the war, and indeed those of all of the Peenemünders, must be viewed in this light.

The administrative threat from the SS did not abate after Dornberger secured the engineers' release in early April. Himmler, confident that his

---

[6] For a complete account, see Neufeld, *The Rocket and the Reich*, 213–220.
[7] Dornberger, *V-2*, 207.

ideologically motivated organization was the best equipped to deal with the growing threat to the Reich, continued to expand his administrative empire and had his sites set squarely on the missile program. The presence of the Army, Armaments Ministry, and the SS, with their multiple, intersecting responsibilities for administering different portions of the program, created fertile grounds for confusion and conflict at its uppermost levels, and Himmler's rapacious empire building knew few limitations. Even before his March meeting with von Braun, Himmler had authorized Kammler to subvert Dornberger's control of the program. When the program began its shift to mass production the previous autumn, it appeared to Reich officials that military operations were imminent. The Army bestowed the title of *BzbV Heer* (Army Commissioner for Special Tasks) on Dornberger and gave him overall command of the future operational missile batteries. This arrangement meant that Dornberger had to give up his formal control at Peenemünde, and Zanssen returned to take his spot.[8]

But Dornberger's troubles were just beginning. By December 1943, the commander of a new interservice missile corps, an organization that was ordered into creation by Hitler and given responsibility for firing V-1 cruise missiles and the A-4, forced Dornberger from his command of the operational batteries.[9] Dornberger retained the title of *BzbV Heer*, but his influence was weakening. At the end of May 1944, he attempted to reassert overall operational command of the program, but failed.[10] Dornberger's memoir blames Kammler for quietly and systematically working behind the scenes to undermine his influence and expand SS power.[11]

Dornberger resented the administrative tug of war, but he accepted the necessity of cooperating with the SS and Armaments Ministry. In early May, he attended a meeting in Mittelwerk of key program administrators, including von Braun, Sawatzki (as Kammler's representative), and Rickhey (as member of the Armaments Ministry). The group held a productive discussion of development, slave labor operations, and missile production goals that seemed to highlight the benefits of collaboration between the different branches. Dornberger, happy with the results of their discussions, called the meeting a "model" of cooperative effort, and he encouraged everyone present to continue their work in this light. "Not one against the other," he urged, "but everyone together!"[12] Less than a week later, Speer, who was just recovering from a pulmonary embolism that almost killed him, wrote a memo to high-ranking Army, SS, and Armaments Ministry officials that

[8] Entstehungsgeschichte der Fertigungsstelle Peenemünde, September 28, 1943, RH8/v.1210, *Bundesarchiv/Militärarchiv* (BA/MA).

[9] Neufeld, *The Rocket and the Reich*, 204.

[10] Dornberger to Fromm, May 31, 1944, RH8/v.3730, BA/MA.

[11] Dornberger, *V-2*, 210–211.

[12] Niederschrift über die Besprechung am 6.5.44 im Büro Generaldirektor Rickhey, May 6, 1944, FE 694/b, National Air and Space Museum (NASM).

supported Dornberger's position by suggesting clearly defined boundaries on the activities of the rival institutions battling for control of the program. He placed technical research, development, and testing in the hands of Army Ordnance, where they had been since the program's inception. The A-4 Special Committee, which was responsible to the Armaments Ministry, would be responsible for managing production contracts and guaranteeing production. Mittelwerk GmbH, also nominally an Armaments Ministry organization, was to continue managing actual production. Finally, the SS was left with responsibility for the expansion of the industrial facilities under Kohnstein and for the ongoing construction projects in Sperrgebiet Mittelbau, as well as supplying the labor for the entire production effort.[13] The memo could not have pleased Himmler or Kammler, but it seemed to be an arrangement that everyone else involved in the program could live with.

Speer's influence had declined considerably since his hospitalization, and other events intervened that allowed the SS to assume greater control of the program. After the July 20, 1944 attempt on Hitler's life, in which several high-ranking Army officials – but no one who was directly involved in the A-4 program – had taken part, Himmler and Kammler again made their move. Hitler had scarcely any trust left in the officer corps, and he had A-4 supporter General Friederich Fromm, Commander-in-Chief of the Reserve Army, arrested, jailed, and later shot, allegedly for cowardice, but really because Fromm failed to distance himself from the conspirators during the attempted coup. Dornberger lost one of his most influential patrons. Himmler, who took over Fromm's command of the Reserve Army, placed Kammler in command of accelerating A-4 deployments. Dornberger was virtually left out in the cold, serving in only a supervisory capacity for technical improvements.[14] This was all too much for him, and he considered a transfer to a command at the front, but von Braun and Ernst Steinhoff talked him out of it.[15]

In order to stave off the full control of the SS, the Armaments Ministry officially reorganized civilian Peenemünde development into a state-owned private corporation and christened it *Elektromechanische Werk, GmbH* [Electromechanical Industries – EW]. The idea had been circulating since the early summer of 1944, but the July plot made it a necessity if Peenemünde was to remain nominally independent of the SS. The base was to be dedicated specifically to development and testing. It would serve as a research and training center for various state and industry projects relating to

---

[13] Speer to Leeb, Kammler, Degenkolb, Rickhey, and Mittelwerk Board of Directors, May 12, 1944, R121, Bd. 405, *Bundesarchiv-Lichterfelde* (BAL).

[14] Neufeld, *The Rocket and the Reich*, 239–240. Hitler was furious with Fromm's peremptory action in executing the July conspirators. Fromm was executed in March 1945. See Ian Kershaw, *Hitler, 1936–1945: Nemesis* (New York: Norton, 2001), 680–682, 689.

[15] Dornberger, *V-2*, 236–237.

rocketry, and its primary task was to continue the development of reliable, mass-produced missiles. The day-to-day business activities, such as purchasing and financing, would be managed by the state.[16] This transition officially took place on August 1 and proceeded smoothly, even if it set up a clumsy managerial situation.[17] Essentially, the company was government owned and staffed. The EW operated, but did not own, the equipment at Peenemünde. Rather, the Army, which administered the base's facilities, owned the equipment. The Army also continued its transportation, maintenance, and security tasks. At the middle and lower levels, the technical organization went largely unchanged, but at the top, development and testing were no longer exclusively under von Braun's direct control.[18] The EW Chief Executive, technically von Braun's boss, was Paul Storch, a Siemens engineer who had previously served as the head of the Subcommittee for Electrical Equipment on the A-4 Special Committee. Storch was an advocate of slave labor to help solve the problems in the A-4 production program and had few major problems cooperating with either the Army or SS bureaucracy.[19] By the end of August, he helped oversee just over 4,000 German employees, both civilian and Army, who staffed the base on Usedom.[20] Though this arrangement was somewhat ungainly, the actual practice of work at Peenemünde changed very little.[21]

The transition to state-owned private industry succeeded in limiting SS influence at Peenemünde, but in the late summer and early autumn of 1944, Dornberger still had to fight Kammler for his professional life. The SS general had been attempting to fully isolate Dornberger by seizing direct control of the newly formed operational missile batteries and cutting him out of any administrative decisions. In the end, however, Kammler overreached, receiving an upbraiding from SS Lieutenant-General (*Obergruppenführer*) Hans Jüttner, Fromm's replacement as Commander-in-Chief of the Reserve Army, for his shabby treatment of Dornberger, who was a longtime Hitler loyalist. The Army General's long service and expertise made him too important of a figure to be isolated from the program, and Kammler compromised by placing Dornberger in charge of training and equipping the new missile troops.[22]

---

[16] Die Aufgaben der Elektromechanischen Werke G.m.b.H, June 28, 1944, RH8/v.1960, BA/MA.

[17] Eberhard Rees Oral History Interview (OHI), NASM.

[18] Dieter Huzel, *From Peenemünde to Canaveral*, 106.

[19] In May 1944, Storch had solved the bottleneck in the production of electrical instruments by bringing slave labor to bear. He also introduced the idea of enslaving the French workers engaged in rudder production by transferring them to Mittelwerk. Niederschrift über die Besprechung am 6.5.44 im Büro Generaldirektor Rickhey, May 6, 1944, FE 694/b, NASM.

[20] Storch to Kammler, August 21, 1944, FE 692/f, NASM.

[21] Huzel, *From Peenemünde to Canaveral*, 106.

[22] For a full description of Kammler's efforts to seize control of the program from Dornberger, see Neufeld, *The Rocket and the Reich*, 241–246.

The personal acrimony that no doubt existed between Kammler and Dornberger as a result of Kammler's aggressive moves was a reflection of the general friction between their two organizations. However, their personal distaste for each other faded in the face of larger technical and wartime considerations. Their compromise solution, which also allowed Dornberger to act as an advisor on missile-related issues, functioned relatively well under the circumstances. Kammler could not manage the missile program without the willing participation and expertise of its members, expertise that Dornberger and others possessed. The two men could at least see eye to eye on the fact that speeding the missile into operations was critically important. Their common professional backgrounds also meant that once they got past their personal enmity, they could at least come to a mutual understanding on technological and bureaucratic issues. In November, Dornberger drew up a set of proposals that articulated the coordination of the roles of personnel in the various bureaucracies. His goal was, as he put it, "the smooth cooperation of all offices in the military and civilian sectors in order to achieve the best possible result without consideration of questions about prestige or competency." Dornberger made it clear that he accepted – however grudgingly – Kammler's leadership "in making decisions about the fundamental questions regarding the A-4." He immediately followed this by placing himself as Kammler's deputy "in all A-4 matters." Dornberger's tasks, as he outlined them, would be to coordinate the work of the civilian and military offices in the program. If they could not come to an agreement under his supervision, then he consented to seeking out Kammler's authority for a final decision. Dornberger, however, would be the day-to-day arbiter for the program. He would work with Army Ordnance, the EW, and Mittelwerk to ensure their smooth functioning and coordination, as well as guide the training of the missile battalions. All technical questions were to be referred to him, and he would delegate them accordingly.[23]

Kammler had already agreed in practice, if not on paper, to Dornberger's ideas. He deferred many of the technical decisions to Dornberger and tended to follow the General's suggestions regarding manpower questions. For example, when Dornberger opined to Kammler at the end of August that the Raderach test facility, near Luftschiffbau Zeppelin in Friederichshafen, had a staff of civilian specialists that was far larger than necessary, Kammler immediately agreed. He ordered, "in agreement with Dr. Dornberger," that the staff should be trimmed and many sent to work at Peenemünde.[24] This type of arrangement between the two Generals, with Dornberger the key administrator and Kammler the individual who had final say in all decisions, was, if not totally satisfactory to either one, at least serviceable. In

---

[23] Walter Dornberger, *Abgrenzung der Arbeitsgebiete und Verantwortlichkeiten auf dem Gebiet des A4-Programmes*, November 11, 1944, RH8/v.1265, BA/MA.

[24] Von Braun to Justrow, August 29, 1944, RH8/v.1960, BA/MA.

December, it was officially agreed upon, formalized, and confirmed by the SS.[25] In the face of organizational conflict and personal abhorrence, the two men managed to form a consensus. This arrangement mirrored the situation at other levels of the program's administrative hierarchy.

Von Braun, for example, had struck up a suitable working relationship with production managers and slave labor officials well before Dornberger and Kammler found themselves at loggerheads in 1944 over who would be the program's ultimate decision maker. The technical director began involving himself in determinations about handling slave labor just after the British bombing raid in 1943, when he discussed evacuating production to sites in the Saar region.[26] As 1943 wore on, he became more and more intimately involved in the planning and deployment of concentration camp labor. In November, as more and more skilled civilian labor was moving from development and test sites into Mittelwerk, von Braun sent a letter to Degenkolb suggesting that slave laborers be substituted for German labor at the test facilities at Lehesten and Redl-Zipf. He thought that a ratio of two prisoners for every one German would be appropriate to operate the facilities.[27] Von Braun did not concern himself with the moral questions around the use of slave labor, only with the issue of how to solve the program's problems. Slave labor was just another means to an end. This mindset helped bring him into agreement with more ideologically motivated managers of the project who also were deeply concerned, for different reasons, with the missile's success. Of course, it is asking too much of him to stand up in protest of National Socialist labor policies, but his behavior throughout 1943 and 1944 reveals a relentless pattern of narrow-minded self-interest and technocratic thinking, which in practice lent the use of concentration camp labor a certain legitimacy by embracing it as a viable solution to manpower problems. Another option open to von Braun would have been to do nothing at all, even to delay or equivocate, but he cared too much about the success of his work to jeopardize its progress by making even a quiet moral stand against slave labor.

Von Braun involved himself in several important decisions about slave labor, even at Mittelwerk, later in 1943 and in 1944. He visited the factory itself in August and October 1943, as well as in January and May 1944.[28] The January meeting offers an important example of how the need for further technological development outlined by Peenemünders could lead

---

[25] Walter Dornberger, Aktenvermerk über die Besprechung am 24. Oktober 1944 bei Generallt. d. Waffen-SS Kammler in Berlin, October 25, 1944; Dornberger, Aktenvermerk über die Besprechung am 3.11.44 bei Generallt. d. Waffen-SS Kammler in Berlin, November 4, 1944, RH8/v.1960, BA/MA. Jüttner to Kammler, December 31, 1944, RH8/v.1265, BA/MA.

[26] Neufeld, *The Rocket and the Reich*, 202.

[27] Von Braun to Degenkolb, November 12, 1943, FE 732, NASM.

[28] Michael Neufeld, "Wernher von Braun, the SS, and Concentration Camp Labor: Questions of Moral, Political, and Criminal Responsibility," *German Studies Review* 25 (2002), 65.

to the increased demand for concentration camp prisoners to work in the factory tunnels. One of the technical details that became clear earlier in the year was that the servomotor assembly, which helped direct the thrust to steer the missile, needed to be strengthened. After Peenemünde engineers developed a method to manufacture improved versions, von Braun presented this information to Rudolph at a conference in Mittelwerk. Accordingly, Rudolph set aside more workspace in the factory to install the necessary machinery. After a short delay for unknown reasons, Rudolph informed von Braun that the work would begin in January 1944. He reported that "the necessary prisoners [to work in the transport kommando] and guards have been ordered from KL Dora."[29] Rudolph anticipated that the arduous transport and setup tasks would take a total of three months to complete.

In May – after his arrest – von Braun attended a meeting at the factory (among others present were Dornberger and Rudolph) in which Sawatzki informed them that he would request an additional 1,800 prisoners for tunnel work to replace those lost during the winter of 1943 to 1944.[30] Von Braun behaved guardedly at this meeting and said little, but his drive returned to full form by August. That month, he wrote to Sawatzki about a French physics professor, Charles Sadron, who was a prisoner in Buchenwald and whom von Braun hoped to bring to Mittelwerk. Von Braun had actually traveled to the camp himself in order to evaluate the skilled labor there, were he met Sadron. While at Buchenwald, von Braun informed Sawatzki that he had arranged for the transport of prisoner labor to Mittelwerk. He also requested that Sadron be given special privileges in Mittelwerk, such as permission to wear civilian clothing to encourage his willingness to perform the necessary work, as von Braun put it.[31] Von Braun, who spoke perfect French, may indeed have identified with the physics expert and had every good intention in attempting to secure some level of humanity for Sadron. However, it was also clear that Sadron possessed certain skills that would help push the program forward, and von Braun recognized this. Von Braun's request to Sawatzki was not based on purely humanitarian considerations, then; utilitarian motives also played an important part.[32] In any case, von Braun's actions show that he had come to a willing acceptance of slave labor and an agreement on the importance of the SS as a labor supplier.

[29] Rudolph to von Braun, January 21, 1944, FE 694/a, NASM.
[30] Niederschrift über die Besprechung am 6.5.44 im Büro Generaldirektor Rickhey, FE 694/b, NASM. See also Neufeld, "Wernher von Braun," 66.
[31] Von Braun to Sawatzki, August 15, 1944, FE 694/a, NASM. See also André Sellier, *A History of the Dora Camp: The Story of the Nazi Slave Labor Camp That Secretly Manufactured V-2 Rockets* (Chicago: Dee, 2003), 105–106.
[32] Neufeld has also noted that if von Braun had arranged transports of slave labor, this "would at least in theory put him in violation of the Nuremberg standard applied to Albert Speer." See Neufeld, "Wernher von Braun," 69.

In accepting their role and their own places in relationship to the SS, Dornberger and von Braun reflected the more readily achieved arrangement between Peenemünde transfers to Mittelwerk and Himmler's representatives. Dornberger's position was impacted much more forcefully and directly by the growing strength of the SS, but he eventually settled into an uneasy but sensible relationship with Kammler. Von Braun's work was affected less by the SS, and his narrow, self-interested technical vision enabled him to readily adapt to the challenges and benefits of the strong presence of the SS in the missile program. Von Braun had no love for Kammler, but his relationship with other elements of the SS was cemented before Kammler's turf war with Dornberger. It survived these stresses because of von Braun's extraordinary dedication, but also because of his precarious political position after his arrest in March 1944. Both men, like all of the Peenemünders beneath them, continued to work exceptionally hard on behalf of the Nazi regime.

ADVANCES AT PEENEMÜNDE, 1944–1945

Despite the challenges and conflicts posed by the intervention of the SS in the program, missile specialists remained committed to the success of their work and defense of their nation. In late 1943 and 1944, when concentration camp prisoners were dying by the thousands in the service of their work, Peenemünders remained committed to the goals of their project. The program's managers, engineers, technicians, and other specialists all worked to iron out seemingly intransigent technical problems, searched for improvements that would boost performance, and thought creatively about new ways to exploit their technology. This was a period of phenomenal technological ingenuity, as the areas of fuel consumption, accuracy, range, speed, destructive capability, and even the effort to limit the consumption of raw materials experienced major theoretical, if not concrete, advancements. Like many armaments specialists in Germany, the Peenemünders worked desperately hard until circumstances forced them to shut down their work in the last months of the war. Their behavior is a powerful indicator of the hold that Peenemünde's central mission held over them.

   Their attitudes toward work in this period reflected a confrontation with the realities of war that had not existed before the August 1943 bombing raid. A year after the British raid, the American Eighth Air Force attacked Peenemünde on three separate occasions, on July 18, August 2, and August 25, heavily damaging test stands and killing a few dozen people.[33] No longer were the Peenemünders blissfully secluded from the war's effects. Many feared for their safety, as the bombing raids jarred their sense of isolation and security. The increasing shortage of raw materials slowed the pace

---

[33] Neufeld, *The Rocket and the Reich*, 247–248.

of development and production, never mind the construction of necessary bomb shelters and the repair of important buildings. Bombing raids became a source of frequent discussion and more than a little anxiety among the employees.[34] A sense of urgency began to pervade the station, and many Peenemünders naturally responded to the increased privation with anger and even a renewed dedication. Paul Figge, a production specialist with a number of ties to the personnel at Peenemünde, stated after the war, with only some exaggeration, that "The bombings hardly affected progress on the A-4 program, because our enthusiasm still remained high to accomplish the goal. So actually, the more difficult the conditions became, the more the enthusiasm grew to finish what we had begun."[35] Figge's comment overstates the attitudes of most Peenemünders ("enthusiastic" was probably not how they would have described themselves after suffering multiple bombing raids), but the employees at the base certainly were determined to complete their work in the face of enormous difficulties. Their impressive efforts over the last year of the war bear this out.

After Dornberger secured his freedom, von Braun tackled his work with nearly limitless energy. He needed all of his technical know-how and administrative expertise because developers at Peenemünde made thousands of changes to the A-4's design over the course of 1944. Not all of the missile's technical bugs had been worked out before mass production began at Mittelwerk, and during the spring and summer of 1944, developers struggled to solve a number of problems that still plagued the completed assemblies. Nevertheless, 1944 proved to be an extraordinarily innovative year in the missile program. Even late into the year, Peenemünde specialists systematically and painstakingly made great efforts to improve the performance of the missile. While overcoming the lingering technical difficulties such as faulty launches and the propensity of the missile to break up during reentry (so-called air bursts), developers also sought improvements in the A-4's value as a weapon.[36]

This effort took place on two fronts. On one hand, engineers and scientists continually honed and modified the missile's design in order to improve its range, accuracy, and destructive power, all while attempting to curb the massive consumption of raw materials involved in its production. On the other hand, they also attempted to increase its operational flexibility, designing different methods of deployment that ran from the practical to the fanciful. Even though Kammler continually urged the development specialists

---

[34] Huzel, *From Peenemünde to Canaveral*, 110.

[35] Quoted in Donald E. Tartar, "Peenemünde and Los Alamos: Two Studies," *History of Technology* 14 (1992), 163.

[36] Air bursts resulted when the missile's metal skin heated up and tore off during reentry, causing the missile to break up or explode in flight. See Neufeld, *The Rocket and the Reich*, 220–230.

to make advancements on the A-4, particularly its range, they needed no coercion from the SS, OKH, or OKW to press on with improvements in the missile itself or the equipment associated with it. Peenemünders took the initiative themselves in order to continue its development and utility as a weapon system.

The spring of 1944 through the winter of 1945 was an intense period of testing parts and assemblies manufactured by subsidiary firms as well as mass-produced missiles that came out of Mittelwerk. Launch problems had mostly been solved by the spring of 1944, but missiles continued to go awry or break up in flight. Guide-beam receptors, turbopumps, electrical systems, valves, fittings, tail assemblies, and steering machinery all went through extensive testing and modification.[37] Test engineers at Peenemünde carried out over sixty launches between the end of August and December 1944. Their efforts resulted in improvements in instrumentation and guidance capabilities, and they also led them to the possibility of attempting to use alternate fuel in the missile.[38] Tinkering with the missile was a constant process. Although the A-4 was seeing heavy military use by late 1944, in the Peenemünders' view, it was far from a perfected weapon. Between December 1944 and January 1945, missile specialists were making design changes as much as three times per day.[39] All of these changes began to pile up, and by the end of the war, Peenemünde engineers had made approximately 65,000 modifications to the A-4 design.[40] A postwar U.S. intelligence analysis described this as "A frantic hope to pull a guided missile rabbit out of the hat while losing its shirt."[41]

The missile was an incredibly complex and revolutionary piece of technology. It did not lend itself to easy transfer from experimental production to serial production because the experimental designs were far too complicated for mass production purposes. Alterations were also not easily incorporated into work on the assembly line. The process for making changes in design details was extraordinarily confused in late 1943 and well into 1944, and there were no coherent processes for making modifications either in subsidiary firms or in general assembly. Development workers, many of whom were unfamiliar with the demands inherent in switching from experimental production to mass production, who labored under great pressure, and who were also eager to complete the work, were contacting subsidiary

[37] See the collection of test reports, April to June 1944, FE 695.
[38] See the collection of test reports, August to December 1944, FE 723, NASM.
[39] Bauanweisungen für A-4 Gerät 12/W, Dez. '44 – Jan. '45, GD638.0.18, Deutsches Museum (DM).
[40] "Production and Distribution of German A-4 (V-2) Rockets," Box 772, Entry 1, RG 18, Records of the Army Air Forces, Bulky Decimal File, 1946–1947, National Archives and Records Administration (NARA). See also Neufeld, *The Rocket and the Reich*, 224.
[41] "Facilities for Guided Missiles Research and Development, Volume II," Box 553, Entry 1, RG 18, Records of the Army Air Forces, NARA.

firms and ordering changes to parts without informing production engineers.[42] Changes ordered by developers were coming at such a pace that many subsidiary firms had trouble meeting orders for new parts. Worse, when new parts from subsidiary firms arrived at Mittelwerk, they were sometimes incompatible with each other or the larger assemblies because of failed coordination between the three groups.[43] Making matters worse was the fact that throughout the early and middle part of 1944, efforts to design simplified parts were often unsuccessful because, according to one explanation, the development workers had unique skills that came from years of missile work. Employees of subsidiary firms, who did not have as much missile development experience, had a great deal of trouble matching the Peenemünders' skill.[44] The effort to manufacture the A-4 was becoming chaotic.

Von Braun thrust himself squarely into this fray in an effort to coordinate the frenetic development activity with mass production, a task that became more complicated with every change made to the missile's design. He worked hard to increase efficiency in development, in the manufacture of new batch runs, and in the delivery of the proper materials to various factories as well as the Mittelwerk. Von Braun also traveled to many subsidiary firms to examine production, harangued other engineers about the best way to go about making changes in production drawings, and tried to improve and standardize the ways in which developers, engineers in subsidiary firms, and production people communicated.[45] He expressly forbade development people from making changes to parts without first making corresponding changes in production drawings and sending them to his office for approval

[42] Lindenberg to Riedel III, May 14, 1944, FE 732, NASM.
[43] Friederich to Neuhaus, August 1, 1944, FE 694/a, NASM.
[44] Konrad Dannenberg OHI, NASM. According to Dannenberg, the reason for this lay in the unique skills that many Peenemünders picked up on the job. During the research and development stages, technical specialists used a number of insider's tricks to develop something completely new. There was no practical way to easily transfer these informal changes to the impersonal, formal production drawings. This was in fact a fundamental problem with the Peenemünders, who in many ways were akin to master craftsmen in that they incessantly tinkered with parts that they were familiar with. Such an approach causes fundamental problems in a mass production environment.
[45] Part of the effort at standardization consisted of forms, developed by von Braun's office, that proposed changes in parts that were manufactured at subsidiary firms. The part to be changed would be assigned a particular cataloging number and the priority grade of the part would be indicated on the form. In addition, the new form also had a box in which the firm, work group, and engineer who proposed the changes would be indicated, providing a measure of accountability for the work and facilitating communication between the correct people. Finally, a part of the form would provide a space for a detailed explanation of the reason for the change, the importance of the change, and when it could be ready for production. See a large set of these forms in FE 732, NASM.

before they went to the subsidiary firms.[46] If parts producers failed to keep up with the necessary changes, Storch and von Braun interceded forcefully to bring them back into line and remained involved until they were sure that the firm could make timely deliveries.[47] Von Braun also directly intervened on a number of occasions in order to quickly cut through bureaucratic Gordian knots, and he sent his representatives to problematic firms in order to ensure compliance with his directives. The Heinkel factory in the Tyrol, for example, manufactured turbopumps and other specialized hardware for the missile's internal assembly. Most, von Braun discovered, were very poorly constructed. He wrote a stern letter to the factory, complaining that "The output of your firm is simply not useable." To clear up this problem, von Braun informed Heinkel's plant manager that he was sending a deputy to the factory and ordered the Heinkel manager to support his representative's efforts "with all means possible."[48] He meant to bring this problem under control quickly and had no qualms about stepping on the toes of others to do it.

Von Braun's arrest, while no doubt forcing him to speak and act carefully, did little to dampen his enthusiasm for his work. The development chief used his authority to directly and effectively intervene in important parts production issues. He deployed his considerable administrative muscle to sort out the myriad of problems that were caused both by an immature weapon system that was rushed into mass production and an unwieldy administrative system that was not ready to handle the burdens that came with such a rash move. His and others' efforts to coordinate the activities of the developers with those who manufactured the missile paid off handsomely by September 1944.

In that month, Mittelwerk began churning out missiles at the rate of 600 to 700 per month through March 1945.[49] The developers' activity, however, did not center merely on ensuring that the missile functioned at a basic standard of performance. Because of their unique expertise, Peenemünde specialists were also central figures in the process of training the first cadres of missile troops who would conduct operations against Allied targets. Peenemünders wrote the specific handling and transportation instructions for the new troops and responded to inquiries from members of the military involved with support and supply activities for the missile battalions.[50]

---

[46] Von Braun Circular, July 10, 1944, FE 694/a, NASM; Von Braun to Steinhoff, August 15, 1944, FE 694/a, NASM.

[47] Huzel to Friederich, July 11, 1944, FE 694/a, NASM.

[48] Von Braun to Bäderich, April 1, 1944, FE 732, NASM.

[49] Neufeld, *The Rocket and the Reich*, 213.

[50] Vorläufige Transportvorschrift für das Gerät A4, December 1943, GD 639.4.8, DM. Beantwortung der Fragebogen für: Feldspeicherpersonal, FR-Gefechtstaffel, Treibstoffkolonne, January 12, 1944, RH8/v.1265, BA/MA.

When asked by operations officers about night launches, for example, the engineers held experimental shots at night to see if lights used by artillery-men could substitute for natural light. The results were good, and the test engineers recommended outfitting the launch batteries with lights for 'round the clock operations.[51] Earlier that summer, VkN members began training some missile troops at Peenemünde. Others were also a part of the teams of so-called technical storm troops that operated with the batteries during early firing operations. V-1 operations, which began in June, showed that the troops firing the cruise missiles were not prepared to deal with technical difficulties as they came up at the front. Dornberger first ordered missile specialists to assist the troops in July in an effort to avoid a similar prob-lem with A-4 (dubbed "V-2" by Goebbels in November 1944) operations. Dornberger ordered that "as many expert engineers as possible" were to be sent to the firing positions. He specified that the engineers sent to the launch sites needed to be thoroughly familiar with the missile, especially with its on-board electronics and steering, engine operation, and ground support equipment. They should also, according to Dornberger, have a "can do" attitude.[52] The Peenemünde engineers were obviously perfectly suited to the task. Dornberger meant to draw his storm troops from specialists at Use-dom, in the Ordnance division, and at Mittelwerk. By the middle of August, he was beginning to assemble a list of both civilian and military personnel from Peenemünde and Mittelwerk who were to serve with the launch bat-talions.[53] Peenemünde personnel went on to work successfully in their tasks of training and accompanying the launching troops throughout 1944 and 1945.[54]

In the course of their work, Peenemünders also suggested changes that went beyond establishing a minimally reliable level of functionality and actually improved upon the finished product. They did not merely allow themselves to be carried along by the inertia of their project, nor did they need orders to motivate their work. Rather, they drove the work forward in a way that they hoped would offer a solution to Germany's worsening military bind. Though V-2 operations began in earnest in September 1944, the Peenemünde engineers never stopped tinkering with the missile. Much of the work in late 1944 and 1945 pushed the theoretical and practical boundaries of rocket engineering into entirely new territory. Some of the ideas were wholly fanciful and not technologically feasible at that juncture, but the point remains that the Peenemünders' imagination, creativity, and

[51] Arbeiten am Gerät A4 bei Nacht, January 19, 1944, RH8/v.1265, BA/MA.
[52] Walter Dornberger, "Technische Stosstrupps," July 21, 1944, RH8/v.1960, BA/MA.
[53] Magnus von Braun to Wa Prüf 10, August 10, 1944, RH8/v.1960, BA/MA.
[54] Heinz Krause Statement, Gericht Rep. 299, Nr. 588, Nordrhein-Wesfälisches *Hauptstaat-sarchiv Düsseldorf, Zweigarchiv Schloss Kalkum* (HStaD-ZA Kalkum); Leonhard Specht Statement, Gericht Rep. 299, Nr. 588, HStaD-ZA Kalkum.

work ethic were on full display between the summer of 1944 and the winter of 1945.

Individual engineers and their teams worked on specific packages of problems based on their areas of expertise. This not only led to a strong familiarity with the issues involved but also helped make the mechanical difficulties into a very personal problem for the engineers. Because engineers at the base became so thoroughly familiar with the design and operation of particular sections of the weapon, they were often able to come up with a variety of ways in which to improve the missile's performance. Except for general inquiries regarding the range and power of the missile, regime authorities did not dictate the direction of research on the V-2 in this period. Instead, specialists at Peenemünde took the initiative and guided the research in directions that they felt were militarily the most appropriate. Peenemünde administrators had long encouraged them to think flexibly about problems and improvements and to bring any potential solutions to their superiors. This they did, and those who contributed often and in key areas were handsomely rewarded. This, of course, meant that senior administrators such as von Braun, who already found himself inundated with work, had even more tasks to perform and projects to guide. Senior-level Peenemünders remained fully prepared to dedicate themselves to the war effort. Their subordinates were also ready to do their part, and many came forward with a number of imaginative advancements.

There is no doubt that developing technology with clear military applications was important to the employees at Peenemünde. Many of the experiments they carried out in the last twelve months of the war betrayed a need to produce a more militarily effective weapon. Such a weapon was exactly what the increasingly desperate regime wanted. In May 1944, Hitler inquired to Speer about the possibility of increasing the explosive effect of the missile by using liquid nitrogen in the warhead (a technical absurdity, given that nitrogen is not combustible under most circumstances).[55] The question did not trickle down to the shop floor at Peenemünde, but specialists on Usedom did conduct important experiments – again, with little prodding from regime authorities – in order to seek the same increase in explosive capability. For example, throughout 1944, engineers at Peenemünde and the Heidelager test range carried out a series of different experimental detonations of the missile with hollow charges mounted outside the engine block. The tests showed that it was possible to exploit the liquid oxygen and fuel that was unconsumed after engine cutoff to increase the effect of the detonation. They accomplished this by using a shaped charge with a hollow cavity that directed the explosion into the engine and ignited the highly flammable leftover fuels. Engineers completed the testing and reported their

---

[55] Speer Desk Calendar, Roll 192, T-73, NARA.

findings to Kammler, Dornberger, and Army Ordnance later in the year.[56] Though these test results and others like them arrived too late in the war to see their implementation in mass production, they are indicative of the Peenemünders' self-mobilization in armaments development during the closing year of the war.

One way to understand the quality and amount of inventive work that went on in this period is to examine the issue of patenting in Peenemünde. It serves as a key indicator of the specialists' willingness and efforts to advance the technology. In the first place, the effort to earn new patents on their technology reflected the Peenemünders' ongoing desire to maintain their professional ideals, even very late in the war. Second, patent applications filed at Peenemünde help elucidate the underlying technological concerns of the development engineers and also illustrate the rapid technological development that took place in the last years of the war. Not only the patent awards, but also the applications themselves, were an important way for engineers at Peenemünde to show their dedication and hard work done in the name of missile development. This was to become a key issue, as civilians at the base grew increasingly concerned about the status of their draft exemptions in the face of increased conscription into the Army and *Volkssturm* militia. Patenting activity, then, served a number of goals at once.

Technology policy in National Socialist Germany, ideologically reactionary in so many ways, was governed, ironically, by a rather liberal patent system. The Reich Patent Code, promulgated in 1936, borrowed extensively from liberal models that favored individual inventors over corporate interests. The new law was based on a smattering of ambiguous passages in *Mein Kampf* and eliminated the idea that corporations, with their excellent resources and deep pockets, were the fountainheads of invention. The new patent law instead placed individuals at the center of the inventing process. The law also forced corporations to grant appropriate compensation, calculated by the state, to individuals whose ideas were put to use. Refined in 1942, the law was a remarkable success, and for a brief period the number of patents filed in Germany outstripped those filed in the far more populous United States.[57]

The Peenemünders took full advantage of this progressive arrangement. Between 1939 and 1945, Peenemünders filed at least 114 patent applications

---

[56] Engineer Hoffe, "Sonderlaborierung im A4-Gerät mit hohlladungen (Sprengversuch)," December 17, 1945, GD 633.20.11, DM.

[57] Kees Gispen, *Poems in Steel: National Socialism and the Politics of Inventing From Weimar to Bonn* (New York: Berghahn Books, 2002), 177–250. Nazi rhetoric about the value of inventors and technology helped garner support among engineers and technicians in the interwar period. After the war, the Federal Republic removed the racially loaded terms from the law and made it the model for a new patent law in 1957. Gispen points out that, in this case, the regime left "a positive legacy . . . in the politics of inventing" (p. 8).

with the Reich Patent Office (*Reichspatentamt* – RPA). Out of these, the RPA recorded only sixteen applications between 1939 and 1941. After 1941, this number spiked dramatically, receding only in 1945. The increase in applications was geometric. In 1942, thirteen applications from Peenemünde appeared in the RPA. In 1943, this rose to twenty-six, then again to fifty-four in 1944. In 1945, with the abandonment of Peenemünde and the final defeat of Nazi Germany, the number of applications dropped to a mere five.[58]

These numbers reflect several factors. First, the state of missile technology before 1939 was quite primitive. Back in 1936, the Army patented the A-3 test model secretly, but it did not yet award individual developers. The next year, in 1937, the Army bought off Rudolph Nebel by paying him and Klaus Riedel RM 75,000 for a nearly worthless rocketry patent they applied for in 1931 and were awarded in 1936. The patent purchase also cleared the legal ground for Ordnance to continue its work. In any case, by the end of 1941, the specialists at Peenemünde had developed the basic technology that would be the key to the success of their endeavor. The most fundamental problems of missile development in guidance, propulsion, and aerodynamics had been solved.[59] Even so, the number of patent applications did not rise sharply until 1943. By the end of that year, as more instruments and parts began to show success, the likelihood that they could be patented increased.

The huge technical and scientific strides made between 1937 and 1941, and the relative paucity of patent applications in this period (twenty-one), indicates that technical maturity, though important, was not a central factor in the increased number of applications in the second half of the war. Another critical factor can be found in the rewards that inventors at Peenemünde received for successful patent submissions or, at least, the applicability of their inventions to missile technology. The Peenemünders collected handsome monetary compensation if their work was patented or used on the A-4. Shortly after July 1942, when the Reich government clarified the patent code to emphasize the rewards due to individual inventors, military patent evaluators developed a calculus for remunerating their inventors, which were sometimes small groups or teams of people. They evaluated the position of each individual in the team according to technical training, management position, and key contributions to come up with a performance assessment for individuals involved in the invention. Second, they assigned the patented invention or process a numerical value based on its applicability to work being done. More important were the final two factors in the calculation. Evaluators divided the usefulness of the invention

---

[58] These numbers are culled from the patent applications found in the Deutsches Museum's Peenemünde Archive Reports, largely a collection of scientific and technical documents that, until this point, was of greater use for scientists and engineers than historians. This series holds the most complete collection of purely technical documents that are available.

[59] Neufeld, *The Rocket and the Reich*, 73–109.

into categories ranging from "pure military application" through "equal military and other applications" to "predominantly for other applications." The final step categorized the invention or process with values ranging from "crucial military importance" to "little military importance." The greater the military application and the military importance of an invention, the higher numerical value it was assigned. Evaluators then multiplied all of these numbers together in order to come up with the proper remuneration for the inventor.[60] This method of evaluation put a premium on the military value of technology and encouraged individuals to continue to think creatively about how to improve the state of the *Wehrmacht*'s own technology. In an operation like that at Peenemünde, busy scientists and engineers stood to profit handsomely from such a system.

Monetary gain was not always a primary concern for all Peenemünders, however. On some occasions, they sought only recognition of their accomplishments and the professional satisfaction of a patent award. In May 1943, for example, the RPA awarded propulsion specialists Karl Neubauer and Friederich Wilhelm Dürre with a patent for their work on the cooling jacket for the missile's combustion chamber. Dürre did not want money for the use of his patented ideas, and he was satisfied simply to be named as inventor on the officially issued patent. In the end, Riedel III insisted that both men receive at least RM 150 for their contributions.[61] Other Peenemünders were similarly motivated by such professional technical considerations. Their work, however, was informed by the ever-present necessity of easing the transition to mass production and simplifying assembly, which simultaneously increased the pace at which functional missiles could be turned out at Mittelwerk. Konrad Dannenberg, in his application, which was also for improvements in the cooling jacket, explained that his improvement would ease manufacturing problems by simplifying the combustion chamber's design. "In this way," he wrote, "the means of cooling with only a single [alcohol] intake will make production easier. This applies for the assembly of the entire engine block. Furthermore, fewer possibilities for disturbances exist because of looseness at weldings and couplings. With this, a higher degree of operational reliability can be counted on."[62] Simplified manufacturing and improved technical performance were obvious concerns for men like Dannenberg. Professional motivation then,

---

[60] Formel für die Berechnung der Vergütung (für Wehrmacht u. Werhmacht-ähnliche Betriebe), GD 634.14.45, DM.

[61] Karl Neubauer and Friedrich Dürre Patentanmeldung [Patent Application], "Raketenofenkühlung," May 1, 1943, GD 624.193.3, DM.

[62] Konrad Dannenberg Patent Application, "Raketenofen-Kühlung," August 16, 1944, GD 624.193.2, DM. Alcohol burned at a lower temperature than the gasses in the combustion chamber, so a film of alcohol fuel along the wall of the combustion chamber would provide an insulating layer against the heat and prevent burn-throughs of the combustion chamber.

with the extra incentive provided by monetary rewards, was enough for most Peenemünders.

The missile specialists did not forget the purpose of their work. They directed much of their effort toward increasing the missile's destructiveness and improving its flight characteristics. A typical example is the prodigious Konrad Dannenberg's patent application for a "Process to Hinder Explosions."[63] In the summer of 1944, Peenemünders were bedeviled by the problem of air bursts as inbound missiles broke up upon reentry to the atmosphere. Dannenberg reasoned that when missiles reentered the atmosphere, the volatile fuels remaining in the tank exploded because of heat friction. His solution was to drain the remaining liquid oxygen into the fuel tank, and the liquid-oxygen/fuel mixture would become more stable by gelling. Dannenberg argued that his idea improved flight stability by moving the inbound missile's center of gravity toward the rear while decreasing the hypervolatility of the propellants. Importantly, he also held that this was a valuable advance because the fuel, although more stable, still increased the punch offered by remaining propellants upon impact detonation.[64] Von Braun and Eberhard Rees were thrilled with the idea and, after determining how they could incorporate Dannenberg's ideas, forwarded the application to the RPA.[65] Dannenberg was eventually notified that his process would be put to use and that he could expect compensation for it as soon as the patent was granted.[66] In addition to concerns about upholding engineering's professional standards, then, military considerations and the effort to help bring about victory were also of paramount importance for development engineers in the project. They were not only interested in spaceflight; rather, Peenemünde engineers took the initiative and made improvements in the missile's destructive capability themselves, enhancing the military effectiveness of the weapon.

---

[63] Dannenberg's background is a common one among the Peenemünders. Born in 1912 just south of Leipzig, his interest in rocketry first emerged after he saw Max Valier speak and he witnessed a demonstration of Opel's rocket cars in Hanover. He joined the Nazi party in 1932 and completed the work toward his Diploma-Engineer degree at the Technical University of Hanover in 1936. In 1939, he was drafted into the reserves, serving in France before being assigned as a civilian reservist with the VkN. See Konrad Dannenberg OHI, NASM. See also Joint Intelligence Objectives Agency (JIOA) Form. No. 1, Biographical and Professional Data, Exploitation of German and Austrian Scientists, Box 28, Entry 1B, RG 330, Records of the Secretary of Defense, JIOA Foreign Scientist Case Files, File "Dannenberg, Konrad," NARA.

[64] Konrad Dannenberg Patent Application, "Verfahren zur Verhindern von Explosionen," July 8, 1944, GD 634.11.12, DM. Dannenberg, incidentally, was half-right. Heat friction was the problem, but inbound missiles were exploding because the heat of reentry weakened the welds on the missile's skin, allowing the skin to tear away.

[65] Von Braun and Rees to the RPA, September 21, 1944, GD 634.11.12, DM.

[66] Schilling and Rees to Dannenberger, December 15, 1944, GD 634.11.12, DM. Because of its submission so late in the war, the RPA never made a decision on Dannenberg's application.

These military concerns resulted in a number of interesting and forward-thinking experiments, but given the state of missile technology and availability of resources in 1944 and 1945, many of their theoretical advances were difficult, if not impossible, to fulfill. One relatively simple concept they attempted was putting wings on an A-4 missile to expand its range, a concept first given attention in 1939. They eventually shelved the idea for this missile, with the code name A-9, because of cost and priority problems. Responding to pressure from military authorities in the middle of 1944, the Peenemünders revived the project and rechristened it the A-4b. By September and October, the Peenemünders had developed test missiles and were preparing launch experiments, which they carried out with very limited success in December 1944 and early January 1945. The project ended shortly thereafter.[67]

Military necessity partly drove this work, as Allied forces steadily regained ground in Western Europe throughout the summer, forcing the operational missile batteries to launch from greater and greater ranges. Wernher Dahm, a test engineer in the Projects Office at Peenemünde, also pointed to another concern that had long plagued the Peenemünders but became particularly acute at the end of 1944. In an interview given decades after the war, he admitted that the motivation for reviving the A-9 was to show that the Peenemünders were in fact attempting to make dramatic advancements in the missile's range. However, this was done in order to show that, as a group, missile developers were too important to be conscripted into the Army or the *Volkssturm*. According to Dahm, local authorities around Peenemünde had been clamoring in late 1944 to draft the specialists on Usedom, and employees at the base were becoming increasingly concerned about just such an event.[68] This is an entirely plausible argument and probably true, but it must also be seen in the larger context of the nearly manic pace of development and design taking place at Peenemünde in the second half of 1944.

Other projects reflect the Peenemünde specialists' strong motivation to create powerful weapons in the service of the Nazi state. Perhaps the second most important development project after the A-4 was the Wasserfall anti-aircraft missile, which had its origins in plans begun in 1941. The crash status of the ballistic missile project, however, meant that most of Peenemünde's resources were not dedicated to Wasserfall until the second half of the war. The Allied bombing campaign and slow but steady seizure of air superiority made Air Ministry officials increasingly desperate to develop a weapon to reverse Germany's losses in the air. Developers hoped to have it operational

[67] Heinz-Dieter Hölsken, *Die V-Waffen: Entstehung, Propaganda, Kriegseinsatz* (Stuttgart: Deutsche Verlags-Anstalt, 1984), 135–145, 200. See also Neufeld, *The Rocket and the Reich*, 251.

[68] Wernher Dahm OHI, NASM.

by May 1946.[69] Despite major efforts by engineers at Peenemünde East and West – it was von Braun's other major project of the war – Wasserfall failed because of extremely complex problems of fuel supply, guidance, and control.[70] As a long-term project, a surface-to-air missile was a sensible idea, but in a nation so desperately short of supplies, manpower, and time, there was no way for Wasserfall to be completed quickly. But the regime's fascination with war-winning technology had entrenched the project into the bureaucracy, and the specialists' dedication to its success meant that the project would continue despite its short-term uselessness.

Peenemünde developers also carried out smaller scale but equally important tasks in the last eighteen months of the war. Guidance, steering, and fuel-injection improvements for the A-4 all steadily emerged over the course of late 1944.[71] In addition, by the end of November, fuel was in drastically short supply across the Reich, a shortage possibly made even worse by the early preparations for Hitler's Ardennes offensive in the West. Development engineers took a number of steps to get around this deficiency. In one case, chemical engineers Tschinkel and Rössler came up with a process for using lignin as a dilutant in fuel. Lignin was a cheap and abundant by-product of the cellulose industry that, through various chemical processes, was fully soluble and burned fairly efficiently.[72] Their idea was never utilized on the A-4, but even so, it was a clear attempt to alleviate fuel shortages and make a substantial impact on Germany's precarious supply problem, especially as it impacted the missile. In addition, the problems created by fuel shortages mandated that a standard fuel combination be used for all new missile projects. By this point, Peenemünde engineers were directly or indirectly involved with several different anti-aircraft missile projects: Wasserfall, Taifun (designed at Peenemünde), *Rheintochter* (Rhine Maiden – a solid-fuel missile developed and manufactured by Rheinmetall-Borsig), *Schmetterling* (Butterfly – designed by Henschel), and Enzian (a wooden, unmanned, rocket-powered interceptor). All of these used different types or combinations of fuel. Peenemünde engineers conferred about this problem and, after a number of experiments, came up with a single mixture of fuel that would be used for all liquid-fuel missile projects, in theory rationalizing and standardizing future supply needs. There is, however, no evidence that this change was implemented before the end of the war.[73] Smaller projects of this sort are reflective of a willingness by the Peenemünders to continue their

---

[69] "Group 2 Targets in Nordhausen Area," Box 678, Entry 82, RG 319, Records of the Army Staff, "P" Files, NARA.

[70] Neufeld, *The Rocket and the Reich*, 230–237.

[71] Ibid., 251–252.

[72] Chemical Engineers Tschinkel and Rössler Patentanmeldung [Patent Application], November 7, 1944, GD 624.624.4, DM. The two engineers never heard back from the RPA about whether or not they would be awarded a patent.

[73] Von Braun and Riedel to Halder, November 18, 1944, GD 624.623.14, DM.

hard work by improvising, adapting to changing circumstances, and over-coming difficulties imposed by the war's increasingly bleak circumstance. They did not fold in the face of military adversity.

Other projects were less technically mundane as well as a reflection of the professional commitment and technological desperation exhibited by Peenemünde engineers in the war's last year. One proposal, written jointly by Peenemünde development engineers and Luftwaffe officials, called for the rapid deployment of the A-4 by using large Messerschmitt Me-323 "Gigant" transport planes, which had six engines and a carrying capacity of twelve tons. First circulated in March 1944, this proposal argued that it was impos-sible to set up forward launch areas because it took too long for the support equipment to get to the location, deploy, and launch. Rapid deployment to these areas was for all intents and purposes impossible. Airlift to specific locations would solve this problem, and the proposal included technical requirements, manpower needs, and equipment quantities for large opera-tions.[74]

This plan had virtually no grounding in reality. To be sure, the Me-323 could carry the missile, but to have all of the necessary ground equipment transported with even one A-4 required a small fleet of transport aircraft. An astounding 123 planes were required to move just one complete battery of A-4s.[75] By March 1944, only about seventy Me-323s were ready for operation, with the underwhelming number of six to eight new planes per month being added to the fleet. In any event, air transport to forward areas in the middle of 1944 would have been a suicide mission. Allied warplanes were rapidly gaining control of the skies and could operate relatively freely over much of Western Europe. Even if the lift capacity for a battery of A-4s existed, any mission of this sort would have rendered the transport fleet completely impotent because of Allied air superiority.

Other ideas were impressively farsighted, but even more farfetched, given the realities of the war.[76] The embodiment of both was the idea for a submarine-launched ballistic missile, known first as Project Swimming Vest [*Projekt Schwimmweste*] and then as Test Stand XII. As early as 1942, the

---

[74] Vorschlag für einen Schnelleinsatz des Sondergerätes mit Lufttransport der Me323 "Gigant," March 1, 1944, GD 639.4.6, DM.

[75] Ibid.

[76] One important project that was envisioned as early as 1936, but shelved in 1942, was the A-10 intercontinental ballistic missile. This was a two-stage missile with a range that would enable it to reach New York. It was beset by a number of scientific and technical problems and remained on the drawing board until 1942, when it was set aside. According to Peenemünde design engineer Wernher Dahm, "It might have worked in the long run, but the technology was not available to make it a really useful project." See Wernher Dahm OHI, NASM. Though it was not a part of the spasm of desperate ideas in the closing stages of the war, it is indicative of the Peenemünders' willingness and ability to think about developing increasingly complex weaponry.

Kriegsmarine had taken an interest in installing unguided artillery rockets on its ships and submarines. Its Commander-in-Chief, Admiral Karl Doenitz, also met with von Braun in August of that year to discuss the use of larger missiles as antiship weapons. Despite the major technological complexities of ship-to-ship missile operations, von Braun was interested and promised to consult Dornberger about it.[77] Dornberger also thought that the idea had merit and proposed that Peenemünde developers work with *Kriegsmarine* engineers to study what would be needed to develop such a weapon as long as no resources were diverted from the A-4 effort, which, as of that point, had not experienced a successful launch.[78] Very little came of it, as the Peenemünders were far too busy with the A-4 to dedicate any of their resources to such a difficult problem. *Kriegsmarine* officials brought the idea up again in 1943, and there were a few meetings to discuss technical details, but the two groups laid no concrete plans for development.[79]

However, in June 1944, less than two weeks after the Allied invasion at Normandy, a Peenemünde engineer named Sachsenberg approached Riedel III with technical drawings and the outline of a plan for an underwater launch canister to be used in conjunction with U-boat operations. The concept was based around the idea that the United States might reconsider its participation in the war if A-4s began falling on New York City and other urban areas along the Atlantic coast. The missile canister, displacing approximately 500 tons when fully loaded, would be towed behind a U-boat. Once at its launch destination, the carrier's ballast tanks could be flooded in order to raise it into vertical position, which would extend the front of the carrier above the surface of the water. From that position, the doors of the carrier could be opened, the missile prepared for launch, and fired. One U-boat could supposedly tow three launch canisters across the Atlantic at a speed of approximately twelve miles per hour.[80]

The proposal was revolutionary – one of many ideas by the Peenemünders that had far-reaching strategic implications – but given the state of missile and submarine technology in 1944, it was also absurd. The technical difficulties of towing the canisters, erecting them, and launching the missile were enormous; the resources limited. However, in an atmosphere in which German officials and armaments specialists increasingly cast about for solutions to get them out of their predicament, it seemed to have merit. Peenemünde developers embraced the idea and, in September, reported the

---

[77] Wernher Von Braun, "Niederschrift über die Besprechung in Kiel," August 28, 1942, RH8/v.1960, BA/MA.

[78] Dornberger to Goetze, September 21, 1942, RH8/v.1960, BA/MA.

[79] Von Braun to Loewe, May 29, 1943; OKM to Dornberger, June 17, 1943, RH8/v.1254, BA/MA.

[80] Sachsenberg, Aktennotiz über eine Besprechung mit herrn Direktor Diplom. Ing. Riedel, June 15, 1944, RH8/v.1276, BA/MA.

results of preliminary experiments to the division heads of EW.[81] In December, Peenemünde developers met with representatives of Vulkan Docks in Stettin, which was contracted to build a prototype, in order to hammer out the difficulties inherent in such a project. Everyone involved, including many important Peenemünders such as Hans Hüter, Riedel III, and Kurt Debus, took the project very seriously and attempted to solve many pressing technical questions at the meeting.[82] The project was conducted in total secrecy (at this point, the revealing code name "Schwimmweste" was changed to "Test Stand XII"). Even other engineers inside Peenemünde were not to be told of the work if they were not involved in it.[83] Test engineers expected to have the prototype vehicle available to them by March 1945 and requested that the building contracts be handled expeditiously.[84] Though the submarine-launched ballistic missile would go on to become one of the twentieth century's most fearsome weapons systems, in 1944–1945, the idea of a delivery platform such as "Schwimmweste" was too ambitious in its strategic concept, technical demands, and tactical applications. It was a reflection of the desperation with which Peenemünde developers carried out their work in the last year of the war.

Thus, missile designers at Peenemünde did not just satisfy their own professional standards with their work, nor did they work merely to meet demands of the state or Nazi Party, as many of them might argue. Rather, they strove to push the technology to its performance and destructive limits. The Nazi regime had long benefited from the Peenemünders' ability to direct their own development activities, and this approach continued to pay dividends in the closing months of the war. The missile specialists mobilized nearly all of their efforts in support of the regime. Some began to feel an increased disillusionment with the idea of a victorious finish to the war – on a report underlining the importance of missiles for breaking Allied air superiority and "therefore the achievement of final victory," von Braun sarcastically scrawled, "Final victory, well, well!" – but this did not dampen their enthusiasm.[85] The Peenemünders absolutely buried themselves in their work, making major theoretical and practical strides in the field of missile technology. For this, they were amply compensated in a number of ways.

Von Braun had a long track record of looking after people who worked hard for the sake of the missile program. This was no different after his arrest. In addition to the compensation given employees whose inventions

[81] Vorversuche für Projekt Schwimmweste, September 11, 1944, RH8/v.1276, BA/MA.
[82] It is perhaps revealing that Riedel III and Debus were still dedicated Nazi party members.
[83] Niederschrift über die Besprechung vom 9.12.44 bei Wa Prüf (BuM) 10, RH8/v.1276, BA/MA.
[84] Aktenvermerk über die Besprechung am 25.1.45 in K., RH8/v.1307, BA/MA.
[85] Neufeld, *The Rocket and the Reich*, 247.

FIGURE 13. Civilian specialists work on a *Meilerwagen*, a truck trailer designed to make the A-4 missile road mobile. [National Air and Space Museum, Smithsonian Institution (SI 78-2955)]

were used at Peenemünde, he tried to make other rewards available as well. For example, in early April 1944, the technical director attempted, though Heinz Kunze, to procure extra food rations for those "intellectually creative workers" in armaments industries. The group from Peenemünde that he recommended included engineers, scientists, technicians, and secretaries, all of whom "performed a great service in the area of development and serial production. ... They have worked long hours, day and night, Sundays and holidays, foregoing free time and have exhibited exemplary lives in their private activities."[86] Von Braun wanted to forward Kunze a list of new nominees every six months. Kunze promptly informed von Braun that this ration program had been cancelled, but von Braun's note to Kunze underlined the extraordinarily hard work that Peenemünde specialists had been putting in as well as von Braun's own efforts to reward them for their activities.[87]

Intellectually creative, hardworking Peenemünders received other forms of rewards as well. In late 1944, many developers at Peenemünde began

[86] Von Braun to Kunze, April 5, 1944, FE 732, NASM.
[87] Kunze to Von Braun, April 21, 1944, FE 732, NASM.

receiving official, nonremunerative rewards in addition to the money that
was their due if their ideas were to be patented. Some individuals won
prizes for technical improvements that they made on the missile. Technician
Bruno Helm, for example, received a prize for improvements he made in
sealing missile combustion chambers.[88] Promotions, titles, and medals were
all distributed in the closing months at Peenemünde. Engineers Dannenberg,
Hackh, Heimburg, Tessmann, and Martin were awarded the title of *Oberin-
genieure*, and all of the "authority of leadership that comes with this title,"
by von Braun and Storch in October 1944.[89] Administrators at Peenemünde
recommended many of their employees for the War Service Cross, either first
or second class, in the same period. Friedrich Dürre, an engineer who also
helped with security measures at the base, was recommended for this award
because he "has fully proven his worth" in teaching counterintelligence
measures to employees of EW. Richard Lochman, in charge of organizing transportation at Peenemünde, was recommended for the same award
because of his hard work and skill in carrying out his duties.[90] None of these
rewards were the result of political cronyism on the part of management or
employees. Rather, they reflected the hard work and long hours put in by
Peenemünde specialists to push forward their project and do all they could
to contribute to their nation's efforts in the war.

Finally, throughout all of their activities over 1944 and 1945, the senior
Peenemünders and most rank-and-file employees were clear about the results
when their missile crashed into cities in England, Belgium, and France. The
first reports of A-4 operations came to Peenemünde in September 1944
after two sets of launches, one against Paris and another later in the day
against London.[91] Dieter Huzel recounted how the reports electrified the
Peenemünders.

The news had arrived there [von Braun's office] also, and the room was rapidly
filling as staff engineers drifted in. A dozen excited conversations were going at once.
Von Braun cut in on the enthusiasm with a sober dose of reality. This was not the
final payoff – far from it. The V-2 was not fully developed. Many specific problems
remained to be overcome, despite the exaggerated propaganda of the Hitler government.[92]

A concerted propaganda campaign for the V-2 did not actually begin until
November, but the Peenemünders were nevertheless aware of missile operations months before. Foreign press reports about the V-2 campaign emerged

[88] Bruno Helm Basic Personnel Record, Box 703, Entry 179, RG 165, File "Boston," NARA.
[89] Von Braun and Storch to Dannenberg, Hackh, Heimburg, Tessmann, and Martin, October
    15, 1944, RH8/v.1941, BA/MA.
[90] Storch to Rees, December 26, 1944, RH8/v.1941, BA/MA.
[91] See Michael Neufeld, *Von Braun: Dreamer of Space, Engineer of War* (New York: Knopf,
    2007).
[92] Huzel, *From Peenemünde to Canaveral*, 119.

steadily in late 1944. These reports also arrived at Peenemünde, ostensibly for intelligence reasons, in order to evaluate the results of the attacks. They included photos showing buildings reduced to rubble and massive craters left in urban areas.[93] Perhaps understandably, Peenemünders had little remorse for the victims of their weapon. Auguste-Elfriede Friede, one of von Braun's secretaries, recalled after the war that "We worked from the standpoint that war is war, and when their bombs stopped falling on the populations of Hamburg, Cologne, and other areas, things would change."[94] Others were even more bitter about the Allied bombing campaign. They were proud to strike a blow for their country: "When the first V-2 hit London, we had champagne. And why not? We were at war, and although we weren't Nazis, we still had a Fatherland to fight for."[95] Most Peenemünders felt no obligation to reign in their efforts to maximize this response. To the extent that they did reflect on their work, they did so in the context of a war in which neither side granted any quarter. Virtually no one felt any guilt about their work or the use of the missile against civilian targets in England and elsewhere.

The feelings emerged in part because the virtually unimpeded destruction by air attacks in this period produced a siege mentality that penetrated well into the consciousness of individual Germans. This state of mind brought German citizens closer together through shared physical and psychological stress. Citizens in Berlin, Schweinfurt, Essen, Dresden, and Peenemünde could all relate to each other on the basis of the shared suffering that they experienced. This reinforced the idea of a national community throughout the country.[96] The Peenemünders were not immune to this feeling. In his memoir, Huzel recalled the "universal expression of *Kameraderie* which these declining days had so brought about."[97] Indeed, the true meaning of the *Volksgemeinschaft* was made abundantly clear in the last years of the war, when class barriers utterly vanished in the rubble of Germany's bombed-out cities.[98] Despite this destruction, Germans proved ever more

---

[93] See collected foreign news reports, FE 688, NASM.

[94] Auguste-Elfriede Friede Statement, "Peenemünde: Schatten eines Mythos." Dieter Huzel remembered being stunned at the magnitude of destruction on a trip from Peenemünde to Bavaria. He wrote that "In the isolation of Peenemünde, I had not realized the extent of damage from Allied air attack. Practically every city or town of any size I passed through showed the marks of heavy bombings, particularly in the immediate vicinity of railroad stations and along the tracks." See Huzel, *From Peenemünde to Canaveral*, 118.

[95] Quoted in Neufeld, *Von Braun*.

[96] Norbert Frei, "Peoples' Community and War: Hitler's Popular Support," in Hans Mommsen, ed., *The Third Reich Between Image and Reality: New Perspectives on German History, 1918–1945* (New York: Berg, 2001), 59–75.

[97] Huzel, *From Peenemünde to Canaveral*, 168.

[98] Hans Mommsen, "The Dissolution of the Third Reich: Crisis Management and Collapse, 1943–1945," *GHI Bulletin* 27 (Fall 2000), 3; see http://www.ghi-dc.org/bulletin27Foo/b27mommsen.html.

willing to invest even more in the regime and offer up their services to the nation in its time of need.

Peenemünders felt this siege mentality as well and contributed in their own way to Germany's last, desperate efforts. Their frantic, last-ditch exertions were prosecuted with phenomenal effort under increasingly inadequate conditions. Of course, not every engineer was inspired to work with such desperation in the last year of the Nazi regime and there were gray areas of motivation, but for the most part, the profound dedication to success that the project elicited from the Peenemünders in its early years merged with a nationwide fortress mentality. As the regime became even more desperate, its goals and those of the Peenemünders became even more deeply enmeshed. For their part, the Peenemünde specialists continued to push the technical limits of missile development even as time and resources dwindled away.

TWILIGHT OVER PEENEMÜNDE

By January 1945, an untenable war situation had became an unwinnable one. On the Eastern Front, a huge Soviet winter offensive swept aside German defenders between the Oder and Vistula Rivers, completing the replacement of one dictatorship with another and finally pushing into Germany proper. By February, Soviet generals were making preparations to cross the Oder River and descend on Berlin, barely fifty miles away. In the West, British and American forces crushed Germany's last, desperate gamble in the Ardennes, and they were poised to advance across the Rhine River into Germany. The war was irrevocably lost and the outlook for the Peenemünde missile base, less than 100 miles from the Soviet troops encamped on the Oder, was hopeless. Evacuations began in early February, and by early March the formerly bustling and dynamic missile development center was little more than a ghost town. The final collapse of the program was not far behind.

The staff at Peenemünde East had been shrinking throughout 1944. In August, the German staff at the base, including its military personnel, had been reduced to 4,262, while 379 forced Eastern workers and prisoners of war remained there.[99] That number was further reduced when Kammler demanded that personnel be reallocated for military service or for work elsewhere. Storch wrote to Kammler that 342 employees could be given up at

---

[99] Storch to Kammler, August 21, 1944, RH8/v.1960, BA/MA. Of this number of employees, about half were salaried employees, that is, engineers, mathematicians, physicists, technicians, and clerical help. The balance was made up of hourly workers such as joiners, welders, electricians, and mechanics. Of the 4,262 workers, 618 were women, mostly clerical staff. Of the 379 forced laborers, 126 were Eastern workers and 253 were prisoners of war. According to Storch, there were no concentration camp prisoners engaged in missile work at Peenemünde.

Peenemünde without limiting development activities. He also reported that another 78 could be sent east for "East Wall Operations" [*Ostwalleinsatz*], digging trenches to try to stem the onslaught of the Red Army. A meager 21 employees, according to Storch, had volunteered for duties at the Front, and another 34 were being prepared to serve as members of Dornberger's technical storm troops, troubleshooting launch operations with the missile batteries. Storch also noted that as more projects neared completion, the base at Peenemünde could give up even more people for other activities.[100] By October, however, development administrators were beginning to feel the sting of these losses. Von Braun wrote to Army Brigadier General Josef Rossmann, who took command of the Ordnance section responsible for liquid-fueled missile development earlier in the year, to complain that the base's ability to keep up its output was declining "as we give up workers for the war or they are replaced by women." Most personnel were being assigned to missile batteries, but the use of the batteries also depended on the completion of proper ground equipment, a task for which those who were leaving were needed. Most employees were already working at least twelve-hour days. Peenemünde's forfeiture of personnel, argued von Braun, was delaying the development of the equipment while the backlog of missile troops that awaited outfitting grew.[101] The continued development pressure at Peenemünde, made worse by the Allied advances on both fronts, was only increasing as personnel departed Usedom for other projects or operations across the shrinking Reich.

Von Braun was certainly concerned about Peenemünde's decline, but his complaints to Ordnance also stemmed from his concern to keep the facility's employees as safe and comfortable as possible. They had forged strong bonds of friendship and professional camaraderie in the course of their work on the base, and he could not but feel some responsibility for them. By January 1945, von Braun was pushing Storch on the "unfortunate question" of a "separation allowance" (*Trennungsentschädigung*), an issue that he first raised with EW's director in December. He argued that employees who, because of the program's dispersal, had been separated from their families should receive an extra allowance for the difficulties of this separation; he urged Storch to take up the matter with regime authorities.[102] In a separate notice to Storch written on the same day, von Braun asked the director to consider transferring funds from a social insurance program for VkN members into a general insurance fund for all employees. He thought EW could dole out funds to offset other individual economic losses because of

100 Paul Storch, "Berufsaufteilung der zur Abgabe vorgeschlagenene Arbeitskräfte," August 23, 1944, RH8/v.1960, BA/MA.
101 Von Braun to OKH Wa Prüf (BuM) 10, October 6, 1944, RH8/v.1960, BA/MA.
102 Von Braun to Storch, "Trennungsentschädigung," January 13, 1945, RH8/v.1941, BA/MA.

the constant air raids.[103] It is unclear whether or not Storch acted on von Braun's proposals.

During these last days at Peenemünde, rumors and fears of the Soviet Army were rampant while confusion about the military situation grew. Thousands of German refugees from the East marched past Usedom, bringing with them stories of rape and pillaging by Soviet troops. VkN members were ordered to carry guns, and many civilians began receiving terse orders to report to their local *Volkssturm* units for rudimentary military training.[104] The fear of espionage increased as the Soviets steadily advanced in the East, and administrators began laying plans for destroying technical documents and even sensitive testing equipment.[105] At the end of January, when Soviet tanks were rumored to be in the area, Rossmann issued a set of orders outlining procedures over the next few days. He ordered that launch tests be completed as quickly as possible, and that when evacuation did occur, all unimportant documents were to be burned. The rest would be removed to their evacuation location, probably somewhere near the Mittelwerk factory. All missiles not ready for launch tests in the next few days were to be relocated, as were key components and equipment.[106] Rumors of Soviet troops nearby turned out to be false (the focus of the Soviet advance was Berlin, not northern Germany), and the very next day, Rossmann wrote, "The situation has calmed. It was only a few tanks that appeared. ... The situation in Pomerania has not been upset." He went on to order that the launch experiments and static tests would continue to go forward until the instruments "were totally serviceable and will fulfill their purposes without any trouble." All employees at Peenemünde were to remain in place.[107]

Work at Peenemünde continued in its last desperate days. Von Braun, who saw the writing on the wall, still continued to push hard to continue development work and was unafraid to use his connection to Kammler to do so. In January, the firm *Gema Blucherwerk* in Leignitz (Legnica, east of Breslau/Wroclaw) in Poland was conducting work on the A-4's guidance system. When the Soviet advance forced the firm to abandon Leignitz, von

[103] Von Braun to Storch, "Monatsbericht Dezember 1944," January 13, 1945, RH8/v.1941, BA/MA.
[104] Huzel, *From Peenemünde to Canaveral*, 133.
[105] Vorbereitung zur schnellen Vernichtung geheimen Aktenmaterials, January 23, 1945, RH8/v.1941, BA/MA. Officials at the base had been concerned about important documents falling into enemy hands since September 1944. Peenemünde administrators began to worry about surprise attacks by both Soviet and Western Allied forces. See Dieter Huzel Circular, "Sicherstellung von geheimen Aktenmaterial," September 25, 1944, FE 734, NASM; Paul Storch, "Sicherstellung von geheimen Aktenmaterial," October 3, 1944, RH8/v.1294, BA/MA. Secrecy remained a key consideration, even at the end of the program's existence.
[106] Rossmann Order, January 30, 1945, RH8/v.1941, BA/MA.
[107] Ibid.

Braun arranged for its workers to come to Peenemünde to continue their experiments. Managers at Gema, however, told von Braun that Dr. Karl Rottgart, the Director of the electronics firm Telefunken and Chairman of the Development Subcommittee for Radio Measurement, had stopped development and wished to transfer guidance specialists at Gema to other work. Von Braun immediately wrote to Kammler, asking him to intercede directly with Rottgart and order him to send all of the Gema employees and their equipment to Usedom.[108] Even though he was becoming increasingly disenchanted with Nazism, von Braun had pinned his professional existence on the success of the A-4 and, in so doing, coupled his career to the Nazi state. He was not yet ready to give that up, despite his disillusionment. Certainly, von Braun still felt concerned in the wake of his arrest to at least give the impression of loyalty, but he was doing more than just keeping up appearances. Rottgart's decision to move the guidance engineers to other work directly impacted the progress of A-4 work. It offered von Braun a choice between continued maximum effort and allowing the program to decline, with no penalty for the latter. That von Braun chose to go to Kammler is indicative of his limited priorities.

That same day, however, Kammler ordered the final evacuation of the base. Preparations were hastily begun, and within two weeks, employees, equipment, and instruments began moving south, to be relocated to Bleicherode, just to the southwest of Mittelwerk.[109] Most of the personnel made the dangerous journey either by truck or train, while much heavy equipment went by boat.[110] By March, most of the remaining 4,000 Peenemünders left Usedom to join what came to be known as the Central Construction Development Cooperative (*Entwicklungsgemeinschaft Mittelbau*), made up of thirty firms (with approximately 7,000 total employees) such as Henschel, Dornier, Ruhr Steel, and others.[111] About 400 people remained behind at Peenemünde because they refused to leave their homes.[112] For many of those who left the base, it would be the last time they would ever see it.

---

[108] Von Braun to Kammler, January 31, 1945, RH8/v.1265, BA/MA.

[109] Neufeld, *The Rocket and the Reich*, 258–259. See also "Group 2 Targets in the Nordhausen Area," Box 678, Entry 82, RG 319, Records of the Army Staff, "P" Files, NARA.

[110] According to U.S. authorities, about seventy percent of the equipment reached the town of Barby on the Elbe River, where it was to be transported by rail to the Harz. However, destruction of railroad tracks along the route meant that it was impossible to ship most of it to its final destination. Much of the stock was either in Lübeck or Barby when the war ended. Intelligence Report, File "V-2 (A-4) Missile (Germany, WWII), Intelligence Interrogations," NASM.

[111] Huzel, *From Peenemünde to Canaveral*, 139. See also "Group 2 Targets in the Nordhausen Area," Box 678, Entry 82, RG 319, Records of the Army Staff, "P" Files, NARA; Von Braun, Antrag auf Genehmigung und Einstufung eines Bauvorhabens der Entwicklungsgemeinschaft Mittelbau, March 6, 1945, RH8/v.852, BA/MA.

[112] U.S. Army Intelligence Report, File "V-2 (A-4) Missile (Germany, WWII), Intelligence Interrogations," NASM.

After the evacuation, a strange atmosphere of *Götterdämmerung* hung over Peenemünde. In the face of utter collapse, the few remaining specialists did their best to enjoy the benefits of life at the base in its last bleak days. According to Huzel, alcohol reappeared in relative abundance, local cinemas still showed films, and the trains, most empty, still operated. "One of the flak units which included in its personnel a number of women telephone operators sponsored a few dances," Huzel recalled. "These usually ended early since the port wine was sweet and easy to take." For Huzel personally, who took strolls along the beach and watched the waves while listening for the sounds of launch tests that never came, it was as if "The uneasy stillness of a death watch had settled over Peenemünde."[113]

The newly relocated EW set up its headquarters in the town of Bleicherode, not far from Mittelwerk, and its employees were scattered all over Thuringia. Those Peenemünders who arrived safely set themselves to work trying to organize themselves in their new accommodations. Efforts to restart the program began, and its administrators expected to have operations fully reset by July.[114] EW administrators hoped to install a plant in a salt mine in Bleicherode to serve as the primary production and testing facility for missiles. Their idea was to link these tunnels to a larger complex being planned for the area.[115] Rossmann also wrote to Dornberger that a crash program for improvements in the A-4 was in development.[116] In March, his staff reckoned that these modifications could be brought on line by September 1945, and they laid development plans stretching six months into the future.[117] Engineers busied themselves with other tasks as well. The Taifun anti-aircraft missile, a small, unguided weapon initiated by Luftwaffe Lieutenant Klaus Scheufelen and developed at Peenemünde, received much of their attention during this period. This desperation project began in the late summer of 1944, and by March 1945, Peenemünde engineers, now in Bleicherode, sought to clarify what development problems remained and how they could give Mittelwerk the help necessary to bring it quickly into production. However, the test stands built to launch the seventy missiles delivered had not been completed. Two, constructed by concentration camp prisoners, were nearing completion, and two more were still in the planning

[113] Huzel, *From Peenemünde to Canaveral*, 142.
[114] C.I.O.S. (Combined Intelligence Objectives Subcommittee) Report, "Investigation of Rocket Research, Elektromechanische Werke GmbH," Allied Operational and Occupational Headquarters, WWII, Box 93, Entry 13D – General Staff, G-2 Division, RG 331, NARA.
[115] "Group 2 Targets in Nordhausen Area," Box 678, Entry 82, RG 319, Records of the Army Staff, "P" Files, NARA.
[116] Rossmann to Dornberger, February 27, 1945, RH8/v.1307, BA/MA.
[117] Schneider, "Niederschrift über die Entwicklungsvorbesprechen bei B.z.b.V. Heer am 20. und 21. März 1945; Graphische Darstellung der Entwicklungsvorhaben gemäss Entwicklungsbesprechung von 21. u. 21.3.45," RH8/v.1307, BA/MA.

stages and would be located somewhere in the vicinity of Mittelwerk. In addition, EW managers agreed to subordinate a number of engineers to a test engineer from the Luftwaffe in order to streamline the development process as much as possible.[118] Von Braun had little faith in the comparatively primitive Taifun, but the industrious work of others continued to advance this desperate project, which never had any hope of breaking the "terror bombing."[119] Despite von Braun's misgivings, the military placed an order for 2 million missiles, which were to be assembled in Mittelwerk.[120]

In early March, von Braun drafted a seven-page proposal for laying out the development and production areas in Sperrgebiet Mittelbau for the V-1, V-2, Schmetterling, and Wasserfall missiles. The outlook was bleak. Because of security considerations, continued testing for the V-2 and Wasserfall could no longer be conducted. The area around Bleicherode was filling with refugees who occupied not only possible factory spaces but also accommodations needed to house factory workers. Von Braun concluded that although some expansion of the underground facilities was possible, time constraints did not make this a viable option. Rather, he held that above ground construction of machine shops, testing facilities, barracks, and assembly plants was absolutely necessary. Von Braun argued that "All working and living space not already requisitioned" must be augmented by new construction, suggesting also that more space might be made available if they occupied buildings being used by refugees from the East and employees of other factories in the area.[121] It is worth mentioning again that concentration camp prisoners would carry out all new construction. Huzel, who joined von Braun at Bleicherode in March and received personal instructions from the technical director to help reestablish plant operations, recalled that "even in the face of such hopelessness, I observed, von Braun's agile mind continued to function and to plan ahead. ... We had no sensible choice but to continue working."[122] The narrowed sense of responsibility engendered by the years at Peenemünde had cut off any option but to forge ahead as best as possible.

In Mittelbau-Dora, the worsening war situation had an even more dramatic effect. In January 1945, thousands of evacuees from Auschwitz, Gross-Rosen, and other camps in the East began arriving, worsening a food

[118] Aktenvermerk über Taifun – Besprechung am 24.3.45 bei den Elektromechanischen Werken GmbH, RH8/v.1941, BA/MA. Present at this meeting were a number of middle- and upper-level engineers, including Konrad Dannenberg, Hans Hüter, and Klaus Scheufelen, as well as Storch, Dornberger, von Braun, and Sawatzki.

[119] Von Braun dismissed the Taifun in early December 1944, telling his shop managers to treat it as "filler work." See Neufeld, *The Rocket and the Reich*, 255.

[120] "Group 2 Targets in Nordhausen Area," Box 678, Entry 82, RG 319, Records of the Army Staff, "P" Files, NARA.

[121] Von Braun, Antrag auf Genehmigung und Einstufung eines Bauvorhabens der Entwicklungsgemeinschaft Mittelbau, March 6, 1945, RH8/v.852, BA/MA.

[122] Huzel, *From Peenemünde to Canaveral*, 147.

Missiles for the Fatherland

situation that was already tenuous at best.[123] The population of prisoners in the entire complex of camps skyrocketed from approximately 27,000 in November 1944 to over 40,000 in March 1945.[124] The food supply, already stretched to the breaking point just to feed the German population, could not handle the strain of the added prisoners, and deaths due to starvation soared. Between the end of December and the beginning of March, over 5,300 people in the complex perished, 1,090 of whom, or just under twenty percent, lived in Dora.[125] Among this number are the prisoners who died because of mass hangings that took place in March under the orders of the new camp commandant, Richard Baer. The executions, 162 in all, took place both inside Mittelwerk and in Dora. The bodies in Mittelwerk were left dangling for twenty-four hours so that all the factory's employees could view them.[126] Despite these horrific circumstances, the factory continued to turn out missiles, with at least 362 A-4s emerging from the tunnels in March.[127] In the fifteen months since August 1943, the missile program's administrators, along with their partners in the Armaments Ministry and SS, expanded the tunnel complex under Kohnstein, relocated and installed a huge factory, and pumped out 5,789 V-2 missiles. It is a feat that boggles the mind both for its technological accomplishment and the horrific brutality with which it was achieved.

At the beginning of April, Kammler ordered that nearly 500 people in the missile program evacuate Bleicherode to Oberammergau, in Bavaria. Allied armies had collapsed the Western Front and were streaming into Germany. The core of the Peenemünde group, including von Braun (who had broken his arm in a serious car accident – his driver fell asleep at the wheel and his car went over an embankment), Dornberger, and many others, proceeded south by car and train. Some thought their move would only be temporary. A U.S. intelligence report filed shortly after the war pointed out the profound impact that years of Nazi propaganda about Germany's final victory had on these engineers and their lasting faith in the Nazi regime. Investigators explained that when the Peenemünders arrived in Bavaria, "It was thought

---

[123] See transport lists on Roll 18, RG 04.006M, Nazi Concentration Camp Records and Roll 161, 196.A.0342, National Archives Captured German Records Collection, both at United States Holocaust Memorial Museum.

[124] Manfred Bornemann and Martin Broszat, "Das KL Dora-Mittelbau," in *Studien zur Geschichte der Konzentrationslager, Schriftenreihe der Vierteljahreshefte für Zeitgeschichte* 21 (Stuttgart: Deutsche Verlags-Anstalt, 1970), 191–194.

[125] Wincenty Hein, "Lagerstärke in KL Dora," U.S.A. vs. Kurt Andrae, et al., Roll, M-1079, NARA.

[126] Erich Ball Testimony, U.S.A vs. Kurt Andrae, et al., Roll 1, M-1079, NARA. Baer arrived from Auschwitz on February 1 to replace Förschner, who assumed command at Kaufering, a subcamp of Dachau. See Jens-Christian Wagner, *Produktion des Todes: Das KZ Mittelbau-Dora* (Göttingen: Wallstein Verlag, 2001), 200.

[127] Neufeld, *The Rocket and the Reich*, 263. According to Neufeld, documentation exists only for missiles shipped up to March 18.

by some that they would enjoy somewhat of a vacation until the Wehrmacht drove the Allies back across the German border, at which time the research people would return to their work."[128] Most of the group probably realized that the end was near, but some still held on to a shred of hope that the regime would survive and that they could resume their work in short order. Even in the face of total collapse, some Peenemünders still held on to their belief, born of years of Nazi propagandizing, in a final, miraculous victory.

This intelligence also forces a reevaluation of another act during the last weeks of the war. In early April, Dieter Huzel and Bernhard Tessmann, both longtime Peenemünders, buried Peenemünde's most important documents in a mine northwest of Mittelwerk. Huzel and others' plausible claim is that they did this so that they could use it as a bargaining chip in their dealings with the Allies at the end of the war. However, in light of the U.S. intelligence report indicating that some engineers were still convinced of the *Wehrmacht*'s ability to hurl the Allies back across the Rhine, their view of the inevitability of Allied takeover, trumpeted after the war's conclusion, must be questioned. It is perhaps equally plausible that yet another reason to keep the documents at arm's length from the Americans and British was based on the chance, however slim, that Germany might still be victorious. In any case, Huzel put himself in grave danger by carrying out this task. He spent an anxious number of days dodging Allied soldiers and airplanes in his quest to hide the documents, retrieve his fiancée from Berlin, and then drive to southern Bavaria to reunite with his colleagues.[129]

The Peenemünders who went south – 300 others remained behind in Thuringia – had meanwhile spread themselves out in the hotels and resorts outside of Oberammergau. They did not work, and most merely waited for the war to end and to surrender. One aerodynamicist, Hermann Steuding, a Nazi Party member who was distraught over the prospect of having his skills put to use either by the western Allies or the Soviet Union, disappeared and was rumored to have committed suicide.[130] On May 2, Magnus von Braun, at the bidding of Dornberger and his older brother, pedaled his bicycle down the mountain and surrendered himself and the program's technical leadership to the U.S. Army.[131] The German missile program, with

---

[128] C.I.O.S. Report, "Investigation of Rocket Research," Box 93, Entry 13D, RG 331, Records of the Army Staff, NARA.

[129] Huzel remembers that road blocks were set up all over Germany. However, even at this late date, secrecy provided the Peenemünders with the privilege of passing them by with no trouble. "Everywhere little units were setting up road blocks, defense establishments, vain efforts to hold back the inevitable," he wrote. "Always our 'Secret Material' pass got us through. *Destination: Classified*, it proclaimed." See Huzel, *From Peenemünde to Canaveral*, 168. On his adventures across Germany, see 148–180.

[130] Frederick Ordway and Mitchell Sharpe, *The Rocket Team: From the V-2 to the Saturn Moon Rocket* (Cambridge, MA: MIT Press, 1982), 268.

[131] Neufeld, *The Rocket and the Reich*, 265.

FIGURE 14. Walter Dornberger and Wernher von Braun in Reutte, Austria, just after their surrender, May 3, 1945. Von Braun broke his arm in an auto accident about seven weeks earlier when his driver fell asleep at the wheel. Between Dornberger and von Braun is Dornberger's adjutant, Herbert Axster. To von Braun's immediate left is Hans Lindenberg, the head of the Quality Control division at Mittelwerk. [National Archives and Records Administration]

its modest roots in the Weimar rocket societies, its meteoric rise under Army supervision, its culmination at the world's most advanced missile research facility, and its descent into barbarism, had finally come to an end. The surrender of the leading technical experts to the United States signified the end of the missile base's existence, but its work, as well as its culture, would be perpetuated in the years afterward by those Germans who came to America. It was a system and a culture that worked, both technically and politically, and it has a legacy that stretches into the twenty-first century.

Conclusion
Engineering Consent at Peenemünde

A screaming comes across the sky. It has happened before, but there is nothing to compare it to now.

Thomas Pynchon, *Gravity's Rainbow*[1]

In the summer of 1945, when the former Peenemünders were awaiting transfer to the United States, a team of interrogators from the U.S. Third Army was assigned to screen the specialists for potential security risks before their departure. The interrogators quickly discovered that these Germans in their midst could not be evaluated as individual experts who might or might not be considered reliable from a political standpoint. Rather, they were, in the words of one investigator, "a closely knit research enterprise, firmly controlled and carefully chosen by Dr. Dornberger and Professor von Braun." The investigators also noted in their discussions with the former Peenemünders how Nazi ideology had colored opinions about Germany's supposed victimization and the service that their nation performed in defending the West from communism. The report also noted something even larger and more influential on the Peenemünders' outlook:

The cohesion of the group and their persistence in ideas ranging from German patriotism to Nation-Socialism [sic] is explained by a number of factors. The lived a secluded life on the island of Usedom in which they were not excessively bothered by the party. They were an Army concern, a closed corporation, carefully supervised by the Abwehr in matters of choice of personnel and security. They were enthusiastic technicians with the mission, according to Goebbels, of saving Germany. As a team they were granted all the financial support, materials, and personnel they required, within the means of the German war machine. Continuance of the work depended on continued conduct of the war. At a time when the generals were dissatisfied with the party rule to the extent of attempting to overthrow it, Peenemünde was out of touch and sympathy with such developments – not for love of the party necessarily but because their work and the war were one.[2]

---

[1] Thomas Pynchon, *Gravity's Rainbow* (New York: Viking Press, 1973), 1.
[2] Osborne to Army Chief of Staff, G-2, USFET [U.S. Forces, European Theater], Appendix A, Walter Jessel, Special Screening Report, October 29, 1945, Box 8, RG 260, OMGUS/FIAT, Folder 47.94, National Archives and Records Administration (NARA).

The Army's investigators had discovered something that would be largely forgotten or ignored by chroniclers of the German liquid-fueled missile program in the years after the war. The postwar apologetics and denials offered by the Peenemünders focused on a supposed distaste for both the regime and the purpose of their work. According to this logic, the missile specialists were apolitical technocrats, unhappy with the war and forced by the regime to use slave labor. Former Peenemünders claimed that the only group engaged in oppressing the concentration camp slaves was the SS, whereas they themselves made every effort to ease the prisoners' suffering. The truth is something else entirely.

The dynamic noted in the Army's intelligence report came about because of the specialists' socialization into the secret world of the Peenemünde missile base. Half military facility and half technological Shangri-la, Peenemünde created a cultural environment in which the needs of the regime and the needs of the missile specialists were inseparably intertwined. The Army's construction of the base carved a space in which its employees and their families could live and work, but also in which their activities could be closely regulated. The idea was to create a closely knit community out of which a revolutionary weapon might spring. To design such a weapon required the deep commitment of a huge group of civilian laborers made up of engineers, scientists, technicians, craftsmen, secretaries, and assistants. Through a complex combination of secrecy, regulation, professionalism, and reward, the system came to function so that individuals subject to its rules adhered to them and reproduced them automatically. The result was a group of specialists who all came to instinctively identify with the goals of their work. The deeply self-interested Peenemünde employees became the model of a compliant citizenry in which employees led a pleasant life while conducting interesting work and enjoying each other's company. This led them to do their level best to protect the regime that made their situation possible.

The dynamism of the German ballistic missile program was rooted in the active identification of the Peenemünde specialists themselves with the objectives of their work. The Peenemünders understood that they could rely on the best efforts of their colleagues, even those that they did not know, because for most of the war, none of them doubted the value of what they were doing. This in turn promoted trust and understanding, enabling them to rely on each other to carry out their tasks to the best of their abilities. A large institution like Peenemünde, with its population of thousands, would not have seen the success it did if large-scale dissent seeped into the fabric of the work. To be sure, the size and complexity of the facility opened up numerous opportunities for disagreement, which might have been registered by such nondescript actions as bureaucratic inefficiency, work slowdowns, or unwillingness to put in overtime – a constant demand, given wartime pressure. That virtually none of this took place among the missile development specialists is a testament to their profound dedication to their tasks and belief in the work.

Every facet of the Peenemünders' world was suffused with National Socialist ideological messages and thoroughly imbued with deep secrecy. This was a poisonous combination. Secrecy was not the single overriding factor in decisions made at Peenemünde, but it did provide a framework for those decisions. It must be taken at least as an important factor in the complex cultural dynamic at the base, and its influence in other historical circumstances must be investigated as well. Objects such as stamps and signed declarations, along with the myriad of other secrecy regulations and activities, were in fact daily indicators of inclusion in a strictly limited club that only a small group of professional elites had entrance to. Activities conducted within this community were expected to stay there, and outside influences were explicitly cut off. At the same time, the security measures were also projections of state power into the daily world of the Peenemünders, reminding them of both the importance of their work and the presence of the state's watchful eye. All of this resulted in an important dynamic that offered the Peenemünders a sense of elitism and privilege while narrowing their political, moral, and ethical choices and restricting contravening views.

Of course, secrecy is a complex phenomenon. Its presence does not necessarily equate with moral depravity or the formation of ill-informed assumptions. A comparison with another "big research" undertaking of World War II, the Manhattan Project, specifically the Los Alamos research laboratory, is highly instructive for many different reasons, one of which is that it offers the chance to examine the physicists' relationship with secrecy. The base at Los Alamos was a military facility under the overall control of Brigadier General Leslie Groves, but it was staffed with civilian experts who were managed by scientist J. Robert Oppenheimer, who both charmed and inspired his Los Alamos colleagues. The administration of the Manhattan Engineering District centralized its atomic bomb researchers on top of an isolated, dusty, windswept mesa at Los Alamos in early 1943. Its remote location made for excellent security and easy monitoring of individuals coming and going from the base. Moreover, it contributed to the sense that it was an isolated cloister (Oppenheimer once referred to it as a "monk's colony") that was populated by like-minded physicists and engineers who were dedicated to the goal of producing an atomic bomb.[3] Upon arriving at Los Alamos, all new employees received a series of lectures that indoctrinated them into security measures and briefed them on the present state of the work. As at Peenemünde, the secrecy around the project and the chance to be let in on it was a source of excitement. Physicist L. D. P. King clearly recalled the

[3] Peter Bacon Hales, *Atomic Spaces: Living on the Manhattan Project* (Chicago: University of Illinois Press, 1997), 42.

great anticipation of "going to a secret new place."[4] Karan McKibben, whose father worked at the laboratory, wrote that "The number of fences behind which our fathers disappeared every work day added an aura of intrigue to their already mysterious work in sundry, odd shaped buildings."[5] As at Peenemünde, secrecy created an element of elitism, privilege, and value of the work among those who were privy to the activities at Los Alamos.

In the laboratories, Los Alamos physicists found great professional satisfaction coupled with extraordinary military pressure. The work, so advanced as to be alien to outsiders, was conducted in an atmosphere of informality and collegiality that one might come to expect from a small community of super-elite specialists. The language used to express it was utterly foreign to all but the small community of atomic physicists.[6] Social life was also deeply fulfilling. Many laboratory employees took up hiking, skiing, and other recreational activities. Most employees held dinner parties and weekend events, and dances, plays, and skits were popular.[7] Such events served to draw the Los Alamos scientists even closer together. All the while, the employees of the laboratory never forgot that they were there to construct a weapon that, as opposed to the V-2, was so destructive both physically and psychologically that its application would crush the will of its enemies to continue. Fifty years after the conclusion of hostilities, the physicist L. D. P. King expressed his thoughts on life and work at Los Alamos in terms that might just as easily have been repeated by a missile engineer at Peenemünde.

I would like to say that to have been able to work at the Laboratory during those early, vital, and important years was indeed a memorable experience. The excitement of a small frontier community plus the excitement of working on a new frontier of science and technology cannot often be combined. Where else could one have had so many technical developments in so short a time; where else could one culminate the efforts and singleness of purpose of so many famous men but here in those momentous years of 1943, '44, and '45?[8]

Where else but halfway around the globe on a picturesque island on Germany's Baltic coast?

The cultural parallels between Los Alamos and Peenemünde are in some ways striking. What, then, is to be made of the Peenemünders' decisions to

[4] L. D. P. King, "The Development of Nuclear Explosives and Frontier Days at Los Alamos," in John Allred, ed., *Behind Tall Fences: Stories and Experiences About Los Alamos at its Beginning* (Los Alamos, NM: Los Alamos Historical Society, 1996), 62, 64. King arrived at Los Alamos in 1943.

[5] Karan McKibben, "Behind Tall Fences," in Allred, ibid., 179.

[6] Hales, *Atomic Spaces*, dubs it "speaking in tongues." See 243–272.

[7] Arthur Wahl, "Los Alamos, 1943," in Allred, *Behind Tall Fences*, 173. Wahl was a radiochemist at Los Alamos.

[8] L. D. P. King, "The Development of Nuclear Explosives," in Allred, *Behind Tall Fences*, 67.

offer the Nazi regime their full support, seek out the SS to provide slave labor for their production work, directly or indirectly participate in the abuse and murder of concentration camp prisoners, and work with a furious desperation to reverse the tide of the war in its last few months, especially in light of the close parallels between the two institutions? In the first place, Germany's unique historical circumstances in the interwar years were of fundamental importance. Hitler promised and delivered Germany's rebirth, and weapons engineers, who came of age in the conservatively charged atmosphere of the technical universities, owed him a great deal. The Peenemünders in particular were deeply in the Nazis' debt. Moreover, missile specialists on Usedom proved to be intensely self-interested. Like those at Los Alamos, their work kept them off of the front lines and safely tucked away in a comfortable community that for a long time managed to avoid the deprivations of war. They were paid well and received both professional and official recognition. Continued efforts to fulfill the program's goals meant the maintenance of this situation, something nearly all Peenemünders were loath to give up.

Even so, these factors still do not fully explain their descent into moral abomination and the belief that slave labor was the proper course to fulfill the program's objectives. Of more immediate and direct importance was the pervasiveness of National Socialist ideology and rhetoric in which their work was framed. Years of public adulation of Hitler and the Nazis appearing in print and on the radio, numerous speeches about Germany's victimization at the hands of nations east and west, and never-ending grandiloquence about the international conspiracy that forced war upon Germany created a hyper-nationalist, xenophobic atmosphere that was intensified by the increasing violence of the war. Even if they were not dedicated Nazis, the Peenemünders came to see such bromides as unshakeable truths. Like many Germans, they internalized these feelings and turned them into action to defend their nation. The fact that all of this took place in the framework of deepest secrecy made for an even more poisoned environment by reinforcing received assumptions, limiting conceivable alternatives, and even making possible acts that might be expected to never see the light of day. In short, it ensured that there would be no opportunity to carve out a counter or dissenting discourse. The toxic atmosphere that these two factors ushered in, coupled with the real benefits of working where they did, is of fundamental importance in explaining how it was that employees at Peenemünde came to embrace slave labor specifically and the Nazi regime generally.

As Michael Neufeld first pointed out, the rise of National Socialism was an important component in the development of this most radical weapon, supplying the materials, bureaucracy, and finances to push the work forward in the context of aggressive rearmament and then global war. Battles over priority of the weapon system and conflicts over controlling it aside, the Nazi regime provided important human resources and raw materials necessary to carry out the work.[9] However, the will to actually complete their tasks was supplied by the Peenemünders themselves. It was a will that stemmed from a deep identification with the work and with each other. Without it, such a complex technological system as a ballistic missile could not have been invented in so short a time. "Self-mobilization," a term first supplied by Karl-Heinz Ludwig, is an apt description of what the Peenemünders engaged in daily while developing the missile. Throughout their time on Usedom, they worked furiously to complete their Herculean task, not, as they would claim later, primarily because they feared for their lives or wished to explore space, but because they were so profoundly self-interested. Employment on Usedom gave them comfortable accommodations, stimulating work, excellent pay, professional satisfaction, and a vibrant social life. Peenemünders consciously understood that they owed the good circumstances of their lives to their skills and to a Nazi state that valued what they could do for it. In turn, they felt an internal compulsion to work as hard as they could on behalf of the regime that sponsored their work.

In addition, Ludwig found among engineers in Nazi Germany both strong ideological and practical reasons for supporting Hitler and the regime. The same is true in the more specific case of the Peenemünders. The institutional practices that employees found at the base connected them to the nation and the regime in novel ways. The Peenemünders had a clear vision of what was best for both themselves and the nation. This vision and that of the Nazis mutually reinforced each other, and the Peenemünders engaged in very little systematic reflection about the direction that National Socialism was taking them. For them, crucial political issues turned on the quality of social and cultural conditions that made up their lives. Their perceived role in the nation as well as the duty assigned to them by the regime encouraged varying degrees of affinity that were nonetheless long lasting.[10] Some were at least sympathetic to Nazism whereas others were outright supporters, but ideology was only one factor among many in play at the facility. Technological fascination, nationalism, money, and careerism all joined with ideological considerations to play key roles in building a compliant community of consent at Peenemünde.

---

[9] Michael Neufeld, *The Rocket and the Reich: Peenemünde and the Coming of the Ballistic Missile Era* (Cambridge, MA: Harvard University Press, 1995), 276–278.

[10] Indeed, support for the Nazis is no longer to be understood as the result of dislocation, crisis, and collapse. Historians now acknowledge the broad popularity of the Nazis and the strength of the relationship between average Germans and the regime. See Alf Lüdtke, *Eigen-Sinn: Fabrikalltag, Arbeitererfahrungen und Politik vom Kaiserriech bis in den Faschismus* (Hamburg: Ergebnisse Verlag, 1993); Donna Harsch, in *German Social Democracy and the Rise of Nazism* (Chapel Hill, NC: University of North Carolina Press, 1993), shows that the Nazis were more successful than other parties because they mobilized the idea of a new, forward-looking national identity that was not associated with the defeats and losses in the past.

After the war, in late 1945 and early 1946, the U.S. Army transferred some 120 former Peenemünders, the core of the group that previously worked on Usedom, to Fort Bliss, Texas, where it assigned the group to assist in V-2 experiments that were to take place in White Sands, New Mexico and help with Project Hermes, the U.S. Army Ordnance's primary missile program.[11] The former Peenemünders distanced themselves as much as possible from the Nazi regime, telling anyone who asked that they only wanted to build rockets to explore space and were forced by the Nazi party and SS to build missiles using slave labor (if they mentioned the latter point at all). The U.S. Army assisted them in their obfuscation, burying their records as best they could.[12] The only high-ranking individual to be brought before a war crimes tribunal was Georg Rickhey, Mittelwerk's General Director. He convinced the court in 1947 that he was also a pawn in the machinations of the SS and did everything he could to help the prisoners. The court found him not guilty and released him.[13] After this brief and mild embarrassment for the Army, the worst of the incidents were behind the missile specialists until Rudolph came under investigation in the early 1980s. Throughout the Cold War, the former Peenemünders carefully cultivated an image that distanced themselves from their Nazi past and played up their dedication to the U.S. space program that was engaged so heavily in the space race with the Soviet Union.

In 1950, the group found itself transferred to Redstone Arsenal in Huntsville, Alabama. Here, they reestablished the system of "everything under one roof" that had worked so well for them at Peenemünde.[14] The concept was pivotal in the development of the Redstone and Jupiter missiles.[15] This style of organization, which had proven highly efficient in

---

[11] Frederick Ordway and Mitchell Sharpe, *The Rocket Team: From the V-2 to the Saturn Moon Rocket* (Cambridge, MA: MIT Press, 1982), 310–317. Von Braun commented on their time at Fort Bliss and White Sands: "Frankly, we were disappointed with what we found in this country during the first year or so. At Peenemünde, we had been coddled. Here they were counting pennies." See Ordway and Sharpe, 352. The Soviet Union also took part in the intellectual plundering of the German missile program, but the specialists who found themselves launching the V-2 at a site outside of Stalingrad were never fully integrated in the Soviet missile establishment in the way that the Peenemünders who came to the United States were. They were headed by Helmut Gröttrup, who, along with von Braun and others, was arrested by the Gestapo in 1944. See Ordway and Sharpe, 318–343; Irmgard Gröttrup, *Rocket Wife* (London: Andre Deutsch, 1959).

[12] Linda Hunt, *Secret Agenda: The United States Government, Nazi Scientists, and Project Paperclip, 1945 to 1990* (New York: St. Martin's Press, 1991), 41–77. Hunt sees a Pentagon conspiracy to hide their Nazi records, but it is more accurate to say that it was motivated by a combination of technical expedience and Cold War politics.

[13] See the record of the trial, "United States vs. Kurt Andrae, et al.," M-1079, RG 226, NARA.

[14] Neufeld, *The Rocket and the Reich*, 271.

[15] See Ordway and Sharpe, *The Rocket Team*, 363–387. A Redstone rocket carried John Glenn into space.

the war years, also maintained the cohesion of the Peenemünde group; Huntsville came to be affectionately known as "Peenemünde Süd" – Peenemünde South. Only after they were organized in this way and given the proper resources did the U.S. Army's rocket and missile program truly blossom. Shortly after, the Army transferred the group over to NASA when that organization was created in 1958. In the following years, they would play a central role in the U.S. space establishment. Von Braun went on to become the Director of the Marshall Space Center in Huntsville; Arthur Rudolph was the Project Manager of the Saturn V rocket program; and Kurt Debus, the Director of Testing at Peenemünde, became the first Director of the Kennedy Space Center in Florida. The group's enthusiastic pursuit of space exploration contributed to the strengthening of the narrative that they had no love for National Socialism and pursued the work they did at Peenemünde for purely humanitarian and scientific purposes. It was buttressed by a raft of books and articles that celebrated their achievements but ignored the dark side of their records. Despite some necessary corrections, this style of work continues to proliferate.[16] In any case, their group cohesion was an essential part of the successful Apollo missions to the moon and also the growing proliferation of more and more advanced intercontinental ballistic missiles in the Cold War. Through all of their assignments, many of the Peenemünders kept in close touch and spoke warmly of their days on the Baltic coast.

The persistence of the Peenemünders' group cohesion and identification is further evidence of the profoundly formative impact that living and working on Usedom had on their lives. The Peenemünde missile research facility melded a group of individuals into a community of shared conditions and beliefs. The dynamic nature of the project developed out of the initiative that the program's personnel brought to their work, and they would come to define what it meant to be a professional missile developer – the proverbial rocket scientist (an inaccurate but popular characterization). All of this was founded upon a single idea, and within this group, no one questioned the base's central mission of producing missiles for the purposes of defending the Nazi state. This axiom became so powerful during the war that it pushed aside all other considerations and made it possible for the missile specialists to thoroughly enmesh themselves in the structures and practices of the National Socialist regime. Though bureaucratic battles took place over control of the program, the demands of the technology as well as the

---

[16] See Ordway and Sharpe, *The Rocket Team*; Thomas Franklin (pseudonym for Hugh McInnish), *An American in Exile: The Story of Arthur Rudolph* (Huntsville, AL: Christopher Kaylor, 1987); Marsha Freeman, *How We Got to the Moon: The Story of the German Space Pioneers* (Washington, DC: 21st Century Science Associates, 1994); Guido de Maeseneer, *Peenemünde: The Extraordinary Story of Hitler's Secret Weapons V-1 and V-2* (Vancouver: AJ Publishing, 2001)

shared goals between the Peenemünders and more radical elements in the regime ensured that cooperation, not competition, ruled the day. This was what they themselves interpreted as the appropriate behavior of their organization. In the end, this interpretation coupled one of the twentieth century's most impressive technological achievements with a revolting act of criminality. Indeed, in order to continue to enhance our understanding of the hold that National Socialism had over many Germans like the Peenemünders and others, historians must continue to examine the fluid but important combination of nationalist sentiment, political ideology, cultural practices, and collective identities.

Bibliography

A Note on Sources

This book is based primarily on documents found in archival collections in Germany and the United States, but it also relies heavily on oral histories, memoirs, and autobiographies completed by participants in the German liquid-fueled missile program. In the United States, the majority of documents are to be found in the Fort Eustis (FE) Collection in the archives of the Smithsonian National Air and Space Museum (NASM) in Washington, DC. This collection holds sixty-four reels of the technical and administrative correspondence produced at Peenemünde during the war. In addition, the Air and Space Museum archives contain transcripts of numerous oral history interviews carried out by Michael Neufeld and others. The interviews with the Peenemünders cover everything from technical development to personal anecdotes.

The National Archives and Records Administration in College Park, Maryland (NARA) holds a much smaller, though still significant, collection of documents pertaining to Peenemünde and Mittelbau-Dora. Of particular importance is the record of the U.S. Army Trial of Dora defendants, U.S.A. vs. Kurt Andrae, et al., located in the Captured German Documents Collection. Moreover, records relating to Project Paperclip, the program that brought German scientists and engineers to the United States, are located in numerous record groups in the archive. Researchers should consult with archivists and finding aids to gain a complete picture of this large but scattered group of documents.

The United States Holocaust Memorial Museum (USHMM) archive in Washington, DC contains several document collections that include videotaped or transcribed testimony given by former Dora prisoners. In addition, important documents such as transport lists and correspondence about slave labor reside in the museum's archive. I also consulted videotaped Holocaust survivor testimonies held at the Fortunoff Video Archive at Yale University in New Haven, Connecticut.

Huntsville, Alabama, the current home of many surviving Peenemünders, has two archives that focus on the postwar period but can still offer contributions to a study of Peenemünde in the war period. The Willy Ley Collection at the University of Alabama, Huntsville (UAH) holds the videotaped oral histories of former Peenemünders that were conducted in the early 1980s by UAH sociologist Donald Tarter. The Space and Rocket Center holds most of the Wernher von Braun papers. The collection is limited for researchers of the Peenemünde period, but it is possible to find some key documents.

In Germany, perhaps the most important collection of documents can be found at the *Bundesarchiv/Militärarchiv* (BA/MA) in Freiburg. Many of the original documents that can be found in the FE collection at NASM are located at the BA/MA, but more importantly a large proportion of the documents missing from the FE microfilm can be found here. The collection at BA/MA is smaller than that at NASM, but no study of Peenemünde is complete without an examination of these records. For this study, the most important files in Freiburg are those located in the records of the *OKH/Heereswaffenamt* (RH8), but the records of the

Reichsminister der Luftfahrt (RL1) are also helpful. A smallish amount of interesting candid photos are also available in the records in RH8.

In Berlin, the *Bundesarchiv-Lichterfelde* (BAL) proved surprisingly important. Many key documents pertaining to the administration of Dora are located in NS-4 (*Konzentrationslager*) and NS-4 Anhang (*Konzentrationslager, u.a. Mittelwerk GmbH*). Even more importantly, the BAL contains the corporate records of the Mittelwerk GmbH and its umbrella firm, Rüstungskontor GmbH. These valuable records were recently transferred from the *Bundesarchiv Koblenz* and can be found in R121 (*Industriebeteiligungsgesellschaft*). Researchers should request this collection ahead of time, as the records are located in the off-site storage facility at Dahlwitz-Hoppegarten.

Though it is a museum rather than an archive, the *Historisches-Technisches Informationszentrum Peenemünde* (HTIZP) holds a wealth of information that was central to this study. There are certainly few documents in the collection that cannot be found elsewhere. Nevertheless, the archive, closed to the public, houses some documents and artifacts, such as *Festzeitschriften* and other pieces, that may not be kept by a large state or federal archives. Moreover, curators at HTIZP have gathered dozens of videotaped interviews of former Peenemünders, and they made them available to me. These videotapes proved to be of surpassing importance, given the reluctance of Peenemünders to hold interviews with those they do not know well. They, along with a collection of rare photographs, are the treasure of the HTIZP collection.

The records of the 1967 West German trial of three Dora defendants were also important. Those in the dock were SS men at Dora, and a large number of civilian engineers were examined by both sides. Many of them had moved from Peenemünde to Dora. This trial was conducted by West German prosecutors, but an East German attorney participated in this effort as well. For this reason, records of the trial can be found at two locations. The primary repository of the trial record is the *Nordrhein-Westfälisches Hauptstaatsarchiv Düsseldorf, Zweigarchiv Schloss Kalkum* (HStaD-ZA Kalkum). However, many trial documents and a great deal of fascinating correspondence from the East German attorney are held by the *Bundesbeauftragte für die Unterlagen des Staatsicherheitsdienstes der ehemaligen Deutschen Demokratischen Republik* (BStU), or more simply, the Stasi Archive. In the late 1950s and early 1960s, the Stasi also attempted to discredit FRG President Heinrich Lübke, who directed construction at Peenemünde during the war. They were unsuccessful in this effort (they eventually engaged in an amateurish effort to manufacture counterfeit documents to make the point), but in the attempt they uncovered a wealth of information about forced labor at Peenemünde. Documents relating to this topic are scattered across numerous files.

Finally, the Deutsches Museum in Munich holds a large and expansive collection of technical documents pertaining to V-2, Wasserfall, and other ballistic missile development at Peenemünde, filed under German Document (GD) numbers. A substantial part of this collection is made up of correspondence pertaining to patenting and the effort to expand the capabilities of ballistic missiles. Most important for this study, however, was the mammoth collection of original photographs held by the Deutsches Museum. The photos depict everything from trial missile launches and buildings on the base to small instruments and technical parts. They make an important contribution to an understanding of the flavor of life at Peenemünde.

Secondary Sources

Adas, Michael, *Machines as the Measure of Men: Science, Technology, and Ideologies of Western Dominance* (Ithaca, NY: Cornell University Press, 1989).

Albrecht, Ulrich, Andreas Heinemann-Gruder, and Arend Wellman, *Die Spezialisten: Deutsche Naturwissenschaftler und Techniker in der Sowjetunion nach 1945* (Berlin: Dietz Verlag, 1992).

Allen, Michael Thad, *The Business of Genocide: The SS, Slave Labor, and the Concentration Camps* (Chapel Hill, NC: University of North Carolina Press, 2002).

_____, "The Banality of Evil Reconsidered: SS Mid-Level Managers of Extermination Through Work," *Central European History* 30 (1997).

_____, "The Puzzle of Nazi Modernism: Modern Technology and Ideological Consensus in an SS Factory at Auschwitz," *Technology and Culture* 37 (July 1996).

Allen, William Sheridan, *The Nazi Seizure of Power: The Experience of a Single German Town, 1922–1945* (New York: Franklin-Watts, 1965).

Allred, John, ed., *Behind Tall Fences: Stories and Experiences About Los Alamos at its Beginning* (Los Alamos, NM: Los Alamos Historical Society, 1996).

Aly, Götz, *Hitler's Volkstaat: Raub, Rassenkrieg und nationaler Sozialismus* (Frankfurt: Fischer Verlag, 2005).

_____, and Susanne Heim, *Vordenker der Vernichtung: Auschwitz und die deutschen Pläne für eine neue europäische Ordnung* (Hamburg: Hoffmann und Campe, 1991).

Arendt, Hannah, *The Origins of Totalitarianism* (New York: Harcourt Brace, 1979).

_____, *Eichmann in Jerusalem: A Report on the Banality of Evil* (New York: Penguin Books, 1963).

Asendorf, Chrisoph, *Super Constellation – Flugzeug und Raumrevolution: die Wirkung der Luftfahrt auf Kunst und Kultur der Moderne* (New York: Springer, 1997).

Bainbridge, William Sims, *The Spaceflight Revolution: A Sociological Study* (New York: Wiley, 1976).

Bandura, Anthony, "Self-Efficacy: Toward a Unifying Theory of Behavior Change," *Psychological Review* 84 (1977).

Barkai, Avraham, *Nazi Economics: Ideology, Theory, and Policy* (New Haven, CT: Yale University Press, 1990).

Barth, Hans, *Hermann Oberth: Leben, Werk, und Auswirkung auf die spätere Raumfahrtentwicklung* (Feucht: Uni-Verlag, 1985).

Bauer, Yehuda, *Rethinking the Holocaust* (New Haven, CT: Yale University Press, 2001).

Benecke, Theodor, and A. W. Quick, eds., *History of German Guided Missile Development: AGARD First Guided Missile Seminar, Munich, Germany, April 1956* (Braunschweig: Appelhaus, 1957).

Béon, Yves, *Planet Dora: A Memoir of the Holocaust and the Birth of the Space Age* (Boulder, CO: Westview Press, 1997).

Bergaust, Erik, *Wernher von Braun* (Washington, DC: National Space Institute, 1976).

Beyerchen, Alan, *Scientists Under Hitler: Politics and the Physics Community in the Third Reich* (New Haven, CT: Yale University Press, 1977).

_____, "What We Now Know About Nazism and Science," *Social Research* 59 (Fall 1992).

Bode, Volkhard, and Gerhard Kaiser, *Raketenspuren: Peenemünde 1936–1994: eine historische Reportage* (Berlin: Links Verlag, 1995).

Boelcke, Willi, A., ed., *Deutschlands Rüstung im Zweiten Weltkrieg: Hitlers Konferenzen mit Albert Speer 1942–1945* (Frankfurt am Main: Athenaion, 1969).

Bok, Sissela, *Secrets: On the Ethics of Concealment and Revelation* (New York: Vintage Books, 1989).

Bornemann, Manfred, *Geheimprojekt Mittelbau: Die Geschichte der deutschen V- Waffen Werke* (Munich: Lehmanns, 1971).

_____, and Martin Broszat, "Das KL Dora-Mittelbau," in *Studien zur Geschichte der Konzentrationslager, Schriftenreihe der Vierteljahreshefte für Zeitgeschichte* 21 (Stuttgart: Deutsche Verlags-Anstalt, 1970).

Bower, Tom, *The Paperclip Conspiracy: The Hunt for Nazi Scientists* (Boston: Little, Brown, 1987).

Breitman, Richard, *Architect of Genocide: Himmler and the Final Solution* (New York: Knopf, 1991).

Broszat, Martin, *Hitler and the Collapse of Weimar Germany* (New York: St. Martin's Press, 1987).

―――, *The Hitler State: The Foundation and Development of the Internal Structure of the Third Reich*, trans. by John W. Hiden (New York: Longman, 1981).

―――, "Soziale Motivation und Führer-Bindung des Nationalsozialismus," *Vierteljahreshefte für Zeitgeschichte* 18 (1970).

Browning, Christopher, *Ordinary Men: Reserve Police Battalion 101 and the Final Solution in Poland* (New York: HarperCollins, 1992).

Burger, Oswald, "Zeppelin und die Rüstungsindustrie am Bodensee," 1999: *Zeitschrift für Sozialgeschichte des 20. Und 21 Jahrhunderts* 1 (1987).

Burleigh, Michael, ed., *Confronting the Nazi Past: New Debates in Modern German History* (New York: St. Martin's Press, 1996).

Burridge, Michael, and Rolf Torstendahl, eds., *Professions in Theory and History: Rethinking the Study of the Professions* (London: Sage, 1990).

Caroll, Bernice A., *A Design for Total War: Arms and Economics in the Third Reich* (The Hague: Mouton, 1968).

Clark, D. B., "The Concept of Community: A Re-examination," *Sociological Review, New Series* 21 (1973), 397–416.

Craig, Gordon, *Germany, 1866–1945* (New York: Oxford University Press, 1978).

Crew, David F., ed., *Nazism and German Society, 1933–1945* (New York: Routledge, 1994).

Deist, Wilhelm, *The Wehrmacht and German Rearmament* (Toronto: University of Toronto Press, 1981).

Dornberger, Walter, "European Rocketry After World War I," *Journal of the British Interplanetary Society* 13 (September 1954).

―――, *Peenemünde: Die Geschichte der V-Waffen* (Frankfurt am Main: Ullstein, 1989).

―――, "The German V-2," *Technology and Culture* 4 (Fall 1963).

―――, "The Lessons of Peenemünde," *Astronautics* 3 (March 1958).

―――, *V-2* (New York: Viking Press, 1954).

Eisfeld, Ranier, *Die unmenschliche Fabrik: V-2 Produktion und KZ "Mittelbau- Dora"* (Nordhausen: KZ Gedenkstätte Mittelbau-Dora, 1992).

―――, *Mondsuchtig: Wernher von Braun und die Gebürt der Raumfahrt aus dem Geist Barbarei* (Reinbek bei Hamburg: Rowohlt Taschenbuch Verlag, 2000).

Engel, Gerhard, *Heeresadjutant bei Hitler 1938–1943: Aufzeichnungen des Majors Engel*, ed. and commentary by Hildegard von Kotze (Stuttgart: Deutsche Verlags-Anstalt, 1974).

Engelmann, Joachim, *Geheim Waffenschmiede Peenemünde* (Friedberg: Podzun-Pallas-Verlag, 1979).

Erichson, Johannes, and Bernhard Hoppe, *Peenemünde: Mythos und Geschichte der Rakete, 1923–1989* (Berlin: Nicolai'sche Verlagsbuchhandlung Beuermann, 2004).

Essers, Ilse, "Max Valier: Ein Vorkämpfer der Weltraumfahrt, 1895–1930," *Technikgeschichte der in Einzeldarstellungen* 5, (Düsseldorf: VDI, 1968).

Farr, Robert M., and Serge Moscovici, eds., *Social Representations* (New York: Cambridge University Press, 1984).

Fleischer, Wolfgang, *Die Heeresversuchsstelle Kummersdorf: Maus, Tiger, Panther, Luchs, Raketen und andere Waffen der Wehrmacht bei der Erprobung* (Wölfersheim-Berstadt: Podzun-Pallas Verlag, 1995).

Franklin, Thomas (pseudonym for Hugh McInnish), *An American in Exile: The Story of Arthur Rudolph* (Huntsville, AL: Christopher Kaylor, 1987).

Freund, Florian, *Arbeitslager Zement: Das Konzentrationslager Ebensee und die Raketenrüstung* (Vienna: Verlag für Gesellschaftskritik, 1989)

―――, and Bertrand Perz, *Das KZ in der Serbenhalle: Zur Kriegsindustrie in Wiener Neustadt* (Vienna: Verlag für Gesellschaftskritik, 1987).

Fritzsche, Peter, *A Nation of Flyers: German Aviation and the Popular Imagination* (Cambridge, MA: Harvard University Press, 1992).

_____, *Germans into Nazis* (Cambridge, MA: Harvard University Press, 1998).

Garlinsk, Joseph, *Hitler's Last Weapons: The Underground War Against the V1 and V2* (London: Friedman, 1978).

Gatzke, Hans, *Stresemann and the Rearmament of Germany* (New York: Norton, 1969).

Geertz, Clifford, *Local Knowledge* (New York: Basic Books, 1983).

Gellately, Robert, *The Gestapo and German Society: Enforcing Racial Policy, 1933–1945* (New York: Oxford University Press, 1990).

_____, *Backing Hitler: Consent and Coercion in Nazi Germany* (New York: Oxford University Press, 2001).

Gersdorff, K. von, "Die Peenemünde Fernrakete A4 ('V2')," in T. H. Benecke, K. H. Ludwig, and J. Hermann, eds., *Flugkorper und Lenkraketen: die Entwicklungsgeschichte der deutschen gelenkten Flugkorper vom Beginn dieses Jahrhunderts bis Heute* (Koblenz: Bernard & Graefe, 1987).

Geyer, Michael, *Deutsche Rüstungspolitik, 1860–1980* (Frankfurt: Suhrkamp, 1984).

Gilbert, G. Nigel, and Michael Mulkay, *Opening Pandora's Box: A Sociological Analysis of Scientists' Discourse* (New York: Cambridge University Press, 1984).

Gimbel, John, "German Scientists, United States Denazification Policy, and the 'Paperclip Conspiracy'," *International History Review* 12 (August 1990).

_____, *Science, Technology, and Reparations: Exploitation and Plunder in Postwar Germany* (Stanford, CA: Stanford University Press, 1990).

Gispen, Kees, *Poems in Steel: National Socialism and the Politics of Inventing from Weimar to Bonn* (New York: Berghahn Books, 2002).

Goerlitz, Walter, *History of the German General Staff, 1657–1945* (New York: Praeger, 1962).

Goode, William J., "Community Within a Community: The Professions," *American Sociological Review* 22 (April 1957).

Gropp, Dorit, *Aussenkommando Laura und Vorwerk Mitte Lehesten: Testbetrieb für V2-Triebwerke* (Bad Münstereifel: Westkreuz Verlag, 1999).

Gröttrup, Irmgard, *Rocket Wife* (London: Deutsch, 1959).

Grüner, Wolf, *Jewish Forced Labor Under the Nazis: Economic Needs and Racial Aims, 1938–1944* (New York: Cambridge University Press, 2006).

Gusterson, Hugh, *Nuclear Rites: A Weapons Laboratory at the End of the Cold War* (Berkeley, CA: University of California Press, 1996).

Hales, Peter Bacon, *Atomic Spaces: Living on the Manhattan Project* (Chicago: University of Illinois Press, 1997), 42.

Hayes, Peter, *Industry and Ideology: IG Farben and the Nazi Era*, 2nd ed. (Cambridge, UK: Cambridge University Press, 2001).

Henneberg, Ilse, ed., *"Niedergefahren zur Hölle – Aufgefahren gen Himmel:" Wernher von Braun und die Produktion der V2-Raketen im KZ-Mittelbau- Dora* (Bremen: Donat Verlag, 2002).

Henschel, Phillip, *Hitler's Rocket Sites* (London: Hale, 1985).

Herbert, Ulrich, *Fremdarbeiter: Die Politik und Praxis des "Ausländer-Einsatzes" in der Kriegswirtschaft des Dritten Reiches* (Bonn: Dietz, 1985)

_____, "Labor as Spoils of Conquest, 1933–1945," in David F. Crew, ed., *Nazism and German Society* (New York: Routledge, 1994), 240–241.

_____, *"Labour and Extermination: Economic Interest and the Primacy of Weltanschauung in National Socialism,"* *Past and Present* 138 (February 1993).

_____, Karin Orth, and Christoph Dieckmann, eds., *Die Konzentrationslager – Entwicklung und Struktur, Bd 2* (Göttingen: Wallstein Verlag, 1998).

Herdt, Gilbert, *Secrecy and Cultural Reality: Utopian Ideologies of the New Guinea Men's House* (Ann Arbor, MI: University of Michigan Press, 2003).

Herf, Jeffrey, *Reactionary Modernism: Technology, Culture, and Politics in Weimar and the Third Reich* (Cambridge, UK: Cambridge University Press, 1984).

_____, *The Jewish Enemy: Nazi Propaganda During World War II and the Holocaust* (Cambridge, MA: The Belknap Press of Harvard University, 2006).

Hochmuth, Peter, *Der illegale Widerstand der Häftlinge des KZ Mittelbau-Dora: Dokumentation* (Schkeuditz: GNN Verlag, 2000).

Hogg, Michael A., and Dominic Abrams, *Social Identifications: A Social Psychology of Intergroup Relations and Group Processes* (London: Routledge, 1988).

Hölsken, Heinz-Dieter, *Die V-Waffen: Entstehung, Propaganda, Kriegseinsatz* (Stuttgart: Deutsche Verlags-Anstalt, 1984).

Homze, Edward L., *Foreign Labor in Nazi Germany* (Princeton, NJ: Princeton University Press, 1967).

Horeis, Heinz, *Rolf Engel – Raketenbauer der ersten Stunde* (Munich: Lehrstuhl für Raumfahrttechnik, 1992).

_____, ed., *Rolf Engel: Rakentenbauer der ersten Stunde* (Munich: Lehrstuhl für Raumfahrttechnik, Technische Universität München, 1992).

Hortleder, Gerd, *Das Gesellschaftsbild des Ingenieurs: Zum politischen Verhalten der technischen Intelligenz in Deutschland* (Frankfurt: Suhrkamp, 1970).

Hüttenberger, Peter, "Nationalsozialistische Polykratie," *Geschichte und Gesellschaft* 2 (1976).

Hunt, Linda, *Secret Agenda: The United States Government, Nazi Scientists, and Project Paperclip, 1945 to 1990* (New York: St. Martin's Press, 1991).

Huzel, Dieter K., *From Peenemünde to Canaveral* (Englewood Cliffs, NJ: Prentice-Hall, 1962).

International Tracing Service, *Verzeichnis der Haftstätten unter dem Reichsführer-SS (1933–1945)* (Bad Arolsen: ITS, 1979).

Irving, David, *The Mare's Nest* (Boston: Little, Brown, 1965).

Jarausch, Konrad, *The Unfree Professions: German Lawyers, Teachers, and Engineers, 1900–1950* (New York: Oxford University Press, 1990).

Kaufmann, Doris, ed., *Geschichte der Kaiser-Wilhelm-Gesellschaft im Nationalsozialismus: Bestansaufnahme und Perspektiven der Forschung* (Göttingen: Wallstein Verlag, 2000).

Kelly, Anita E. ed., *The Psychology of Secrets* (New York: Plenum Press, 2002).

Kershaw, Ian, *Popular Opinion and Political Dissent in the Third Reich: Bavaria, 1933–1945* (Oxford: Oxford University Press, 1983).

_____, *The Nazi Dictatorship: Problems and Perspectives of Interpretation* (New York: Routledge, Chapman and Hall, 1993).

_____, *Hitler: 1889–1936: Hubris* (New York: Norton, 1998).

_____, *Hitler: 1936–1945: Nemesis* (New York: Norton, 2001).

Klee, Ernst, and Otto Merck, *The Birth of the Missile: The Secrets of Peenemünde*, trans. by T. Schoeters (New York: Dutton, 1965).

Klein, Heinrich, *Vom Geschoss zum Feuerpfeil: Der grosse Umbruch der Waffentechnik in Deutschland, 1900–1970* (Neckargemünd: Kurt Vowinckel, 1977).

Koehl, Robert Lewis, *The Black Corps: The Structure and Power Struggles of the Nazi SS* (Madison, WI: University of Wisconsin Press, 1983).

Kogon, Eugen, *Die Stunde der Ingenieure* (Düsseldorf: VDI, 1976).

Krausnick, Helmut, and Martin Broszat, *Anatomy of the SS State*, trans. by Dorothy Lang and Marian Jackson (London: Granada, 1970).

Lasby, Clarence B., *Project Paperclip: German Scientists and the Cold War* (New York: Atheneum, 1971).

Lattas, Andrew *Cultures of Secrecy: Reinventing Race in the Bush Kaliai Cargo Cults* (Madison, WI: University of Wisconsin Press, 1998).

Leeb, Emil, *Aus der Rüstung des Dritten Reiches (Das Heereswaffenamt 1938–1945) Wehrtechnische Monatshefte*, Beiheft 4 (Berlin: Mittler, 1958).

Ley, Willy, "Count von Braun," *Journal of British Interplanetary Society* 6 (June 1947).

_____, *Rockets, Missiles, and Space Travel* (New York: Viking Press, 1961).

Longmate, Norman, *Hitler's Rockets: The Story of the V-2s* (London: Hutchinson, 1985).

Ludwig, Karl-Heinz, "Die deutschen Flakraketen im Zweiten Weltkrieg," *Militärgeschichtliche Mitteilungen* 1 (1969).

———, *Technik und Ingenieure im Dritten Reich* (Düsseldorf: Droste Verlag, 1974).

MacKenzie, Donald, *Inventing Accuracy: A Historical Sociology of Nuclear Missile Guidance* (Cambridge MA: MIT Press, 1990).

Maeseneer, Guido de, *Peenemünde: The Extraordinary Story of Hitler's Secret Weapons V-1 and V-2* (Vancouver, Canada: AJ Publishing, 2001).

Maier, Charles S., *Recasting Bourgeois Europe: Stabilization in France, Germany, and Italy in the Decade After World War I* (Princeton, NJ: Princeton University Press, 1975).

Maier, Helmut, ed., *Rüstungsforschung im Nationalsozialismus: Organisation, Mobilisierung, und Entgrenzung der Technikwissenschaften* (Göttingen: Wallstein Verlag, 2002).

Mallmann, Klaus Michael, "Milieu, Radikalismus und lokale Gesellschaft: Zur Sozialgechichtes des Kommunismus in der Weimarer Republik," *Geschichte und Gesellschaft* 21 (1995).

Marcus, George E., ed., *Technoscientific Imaginaries: Conversations, Profiles, and Memoirs* (Chicago: University of Chicago Press, 1995).

Marionoff, Dimitri, with Palma Wayne, *Einstein – An Intimate Study of a Great Man* (New York: Doubleday, 1944).

McClelland, Charles, *The German Experience of Professionalization: Modern Learned Professions and Their Organizations from the Early Nineteenth Century to the Hitler Era* (New York: Cambridge University Press, 1991).

McDougall, Walter A., *The Heavens and the Earth: A Political History of the Space Age* (New York: Basic Books, 1985).

McGovern, James, *Crossbow and Overcast* (New York: Morrow, 1964).

Mehrtens, Herbert, and Steffen Richter, eds., *Naturwissenschaft, Technik und NS- Ideologie: Beiträge zur Wissenschaftsgeschichte des Dritten Reichs* (Frankfurt: Suhrkamp, 1980).

Michel, Jean, with Louis Nucera, *Dora*, trans. by Jennifer Kidd (New York: Holt, Rinehart & Winston, 1979).

Middlebrook, Martin, *The Peenemünde Raid: The Night of 17–18 August, 1943* (London: Cassell, 1982).

Mierzejewski, Alfred C., *The Collapse of the German War Economy, 1944–1945: Allied Air Power and the German National Railway* (Chapel Hill, NC: University of North Carolina Press, 1988).

Millinger, Heinz, *Über Peenemünde ins All*, available at http://www.urbin.de/46189376/buch_millinger.pdf.

Mommsen, Hans, ed., *The Third Reich Between Vision and Reality: New Perspectives on German History, 1918–1945* (New York: Berg, 2001).

———, "The Dissolution of the Third Reich: Crisis Management and Collapse, 1943–1945," *German Historical Institute Bulletin* 27 (Fall 2000).

———, *Der Nationalsozialismus und die deutsche Gesellschaft* (Reinbek bei Hamburg: Rowohlt, 1991).

Moore, S. F. and Barbara Myerhoff, eds., *Secular Ritual* (Assen: Van Gorkum, 1977).

Mosse, George, *The Crisis of German Ideology: Intellectual Origins of the Third Reich* (New York: Fertig, 1964).

———, *The Nationalization of the Masses* (New York: New American Library, 1975).

Müller, Klaus-Jurgen, *Das Heer und Hitler: Armee und Nationalsozialistisches Regime, 1933–1940*, 2nd ed. (Stuttgart: Deutsche Verlags-Anstalt, 1988).

Müller, Rolf-Dieter, "Kriegführung, Rüstung und Wissenschaft: Zur Rolle des Militärs bei der Steuerung der Kregstechnick unter besonderer Berücksichtigung des Heereswaffenamtes 1935–1942," in Helmut Maier, ed., *Rüstungforschung im Nationalsozialismus: Organisation, Mobilisierung und Entrgrenzung der Technikwissenschaften* (Göttingen: Wallstein Verlag, 2002).

————, "World Power Status Through the Use of Poison Gas? German Preparations for Chemical Warfare, 1919–1945," in Wilhelm Deist, ed., *The German Military in the Age of Total War* (Leamington Spa: Berg, 1965).

Neander, Joachim, *Das Konzentrationslager "Mittelbau" in der Endphase der nationalsozialistischen Diktatur: zur Geschichte des letzten im "Dritten Reich" gegründeten selbständigen Konzentrationslagers unter besonderer Berücksichtigung seiner Auflösungsphase* (Clausthal-Zellerfeld: Papierflieger, 1999).

Nebel, Rudolf, *Die Narren von Tegel: Ein Pioneer der Raumfahrt erzählt* (Düsseldorf: Droste Verlag, 1972).

Neufeld, Michael J., "Rolf Engel vs. the German Army: A Nazi Career in Rocketry and Repression," *History and Technology* 13 (1996).

————, *The Rocket and the Reich: Peenemünde and the Coming of the Ballistic Missile Era* (Cambridge, MA: Harvard University Press, 1995).

————, *Von Braun: Dreamer of Space, Engineer of War* (New York: Knopf, 2007).

————, "Hitler, the V-2, and the Battle for Priority, 1939–1943," *Journal of Military History* 57 (July 1993), 511–538.

————, "Weimar Culture and Futuristic Technology: The Rocketry and Spaceflight Fad in Germany, 1923–1933," *Technology and Culture* 31 (October 1990), 725–752.

————, "Wernher von Braun, the SS, and Concentration Camp Labor: Questions of Moral, Political, and Criminal Responsibility," *German Studies Review* 25 (2002).

Newman, Leonard S., and Ralph Erber, eds., *Understanding Genocide: The Social Psychology of the Holocaust* (New York: Oxford University Press, 2002).

Nuss, Karl, "Einige Aspekte der Zusammenarbeit vom Heereswaffenamt und Rüstungskonzernen vor dem zweiten Weltkrieg," *Zeitschrift für Militärgeschichte* 4 (1965), 433–443.

Oberth, Hermann, *Die Rakete zu den Planetenräumen* (Nuremberg: Uni-Verlag, 1960; original work published 1923).

————, *Wege zur Raumschiffahrt Reprint* (Bucharest: Kriterion, 1974).

Ordway, Frederick I., and Mitchell Sharpe, *The Rocket Team: From the V-2 to the Saturn Moon Rocket* (Cambridge, MA: MIT Press, 1982).

Orth, Karin, *Das System der Nationalsozialistischen Konzentrationslager* (Munich: Pendo Verlag, 2002; original work published 1999).

Overy, Richard, "From 'Uralbomber' to 'Amerikabomber': The Luftwaffe and Strategic Bombing," *Journal of Strategic Studies* 1 (1978), 154–178.

————, *Goering: The Iron Man* (New York: Routledge & Keegan Paul, 1984).

————, "Hitler's War and the German Economy: A Reinterpretation," *Economic History Review* 35 (1982), 272–291.

————, "Mobilization for Total War in Germany 1939–1941," *English Historical Review* 103 (1988).

Päch, Susanne, "Rolf Engel: Fifty Years of Activity in Rocketry and Space Flight," *Spaceflight* 22 (June 1980), 231–236.

Pachaly, Erhard, and Kurt Pelny, *Konzentrationslager Mittelbau-Dora: Zum antifaschistischen Widerstandskampf im KZ Dora 1943 bis 1945* (Berlin: Dietz, 1990).

Paul, Gerhard, and Klaus-Michael Mallmann, *Die Gestapo – Mythos und Realität* (Darmstadt: Wissenschaftliche Buch Gesellschaft, 1995).

Petzina, Dieter, *Autarkiepolitik im Dritten Reich: Der nationalsozialistische Vierjahresplan* (Stuttgart: Deutsche Verlags-Anstalt, 1968).

Peukert, Detlev, *The Weimar Republic: The Crisis of Classical Modernity* (New York: Hill & Wang, 1987).

Reisig, Gerhard, H.R., "Das kongeniale Vermächtnis Hermann Oberths und Wernher von Brauns für die Raumfahrt-Entwicklung," *Astronautik* (1987), 103–106.

————, "Von den Peenemünder 'Aggregaten' zur amerikanischen 'Mondrakete': Die Entwicklung der Apollo-Rakete 'Saturn V' durch das Wernher von Braun Team an Hand der Peenemünder Konzepte," *Astronautik* (1986).

_____, *Raketenforschung in Deutschland: wie die Menschen das All eroberten* (Berlin: Wissenschaft und Technik Verlag, 1999).

Remy, Steven, *The Heidelberg Myth: The Nazification and Denazification of a German University* (Cambridge, MA: Harvard University Press, 2002).

Renneberg, Monika, and Mark Walker, eds., *Science, Technology, and National Socialism* (New York: Cambridge University Press, 1994).

Riedel, Walter J.H., "A Chapter in Rocket History," *Journal of the British Interplanetary Society* 13 (July 1954), 208–212.

Ringer, Fritz, *The Decline of the German Mandarins: The German Academic Community, 1890–1933* (London: Wesleyan University Press, 1983; original work published 1969).

Ruland, Berndt, *Wernher von Braun: Mein Leben für die Raumfahrt* (Offenburg: Burda, 1969).

Rürup, Reinhard, ed., *Topographie des Terrors: Gestapo, SS, und Reichsicherheitshauptamt auf dem "Prinz-Albrecht Gelände" – Eine Dokumentation* (Berlin: Willmuth Arenhövel, 1987).

_____, ed., *Wissenschaft und Gesellschaft: Beiträge zur Geschichte der Technischen Universität Berlin: 1879–1979* (New York: Springer, 1979).

Rüter, C. F., and D. W. de Mildt, *Justiz und NS-Verbrechen: Die Westdeutschen Strafverfahren wegen nationalsozialistischer Tötungsverbrechen 1945–1997: eine systematische Verfahrensbeschreibung, mit Karten und Registern: bearbeitet im Seminarium voor Strafrecht en Strafrechtspleging 'Van Hamel' der Universität Amsterdam* (Munich: Saur, 1998).

Schmidt, Matthias, *Albert Speer: The End of a Myth*, trans. by Joachim Neugroschel (London: Harrap, 1985).

Schneider, Erich, "Technik und Waffenentwicklung im Kriege," in *Bilanz des Zweiten Weltkrieges* (Oldenburg: Gerhard Stalling, 1953).

Schulte, Jan-Erik, *Zwangsarbeit und Vernichtung: das Wirtschaftsimperium der SS: Oswald Pohl und das SS-Wirtschaftsverwaltungs-Hauptamt, 1933–1945* (Paderborn: Schöningh, 2001.

Schulz, Gerhard, *Aufstieg des Nationalsozialismus: Krise und Revolution in Deutschland* (Frankfurt am Main: Propyläen Verlag, 1975).

Segev, Tom, *Die Soldaten des Bösen: zur Geschichte der KZ-Kommandenten* (Reinbek bei Hamburg: Rowohlt Taschenbuch Verlag, 1992).

Seidler, Franz W., *Fritz Todt: Baumeister des Dritten Reiches* (Munich: Herbig, 1986).

Sellier, André, *A History of the Dora Camp: The Story of the Nazi Slave Labor Camp That Secretly Manufactured V-2 Rockets* (Chicago: Dee, 2003).

Siddiqi, Asif, "The Rocket's Red Glare: Technology, Conflict, and Terror in the Soviet Union," *Technology and Culture* 44 (July 2003), 470–501.

Simon, Leslie, *German Research in World War II: An Analysis of the Conduct of Research* (New York: Wiley, 1947).

Simpson, Christopher, *Blowback: America's Recruitment of Nazis and its Effects on the Cold War* (New York: Weidenfeld & Nicholson, 1988).

Sontheimer, Kurt, *Antidemokratisches Denken in der Weimarer Republik* (Munich: Nymphenburger Verlagsbuchhandlung, 1968).

Speer, Albert, *Inside the Third Reich* (New York: Macmillan, 1970).

_____, *Infiltration*, trans. by Joachim Neugroschel (New York: Macmillan, 1981).

Spoerer, Mark, *Zwangsarbeit unter dem Hakenkreuz: Ausländische Zivilarbeiter, Kriegsgefangene und Häftlinge im Dritten Reich und im besetzten Europa* (Stuttgart: Deutsche Verlags-Anstalt, 2001).

_____, "Profitierten Unternehmen von KZ-Arbeit? Eine kritische Analyse der Literatur," *Historische Zeitschrift* 268 (1999).

Stachura, Peter, ed., *The Shaping of the Nazi State* (New York: Barnes & Noble Books, 1978).

Stamm-Kühlmann, Thomas, and Reinhard Wolf, eds., *Raketenrüstung und internationaler Sicherheit von 1942 bis heute* (Stuttgart: Franz Steiner Verlag, 2004).

Steinhoff, Ernst A., "Development of the German A-4 Guidance and Control System 1939–1945: A Memoir," in R. Cargill Hall, ed., *History of Rocketry and Astronautics: Proceedings of the Third Through Sixth History Symposia of the International Academy of Astronautics* (San Diego: Univelt for the American Astronautical Society, 1986).

Stern, Fritz, *The Politics of Cultural Despair* (New York: Anchor Books, 1965).

Stern, J. P., *Hitler: The Führer and the People* (Berkeley, CA: University of California Press, 1975).

Stubno, William J., Jr., "The von Braun Rocket Team Viewed as a Product of German Romanticism," *Journal of the British Interplanetary Society* 35 (1982).

Stuhlinger, Ernst, "Gathering Momentum: Von Braun's Work in the 1940s and 1950s," in Frederick Ordway III and Randy Liebermann, eds., *Blueprint for Space: Science Fiction to Science Fact* (Washington, DC: Smithsonian Institution Press, 1992).

———, and Frederick I. Ordway, *Wernher von Braun: Aufbruch in den Weltraum* (Esslingen: Bechtle Verlag, 1992).

Syon, Guillaume de, *Zeppelin! Germany and the Airship, 1900–1939* (Baltimore, MD: Johns Hopkins University Press, 2002).

Szöllösi-Janze, Margit, and Helmuth Trischler, eds., *Grossforschung in Deutschland* (Frankfurt: Suhrkamp, 1990).

Tartar, Donald E., "Peenemünde and Los Alamos: Two Studies," *History of Technology* 14 (1992).

Tefft, Stanton, ed., *Secrecy: A Cross Cultural Perspective* (New York: Human Sciences Press, 1980).

Torstendahl, Rolf, and Burrage, Michael, eds., *The Formation of Professions: Knowledge, State, Strategy* (London: Sage, 1990).

Trischler, Helmuth, *Luft- und Raumfahrtforschung in Deutschland 1900–1970: Politische Geschichte einer Wissenschaft* (Frankfurt: Campus, 1992).

United States, Army Ordnance, *The Story of Peenemünde, or What Might Have Been* (also known as *Peenemünde East, Through the Eyes of 500 Detained at Garmisch*), Mimeograph, 1945.

Villain, Jacques, "France and the Peenemünde Legacy," Paper IAA-92-0186, presented at the 43rd Congress of the International Astronautical Federation, Washington, DC, 1992.

von Braun, Magnus Freiherr, *Wege durch vier Zeitepochen* (Limburg an der Lahn: Starke, 1965).

von Braun, Wernher, "Reminiscences of German Rocketry," *Journal of the British Interplanetary Society* 15 (May–June 1956).

Wagner, Jens-Christian, *Produktion des Todes: Das KZ Mittelbau-Dora* (Göttingen: Wallstein Verlag, 2001).

Walker, Mark, ed., *Science and Ideology: A Comparative History* (New York: Routledge, 2003).

———, *Nazi Science: Myth, Truth, and the German Atomic Bomb* (Cambridge, MA: Perseus, 1995).

———, *German National Socialism and the Quest for Nuclear Power, 1939–1949* (Cambridge, UK: Cambridge University Press, 1989).

Ward, Bob, ed., *Wernher von Braun: Anekdotisch* (Esslingen: Bechtle Verlag, 1972).

Wegener, Peter, *The Peenemünde Wind Tunnels: A Memoir* (New Haven, CT: Yale University Press, 1996).

Weinman, Martin, ed., *Das nationalsozialistische Lagersystem* (Frankfurt am Main: Zweitausendeins, 1990).

Welch, David, ed., *Nazi Propaganda: The Power and Limitations* (Totowa, NJ: Barnes & Noble Books, 1983).

Weitz, Eric D., *Creating German Communism: From Popular Protests to Socialist State* (Princeton, NJ: Princeton University Press, 1997).

Wildt, Michael, *Generation des Unbedingten: Das Führerkorps des Reichssicherheitshaup-tamtes* (Hamburg: Hamburg Edition, 2002).

Winter, Frank H., *Prelude to the Space Age: The Rocket Societies: 1924–1940* (Washington, DC: Smithsonian Institution Press, 1983).

———, *Rockets into Space* (Cambridge, MA: Harvard University Press, 1990).

Wolff, Kurt H., ed., *The Sociology of Georg Simmel* (Glencoe, IL: The Free Press, 1950).

Young, Anthony, *The Flying Bomb* (New York: Sky Books, 1978).

Index